Genetic Data Analysis II

Genetic Data Analysis II

Methods for Discrete Population Genetic Data

BRUCE S. WEIR

Program in Statistical Genetics
Department of Statistics
North Carolina State University

Sinauer Associates, Inc. Publishers
Sunderland, Massachusetts

The Cover

This book combines the disciplines of statistics and genetics, and central concepts in these fields are brought together in the cover illustration. The bell-shaped curve is that of the normal distribution, upon which much of statistical theory is based. The elements of the curve represent the four DNA base types in DNA sequences, upon which genetic information is ultimately dependent. Successive triplets of letters along the curve represent amino acids, and the one-letter amino acid code for the sequence spells out the title of this book. The curve is laid over a table containing data from which Mendel deduced the existence of genes.

Genetic Data Analysis II

Sinauer Associates Inc., 23 Plumtree Road/P.O. Box 407,
Sunderland, MA 01375 USA FAX: 413-549-1118
email: publish@sinauer.com

Library of Congress Cataloging-in-Publication Data

Weir, B. S. (Bruce S.), 1943-
Genetic data analysis / Bruce S. Weir. – 2nd ed., [rev. and expanded]
p. cm.
Includes bibliographical references and index.
ISBN 0-87893-902-4 (paper)
1. Genetics – Statistical methods. 2. Quantitative genetics. 3. Gene frequency – Statistical methods. I. Title.
QH438.4.S73W45 1996
575.1'072 – dc20 96-17850
CIP

Printed in Canada

5

To Beth

Contents

Preface

I have been both pleased at the reaction to *Genetic Data Analysis*, and chastened by how quickly the data and their analyses have changed. I have tried to respond to these changes in this second edition while keeping the core of the book intact. With the leap in desktop computing power over the past six years there is a move towards computationally intensive methods rather than classic parametric statistics, as witnessed by the popularity of permutation tests. I have incorporated this approach in several places throughout the book, although I have deferred until a later edition a treatment of Monte Carlo Markov chain methods. I have also made a brief reference to the increasing role of Bayesian methods in statistical genetics.

As I noted in the preface to the first edition, writing about analyses of molecular genetic data is like aiming at a moving target. The flood of data on microsatellite markers as a result of PCR technology, for example, is having a substantial impact on population genetics. With the human genome map essentially complete, and a complete human DNA sequence becoming increasingly likely, the explosion of discrete human genetic data seems certain to continue. New technology is also allowing population geneticists to generate data for other organisms almost without limit. In light of this wealth of new data, the description of sequence analyses in Chapter 9 will seem very brief. It should be regarded as only an introduction and could be supplemented for example by Nei's *Molecular Evolutionary Genetics* or Li and Graur's *Fundamentals of Molecular Evolution.* Likewise, a single chapter cannot hope to cover the whole field of phylogeny reconstruction from molecular data. The second edition of *Molecular Systematics*, edited by Hillis, Moritz and Mable is highly recommended for additional coverage.

Although the observation is colored by personal experience, I note that the analysis of population genetic data has received unaccustomed public attention over the past six years as a result of the forensic use of DNA markers. It is my hope that agreement will be reached on the point at which DNA profiles are considered to be so discriminating that quantification of the evidence of matching profiles is not needed. However, I have laid out a basis for providing quantification in a new chapter. It is a pleasure to record here my indebtedness in this area to Ian Evett, John Buckleton and Bruce Budowle.

The past six years have also witnessed a boom in statistical approaches to locating genes on genetic maps. This book is deliberately focused on discrete characters, and the reader will want to refer to Falconer and Mackay's *Quantitative Genetics* (fourth edition) for a review of methods for mapping continuous trait genes. Searches for some human disease genes do employ the methods of this book, however, and a new chapter on linkage addresses some of the issues of map construction and association mapping. The reader is referred to Ott's *Analysis of Human Genetic Linkage* (second edition) for more details.

I hope that the book will continue to fill its primary role of describing the basic analyses of discrete genetic data, with an emphasis on making inferences about associations between alleles and about population structure. Certainly these were the areas that generated the most enquiries and requests for computer programs over the past six years. In that regard, I am pleased to report the replacement of my FORTRAN programs by the powerful and flexible suite of programs written by Paul Lewis and Dmitri Zaykin. The authors have adopted the same data structure used by David Swofford in PAUP, and have generously made it available without charge on the World Wide Web (URL http://www.stat.ncsu.edu and click on "statistical genetics.")

Our Web site will also contain postings about this book. The need for such postings to announce errors or obscurities has been greatly diminished by the advice of some superb reviewers: Jonathan Arnold and his students, Wen-Hsiung Li, Elizabeth Thompson and Jeffrey Thorne. The reviewers deserve no blame for those instances where I failed to heed their good advice. Students in my classes at North Carolina State University have also diligently corrected earlier drafts. Christopher Basten deserves much credit for the improved appearance of the book. He worked very hard to allow the change to the Palatino font that Andrew Sinauer wisely suggested. The growing list of fine books published by Sinauer Associates is a tribute to Andy and his colleagues, although it does not reveal their much-appreciated patience.

I thank Beth for her wonderful support.

Raleigh, North Carolina
April, 1996

Chapter 1

Nature of Discrete Genetic Data

INTRODUCTION

With genetic information affecting more and more aspects of daily life, the need for careful statistical analyses of genetic data has grown. That need became very public when a prominent athlete was accused of a violent crime, in part, because his DNA profile matched that found at the scene of the crime and a prosecutor told a court that the particular profile is found only once in a billion people. Almost as public was the debate that followed the announcement of evidence for an "African Eve" some 200,000 years ago. The debate, on the time of when the woman who carried the ancestor of all human mitochondria was alive, centered on the statistical methods to study sequence evolution. Sequence evolution on a much shorter time scale was involved when statistical methods were used to compare some HIV sequences to help determine if a Florida dentist had infected some of his patients with the AIDS virus.

Genetic data analysis is now widely used in the search for genes affecting human diseases or traits of economic importance in domestic plants and animals. These searches are often based on associations between genetic markers and the traits of interest, and it is a statistical activity to measure and assess these associations. Further statistical analyses may translate associations into distances between marker and trait loci, paving the way for a detailed biochemical study of the trait. Publication of the first complete DNA sequence of a eukaryote points to the success of the Human Genome Project. Assembling the thousands of partial sequences in such enormous projects is a statistical activity, as will be the long-term task of characterizing features of DNA and protein sequences. Statistical procedures are being developed to detect errors in sequence or linkage data. The prediction of protein structure from amino acid sequence features depends on statistical arguments and has far-reaching implications ranging from an understanding of evolution to the design of pharmaceuticals.

Full benefits of the many exciting advances in molecular genetics will be realized only if appropriate statistical analyses are applied to experimental findings. This book is designed to give an introduction to methods needed for current types of genetic data and to lay a foundation for the new methods that will be needed in the future. It is meant to be accessible to geneticists with only a modicum of statistical training, and it is hoped that it will entice more statisticians to take up the challenges being presented by genetic data.

The book treats data collected for discrete characters that can range from morphological traits such as flower color to individual base types in a DNA sequence. Whatever the nature of the data, the analyses to be presented are for observations on discrete heritable units that can be characterized as ***Mendelizing units***. In other words, observations are made on units that are transmitted from parent to offspring, and whose transmission is a random process. At one time such units would have been called genes, but it is no longer desirable to restrict attention to these DNA regions that encode for proteins. Statistical theories can be developed for units that are indeed genes or that depend on physical properties of gene products, or that may have nothing to do with coding regions.

EXAMPLES OF GENETIC DATA

Phenotypic Data

The best known phenotypic data are those given by Mendel for the garden pea *Pisum sativum*. Mendel observed seven characters in a series of crosses and gave the results, from the offspring of the resulting hybrid plants, shown in Table 1.1. These data allowed Mendel to postulate a 3:1 segregation ratio for dominant and recessive phenotypes, although there has been discussion that they fit this rule rather too well.

Codominant Phenotypic Data

As Fisher (1936a) pointed out in discussing Mendel's work, a problem with dominant characters where heterozygotes and dominant homozygotes cannot be distinguished is that parental types cannot be determined with certainty. This issue does not arise with codominant traits such as some blood groups, and a particularly complete data set, published by Race et al. (1949), is shown in Table 1.2.

Table 1.1 Mendel's results for seven dominant characters in *Pisum sativum*.

	Character	Dominant Form		Recessive Form	
		Seed characters			
A	Seed shape	5474	Round	1850	Wrinkled
B	Cotyledon color	6022	Yellow	2001	Green
		Plant characters			
C	Seed coat color	705	Grey-brown	224	White
D	Pod shape	882	Simply inflated	299	Constricted
E	Unripe pod color	428	Green	152	Yellow
F	Flower position	651	Axial	207	Terminal
G	Stem length	787	Long	277	Short

Source: Mendel (1866).

The *MN* blood groups are recognized by agglutination with anti-M or anti-N serum and can be regarded as being controlled by a single locus with two alleles, M and N. The three genotypes can all be distinguished. The *S* blood groups are detected with anti-S serum and are controlled by a locus with alleles S and s. The two genotypes *SS* and *Ss* both agglutinate with anti-S serum and could not be differentiated by Race et al. The *M* and *S* loci are closely linked. In Table 1.2, the notation *MN.S* indicates that the S gene may be located on either the M or N bearing chromosomes, whereas *MSNs*, for example, is used if it is clear that M and S are on one chromosome, and N, s are on the other. Two children linked with an "=" are monozygotic twins, and those linked with a "–" are dizygotic twins.

Allozyme Data

A major impact on population genetics was provided by development of the technique for electrophoretic detection of charge differences for variants among soluble proteins. Protein material from an individual is placed at one end of a slab of gel and allowed to migrate under the action of an electric field along the gel.

Table 1.2 MNS blood group types for families reported by Race et al. (1949).

	Father	Mother	Child 1	Child 2	Child 3	Child 4	Child 5	Child 6
1	*MSMs*	*MsMs*	*MsMs*	*MSMs*	*MSMs*			
2	*MM.S*	*MsMs*	*MSMs*	*MSMs*				
3	*MM.S*	*MM.S*	*MM.S*	*MM.S*	*MM.S*			
4	*MM.S*	*MM.S*	*MM.S*					
5	*MM.S*	*MM.S*	*MM.S*					
6	*MM.S*	*MM.S*	*MM.S*	*MM.S*	*MM.S*			
7	*MsMs*	*MsNs*	*MsNs*	*MsNs*				
8	*MsNs*	*MsMs*	*MsMs*					
9	*MsNs*	*MsMs*	*MsNs*					
10	*MsMs*	*MsNs*	*MsMs=*	*=MsMs*				
11	*MsMs*	*MsNs*	*MsNs*	*MsMs*	*MsNs*	*MsMs*		
12	*MsNs*	*MSMs*	*MSNs–*	*–MsMs*				
13	*MSMs*	*MsNs*	*MsNs*	*MSNs*	*MsMs*			
14	*MsNs*	*MSMs*	*MsMs*	*MSNs*	*MSMs*			
15	*MM.S*	*MsNs*	*MSMs*	*MSMs*	*MSNs*	*MSNs*	*MSNs*	*MSNs*
16	*MsNs*	*MM.S*	*MSMs*	*MSMs=*	*=MSMs*			
17	*MM.S*	*MsNs*	*MSMs*	*MSMs*				
18	*MsNs*	*MM.S*	*MSMs=*	*=MSMs*				
19	*MsNs*	*MM.S*	*MSNs*					
20	*MsNs*	*MM.S*	*MSMs*	*MSMs*				
21	*MsMs*	*MSNs*	*MSMs*	*MsNs*	*MSMs*	*MsNs*		
22	*MsMs*	*MSNs*	*MSMs*	*MSMs*	*MSMs–*	*–MSMs*		
23	*MSNs*	*MSMs*	*MM.S*	*MsNs*				
24	*MSNs*	*MSMs*	*MM.S*	*MsNs*				
25	*MsNS*	*MSMs*	*MsMs*					
26	*MM.S*	*MN.S*	*MM.S*	*MN.S*	*MN.S*			
27	*MM.S*	*MN.S*	*MM.S*	*MN.S*				
28	*MM.S*	*MN.S*	*MM.S*	*MM.S*				
29	*MN.S*	*MM.S*	*MN.S*	*MM.S*				
30	*MM.S*	*MN.S*	*MN.S*					
31	*MM.S*	*MN.S*	*MM.S*	*MN.S*				
32	*MM.S*	*MN.S*	*MN.S*	*MM.S*				
33	*MN.S*	*MM.S*	*MN.S–*	*–MM.S*				
34	*MM.S*	*MN.S*	*MM.S*	*MM.S*				
35	*MM.S*	*MN.S*	*MM.S*	*MM.S*	*MN.S*			
36	*NsNs*	*MsMs*	*MsNs*	*MsNs*				
37	*NsNs*	*MsMs*	*MsNs*	*MsNs*	*MsNs*	*MsNs*	*MsNs*	*MsNs*
38	*NsNs*	*MsMs*	*MsNs*	*MsNs*	*MsNs*			
39	*MsMs*	*NsNs*	*MsNs*					
40	*MSMs*	*NsNs*	*MsNs*	*MsNs*				
41	*MSMs*	*NsNs*	*MSNs*	*MsNs*				
42	*NsNs*	*MSMs*	*MSNs*	*MsNs*	*MSNs*			
43	*NsNs*	*MM.S*	*MSNs*					
44	*NN.S*	*MsMs*	*MsNS*					
45	*NsNs*	*MSMs*	*MsNs*	*MN.S*				
46	*MM.S*	*NN.S*	*MN.S*					

Table 1.2 Continued

	Father	Mother	Child 1	Child 2	Child 3	Child 4	Child 5	Child 6
47	NN.S	MM.S	MN.S=	=MN.S				
48	MsNs	MsNs	MsNs	MsNs				
49	MsNs	MsNs	MsNs	MsMs	NsNs			
50	MsNs	MsNs	NsNs					
51	MsNs	MsNs	MsNs	NsNs				
52	MSNs	MsNs	MSNs	NsNs	MsNs			
53	MsNs	MSNs	NsNs	MSNs				
54	MsNs	MSNs	MSMs	MSNs	NsNs			
55	MSNs	MsNs	NsNs	MSNs	MSMs			
56	MsNs	MSNs	MSMs	MsNs	MSNs			
57	MSNs	MsNs	NsNs	MSNs	MSNs			
58	MSNs	MsNs	NsNs					
59	MsNs	MSNs	MSNs	MsNs	MsNs	MSMs	NsNs	
60	MsNs	MsNS	MsMs					
61	MsNs	MN.S	MN.S–	–MsNs	MN.S			
62	MsNs	MN.S	NSNs					
63	MsNs	MN.S	MSMs	MN.S				
64	MN.S	MsNs	MN.S					
65	MN.S	MsNs	MN.S					
66	MSNs	MSNs	NsNs	MSNs				
67	MSNs	MSNs	MSMS	MSNs	NsNs	MSMS	MsNs	
68	MSNs	MSNs	MSMS	NsNs				
69	MN.S	MN.S	MN.S	NSNs	MsNs	NSNs		
70	MN.S	MN.S	MN.S	MM.S				
71	MN.S	MN.S	MN.S	MN.S				
72	MN.S	MN.S	MN.S	MN.S				
73	NsNs	MsNs	NsNs					
74	MSNs	NsNs	MSNs	NsNs				
75	NsNs	MSNs	MSNs	NsNs				
76	NsNs	MSNs	NsNs	NsNs				
77	MSNs	NsNs	NsNs					
78	NsNs	MSNs	MSNs	NsNs	NsNs			
79	MsNS	NsNs	MsNs	MsNs	NSNs			
80	NsNs	MN.S	NSNs					
81	MN.S	NsNs	NSNs	NSNs				
82	NsNs	MN.S	MSNs	MSNs	MSNs			
83	NSNs	MSNs	MN.S	NsNs				
84	NN.S	MN.S	MN.S	NN.S	MN.S	NN.S		
85	MN.S	NN.S	MN.S	MN.S				
86	MsNs	NSNs	MsNs	NsNs	NSNs			
87	NN.S	MsNs	MsNS					
88	NN.S	MsNs	MsNS					
89	NsNs	NsNs	NsNs	NsNs	NsNs			
90	NsNs	NSNs	NsNs					
91	NSNs	NsNs	NSNs	NsNs	NsNs			
92	NSNs	NsNs	NsNs					
93	NN.S	NsNs	NSNs	NSNs	NSNs			

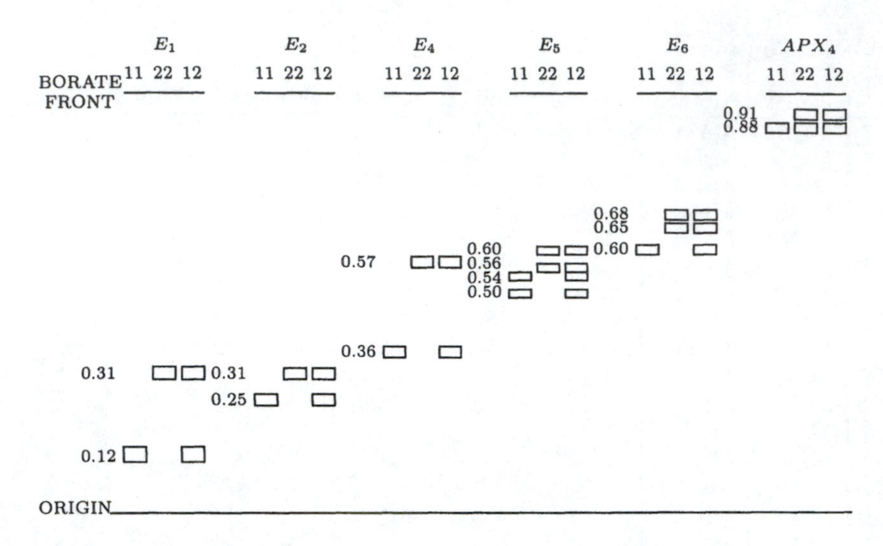

Figure 1.1 Electrophoretic banding pattern for six loci in wild oats. Source: Clegg and Allard (1973).

Migration rates are determined by the size, shape, and charge of the protein. Material from several individuals can be accommodated in distinct lanes on a single gel. A comprehensive discussion of the implications of the finding of large amounts of electrophoretic variation was given by Lewontin (1974). The different allelic forms of a protein detected in this way are called ***allozymes***. Observations could now be made at the level of gene products rather than on the phenotype of an individual. Although only about 30% of proteins are soluble, and although only about 25% of the differences in DNA sequences for the structural genes encoding these proteins lead to charge differences detectable by electrophoresis, the technique remains the easiest and cheapest way of collecting data on many individuals and several loci. Surveys of 10 or so loci in several hundred individuals are often reported.

The electrophoretic banding patterns used by Clegg and Allard (1973) to infer genotypes at six loci in wild oats, *Avena fatua*, are shown in Figure 1.1. All six loci in this study had two alleles, although any particular allele may produce more than one band on the gel. It can be seen from Figure 1.1 that one of the bands for the E_1 locus migrates the same distance as one from the E_2 locus, so that all three genotypes can be scored at either locus only if the other locus is homozygous for the band that does not overlap.

The genetic basis of the banding patterns should be determined before commencing any statistical analysis of electrophoretic data. This is best accomplished by making crosses between individuals with different patterns and checking the segregation ratios in the offspring. Such a procedure was followed by Clegg and Allard, but often it may not be practical. In those cases it is common to argue that banding patterns do indeed correspond to distinct loci by analogy to the situation for the same enzymes in species where genetic tests have been made.

Data from electrophoretic studies are recorded as multilocus genotypes for each of the individuals scored. With horizontal slicing of gels (i.e., slicing along planes parallel to the direction to the lines of movement of the proteins) and application of appropriate stains to each slice, many enzymes can be scored. Genotypes at 11 loci in a sample of the yellow fever mosquito, *Aedes aegypti*, collected in Ghana and made available by J. Powell are shown in Table 1.3. Different alleles at a locus are represented by different numbers, and the pairs of numbers give the genotypes. There is no way to determine from the electrophoretic gel which of a pair of alleles at each locus is paternal and which is maternal. This ambiguity can cause problems when analyses are performed on two or more loci together. Another problem is that not every locus is scored in every individual. Such a lack of balance is a feature of almost all population genetic data sets. Finally, it can be seen in this table that several of the loci have only one allelic type, and so show sample *monomorphism*.

Protein Sequence Data

Electrophoresis works with charge differences among different forms of a protein, but a more complete picture of proteins was provided when it became possible to determine the sequence of amino acids constituting the protein. The variation revealed by this technology allowed more detailed studies of the evolutionary relationships between different species. Figure 1.2 displays a collection of protein sequences for fish protamines (Hunt and Dayhoff 1982). Protamines, or sperm histones, are polypeptides associated with nuclear DNA that replace histones in maturing sperm of vertebrates. The sequences in Figure 1.2 have been aligned to give lowest percentage differences and employ the one-letter code for amino acids in Table 1.4.

Protein sequences are now collected into databases so that they are widely accessible. One such database is the Protein Identification Resource (PIR), accessible from the National Center for Biotechnology Information (World Wide Web URL http://www.ncbi.nlm.nih.gov).

Table 1.3 Multilocus allozyme genotypes in a sample of *Aedes aegypti*.

Indiv.	Sex	*Pgi*	*Pgd*	*Gpd*	*Idh1*	*Idh2*	*Pgm*	*Mdh*	*Hk1*	*Hk2*	*Hk3*	*Hk4*
1	1	11	11	11	12	11	34	12	11	11	11	11
2	1	11	11	11	12	11	22	12	11	11	11	11
3	1	11	11	11	12	11	23	11	11	11	11	11
4	1	11	11	11	11	11	11	11	11	11	11	11
5	1	11	11	11	12	11	34	12	11	11	11	
6	1	11	11	11	11	11	22	11	11	11	11	
7	1	11	11	11	11	11	11	11	11	11	11	
8	1	11	11	11	12	11	24	12	11	11	11	
9	1	11	11	11	11	12	11	11	11	11	11	
10	1	11	11	11	11	11	12	11	11	11	11	11
11	1	11	11	11	12	11	24	11	11	11	11	11
12	1	11	11	11	12	11	24	12	11	11	11	11
13	1	11	11	11	12	11	34	11	11	11	11	11
14	1	11	11	11	11	11	11	11	11	11	11	
15	1	11	11	11	12	11	22	12	11	11	11	
16	1	11	11	11	12	11	34	11	11	11	11	11
17	1	11	11	11	12	11	23	12	11	11	11	11
18	1	11	11	11	12	11	24	11	11	11	11	11
19	1	11	11	11	12	11	22	11	11	11	11	11
20	1	11	11	11	12	11	34	11	11	11	11	11
21	1	11	11	11	11	12	11	11	11	11	11	11
22	1	11	11	11	12	11	23	12	11	11	11	11
23	1	11	11	11	12	11	34	11	11	11	11	11
24	2	11	11	11	11	11	11	11	11	11	11	11
25	2	11	11	11	11	11	34	11	11	11	11	11
26	2	11	11	11	11	11	24	12	11	11	11	11
27	2	11	11	11	11	11	11	11	11	11	11	11
28	2	11	11	11	11	11	34	12	11	11	11	11
29	2	11	11	11	11	11	24	12	11	11	11	11
30	2	11	11	11	11	11	24	11	11	11	11	11
31	2	11	11	11	11	11	23	11	11	11	11	11
32	2	11	11	11	11	11	22	11	11	11	11	11
33	2	11	11	11	11	11	23	11	11	11	11	11
34	2	11	11	11	11	11	11	11	11	11	11	11
35	2	11	11	11	11	11	34	12	11	11	11	11
36	2	11	11	11	11	11	24	11	11	11	11	11
37	2	11	11	11	11	11	34	11	11	11	11	11
38	2	11	11	11	11	11	23	12	11	11	11	11
39	2	11	11	11	11	11	11	11	11	11	11	11
40	2	11	11	11	11	11	23	11	11	11	11	11

Source: J. Powell (personal communication).

Restriction Fragment Data

Information on the DNA sequence itself grew with the use of ***restriction endonucleases***, a class of enzymes that cleaves DNA molecules at specific recog-

```
 1  Tuna Y2       PRRRR--QASRPVRRRRRYRRSTAARRRRRVVRRRR
 2  Tuna Z2       PRRRR--RSSRPVRRRRRYRRSTAARRRRRVVRRRR
 3  Tuna Z1       PRRRR--RSSRPVRRRRRYRRSTAARRRRRVVRRRR
 4  Salmon AII    PRRRRRRSSSRPIRRRR-YRRAS--RRRRRGGRRRR
 5  Trout IB      PRRRRRRSSSRPIRRRR-PRRVS--RRRRRGGRRRR
 6  Salmon AI     PRRRR--SSSRPVRRRRRPR-VSR-RRRRRGGRRRR
 7  Trout IA      PRRRR--SSSRPVRRRRRPRRVSR-RRRRRGGRRRR
 8  Trout II      PRRRR--SSSRPVRRRR-ARRVSR-RRRRRGGRRRR
 9  Herring YII   PRRR-TRRASRPVRRRR-PRRVS--RRRR--ARRRR
10  Herring Z     ARRRRSRRASRPVRRRR-PRRVS--RRRR--ARRRR
11  Herring YI    ARRRRS--SSRPIRRRR-PRRRTT-RRRR-AGRRRR
12  Sturgeon B    ARRRRR--SSRPQRRRRR-RRHG--RRRR--GRR--
13  Sturgeon A    ARRRRRHASTKLKRRRRR-RRHQ--KK----SHK--
```

Figure 1.2 Alignment of fish protamine sequences. Source: Hunt and Dayhoff (1982).

nition sites. Use of such enzymes, coupled with the ability to identify particular regions of the genome with labeled probes, made it possible to score for the presence or absence of several restriction sites in an individual. Unlike observations on proteins or phenotypes, restriction sites can be associated with noncoding as well as coding regions of the genome, although they still meet the criterion of being Mendelizing units. When specific DNA segments are digested with a restriction enzyme, the resulting set of fragment sizes indicates the presence or absence of recognition sites for the enzyme. Varia-

Table 1.4 One-letter and three-letter codes for the 20 amino acids.

	Amino Acid	Codes			Amino Acid	Codes	
1	Alanine	Ala	A	11	Leucine	Leu	L
2	Arginine	Arg	R	12	Lysine	Lys	K
3	Asparagine	Asn	N	13	Methionine	Met	M
4	Aspartic acid	Asp	D	14	Phenylalanine	Phe	F
5	Cysteine	Cys	C	15	Proline	Pro	P
6	Glutamine	Gln	Q	16	Serine	Ser	S
7	Glutamic acid	Glu	E	17	Threonine	Thr	T
8	Glycine	Gly	G	18	Tryptophan	Trp	W
9	Histidine	His	H	19	Tyrosine	Tyr	Y
10	Isoleucine	Ile	I	20	Valine	Val	V

Table 1.5 Some common restriction enzymes and their recognition sequences.

Enzyme	Recognition Sequence
*Bam*HI	GGATCC CCTAGG
*Eco*RI	GAATTC CTTAAG
*Hind*III	AAGCTT TTCGAA
*Taq*I	TCGA AGCT
*Xho*I	CTCGAG GAGCTC

tion in fragment sizes across a sample can also be due to sequence insertions or deletions unrelated to the enzyme. A list of some common restriction enzymes is given in Table 1.5, along with their double-stranded recognition sequences.

Detection of restriction fragment length polymorphisms (RFLP) rests on the ***Southern transfer*** process. The DNA fragments produced by digesting a DNA sample with a restriction enzyme are separated, according to size, by electrophoresis. The resulting collection of fragments form a smear on the gel, from which specific fragments can be detected with a probe. The DNA must first be transferred from the agarose gel onto a medium where this detection can take place. The DNA is ***denatured*** into single-stranded fragments and "blotted" onto a nylon filter, where it is bound in the same positions as on the electrophoresis gel. A probe consisting of a DNA sequence complementary to (part of) the fragment to be detected, and also containing a radioactive label, is applied to the filter. The probe hybridizes to the fragment, and the label is used to expose an X-ray film. The result is a visible band on the film at the same position as that of the original fragment on the gel. The size of the fragment can be inferred from the migration distance on the gel. After the probe is stripped from the filter, another probe can be applied and a second set of fragments detected. This process can be repeated.

Table 1.6 Restriction map variants among *Drosophila melanogaster* lines.

		Restriction Sites				Insertions/Deletions*				
		*Bam*HI	*Hind*III	*Hind*III	*Xho*I	Δa	Δb	Δc	Δd	Δf
Line	*Adh*	−7.1†	−3.0†	+2.7†	+1.2†	20	550	900	180	30
R1	*S*	+	−	−	+	−	−	−	+	−
R2	*S*	+	−	−	+	−	−	−	+	−
M1	*S*	+	−	−	+	−	−	−	−	−
R3	*S*	+	−	−	+	−	−	−	−	−
N1	*S*	+	−	−	−	−	−	−	−	+
N2	*S*	+	−	−	−	−	−	−	−	+
R4	*S*	+	−	−	−	−	−	−	−	−
R5	*S*	+	−	−	−	−	−	−	−	−
R6	*S*	+	−	+	−	−	−	−	−	−
K1	*S*	+	−	+	−	−	−	+	−	−
K2	*S*	+	−	+	+	−	−	−	−	−
K3	*F*	−	−	−	+	−	−	−	−	−
K4	*F*	−	−	−	+	−	−	−	−	−
R8	*F*	−	+	−	+	−	−	−	−	−
M2	*F*	−	+	−	+	−	−	−	−	−
R9	*F*	−	−	−	+	+	−	−	−	−
R10	*F*	−	−	−	+	+	−	−	−	−

Source: Langley et al. (1982).

* a, d, f are deletions; b, c are insertions. Estimated sizes are given in numbers of nucleotides.

† These figures (thousands of nucleotides) indicate the positions of the restriction sites.

Variation at a set of four restriction sites and five sites of insertion or deletion, as well as the two *Adh* allozymes, is shown for a series of lines of the fruit fly *Drosophila melanogaster* in Table 1.6. These data were obtained by Langley et al. (1982) from stocks of *D. melanogaster* made homozygous for independent second chromosomes. A region of about 12,000 nucleotides containing the alcohol dehydrogenase (*Adh*) locus was studied. Apart from the four variable sites, another 20 restriction sites were found to be present in all lines. In Table 1.6, a "+" indicates the presence and a "−" indicates the absence of a restriction site or an insertion or deletion. The two *Adh* allozymes are written as F or S.

The relative positions (about an origin taken at a *Bam*HI site in the *Adh* coding region) of the four restriction sites are given in Table 1.6, illustrating one of the novel features of such data. Not only is information about genetic variation made available at several locations in the genome, but also the physical distance between these locations can generally be determined. Several databases have been established that include information on variable restriction sites, primarily in humans. They include the Genome Data Base (GDB) at the Johns Hopkins University School of Medicine (WWW address http://gdbwww.gdb.org).

Polymerase Chain Reaction

Although the Southern transfer process is widely used, it requires sufficient DNA to produce a detectable band on the film. It is also a lengthy procedure, and requires the use of radioactive chemicals. A process without these disadvantages is the ***polymerase chain reaction*** or PCR (Saiki et al. 1985, Mullis and Faloona 1987). This provides amplification of the DNA between two specific sites in the genome. Two primers are needed, of about 20 bp, complementary to the two sequences defining the boundaries of the region of DNA to be amplified. The DNA is denatured by heat and the primers anneal to the single strands as they are cooled. If *Taq* polymerase has been added, DNA that is complementary to each strand is synthesized from each primer, and the DNA in the region is thereby doubled. Successive cycles of heating and cooling result in repeated doubling of the targeted DNA region, and up to 4×10^6 copies can result from 25 cycles. Each cycle takes about five minutes. PCR can be regarded as a means of *in vitro* cloning, and it does not need purified DNA as a starting point. It is useful for studying very old DNA that may be partially degraded (e.g., Handt et al. 1994).

Length variants in the amplified DNA can be separated according to size on an agarose gel, stained with ethidium bromide, and visualized by illumination with ultraviolet light. Alternatively, fragments generated by PCR can be blotted onto a membrane and probed with a radioactive label, as in the detection of RFLPs. Sequence variants can be identified either by direct DNA sequencing or by the use of allele-specific oligonucleotide (ASO) probes. A survey of the techniques and applications was given by Erlich (1989).

RAPD Loci

A PCR method that has proved useful in generating genetic markers is known as RAPD, for ***random amplified polymorphic DNA*** (Welsh and McClelland 1990, Williams et al. 1990). A single primer is used, but is not chosen to flank a particular region. If the genome contains a sequence complementary to this primer, and also contains the same sequence in inverse orientation within a few kb, the intervening region is amplified by PCR. This region is random in the sense that it is not known ahead of time. Only those individuals with both sequences will result in an amplification product, so RAPD markers are essentially two-allele loci with presence of the two recognition sequences dominant to absence.

VNTR and STR Loci

RFLP and RAPD variants are characterized by the presence or absence of specific DNA sequences that serve as recognition sites either for restriction enzymes or for PCR primers. Another class of variants is due to length variation in a region brought about by varying numbers of copies of short sequences. For ***minisatellites***, or ***variable number of tandem repeats*** (VNTR), the repeat unit is of the order of 10 bp and the number of units can be in the thousands. ***Microsatellites***, or ***short tandem repeats*** (STR), or simple sequence repeats (SSR) have repeat units of only 2 or 3 bp. Provided the length variants can be distinguished, by electrophoresis for example, there can be hundreds or thousands of different genotypes distinguished at a locus. These markers can therefore be used for identification as well as for associations with traits of interest.

DNA Sequence Data

The ultimate step in obtaining genetic data is the determination of the sequence of bases in the DNA molecule. There are four kinds of base, `A`,`T`,`C`, and `G` as shown in Table 1.7 (`U` replaces `T` in RNA molecules), and DNA sequences can be found for coding as well as noncoding regions. Such data have been collected with different objectives in mind. Constructions of phylogenies for a set of species have used a single sequence from each of the species, whereas questions concerning variation within species require several sequences from a species.

Mitochondrial DNA sequences for the *tRNA his* gene are shown in Figure 1.3. These sequences, along with other mitochondrial sequences (Brown et al. 1982), have been used to construct evolutionary histories of man and apes. The complete 16,569 bp DNA sequence for a human mitochondrion was published by Anderson et al. (1981).

Information on the sequences of the *Adh* region in *D. melanogaster* collected from isochromosomal lines derived from 11 populations by Kreitman (1983) is shown in Figure 1.4. Among the 2,721 nucleotides sequenced in the 11 lines, 43 were found to be variable, and it is the bases at these polymorphic sites that are displayed in the figure. The reference sequence shown at the

Table 1.7 The four nitrogenous bases and their types. The same two purines, `A` and `G`, are found in DNA and RNA. The two pyrimidines in DNA are `C` and `T`; in RNA `U` replaces `T`.

Type	Base
Purine	**A**denine
	Guanine
Pyrimidine	**C**ytosine
	Thymine
	Uracil

head of the table consists of the sequence of most common bases among the six lines that had the S allozyme for *Adh*.

As is the case with restriction site data, DNA sequence data contain information about the relative physical locations of the observational units in the genome. Many of the properties of the proteins encoded by DNA sequences depend on the two- and three-dimensional structure of the molecules, and not just on the one-dimensional amino acid sequence. Advances in inferring such structures from nucleotide sequences (e.g., Zuker 1989) are beyond the scope of this book.

DNA sequences are collected into databases in the same way as are protein sequences. The principal databases are GenBank and Genome Sequence Data Base (GSDB) in the United States, the European Molecular Biology Laboratory (EMBL) Nucleotide Sequence Database, and the DNA Data Bank of Japan (DDBJ). These four databases now store 200 million bp, representing human and more than 8,000 other species. The first genome to be sequenced was of the bacteriophage ϕX174, with 5386 bp (Sanger et al. 1977), and the first complete chromosome was number III for yeast *Saccha-*

```
1   GTAAATATAG TTTAACCAAA ACATCAGATT GTGAATCTGA CAACAGAGGC TTACGACCCC TTATTTACC
2   GTAAATATAG TTTAACCAAA ACATCAGATT GTGAATCTGA CAACAGAGGC TCACGACCCC TTATTTACC
3   GTAAATATAG TTTAACCAAA ACATCAGATT GTGAATCTGA TAACAGAGGC TCACAACCCC TTATTTACC
4   GTAAATATAG TTTAACCAAA ACATTAGATT GTGAATCTAA TAATAGGGCC CCACAACCCC TTATTTACC
5   GTAAACATAG TTTAATCAAA ACATTAGATT GTGAATCTAA CAATAGAGGC TCGAAACCTC TTGCTTACC
```

Figure 1.3 Five mitochondrial DNA sequences. 1: human; 2: chimpanzee; 3: gorilla; 4: orangutan; 5: gibbon. Source: Brown et al. (1982).

Reference:	CCGCAATATGGGCGCTACCCCCGGAATCTCCACTAGACAGCCT
Line	
Wa-S	........AT........TT.ACA.TAAC..............
Fl-1S	..C...............TT.ACA.TAAC..............
Af-S	...................................A....T.A
Fr-S	................GT.................A..TA...
Fl-2S	...AG...A.TC..AGGT..................C......
Ja-S	..C............G..............T.T.CAC....T.
Fl-F	..C............G.............GTCTCC.C......
Fr-F	TGCAG...A.TCG..G.............GTCTCC.CG.....
Wa-F	TGCAG...A.TCG..G.............GTCTCC.CG.....
Af-F	TGCAG...A.TCG..G.............GTCTCC.CG.....
Ja-F	TGCAGGGGA....T.G....A...G....GTCTCC.C......

Figure 1.4 Variable nucleotides among 11 *Drosophila Adh* sequences. Source: Kreitman (1983).

romyces cerevisiae, 315 kb (Oliver et al. 1992). The complete genomic sequence for the bacterium *Haemophilus influenzae* (Fleischmann et al. 1995) has been deposited in GSDB.

GENETIC AND STATISTICAL SAMPLING

One of the themes that will recur in this book is that the analysis of genetic data must be based on some theory, or model. The various types of sampling used to generate the data must be kept in mind, for these will affect the properties of estimates and will determine the scope of the inferences that may be drawn from the data. Frequent mention will be made of the distinction between statistical and genetic sampling, and the distinction can be seen most easily from the idealized situation in Figure 1.5.

The data to be analyzed will be for a sample of n individuals taken from a population of size N. Statistical sampling refers to this choice of n individuals from N. Another sample, also of size n, from the same population may give different values of quantities such as gene frequency simply because individuals with different genotypes may be sampled. Most analyses will be based on multinomial sampling, in which every member of the population is equally likely to be chosen at any stage to be a member of the sample. Even though repeated samples are not generally taken, there are statistical theories that predict how much variation would be likely between such samples and this variation is taken into account in making inferences about the

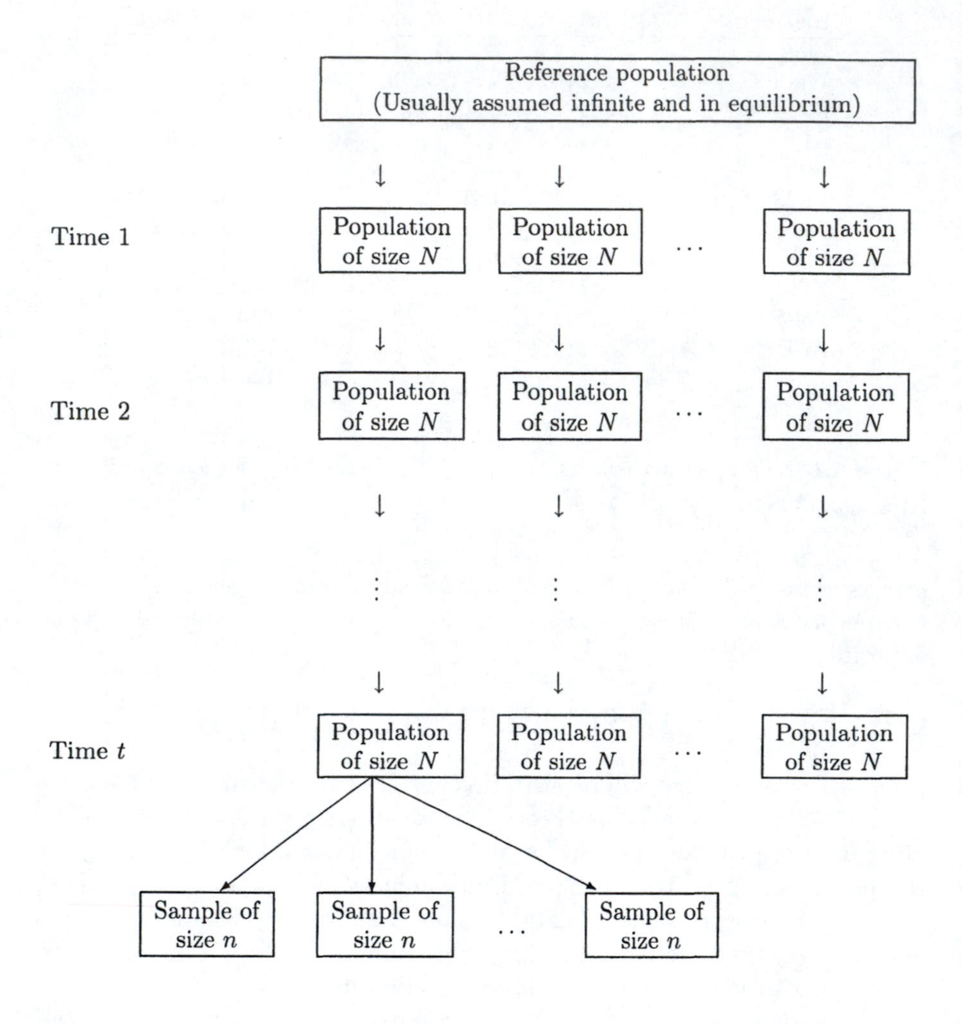

Figure 1.5 Representation of the two processes of genetic and statistical sampling. Genetic sampling causes the populations to differ, whereas statistical sampling causes the samples from one population to differ.

population on the basis of sample data.

For genetic data there is another source of variation to be considered, and this arises from the sampling inherent in the transmission of genetic material from parent to offspring. The population of size N from which a sample is drawn can be regarded as just one of the many replicate populations that could have descended from the same reference, or founder, population. Even if all factors such as population size, mating structure, selection, and mutation forces were kept exactly the same, these replicate populations would differ because different genes may be transmitted at each locus between generations for the different replicates. There is variation between replicate populations because of genetic sampling, and theory is needed to accommodate this variation. A vivid experimental demonstration of the effects of drift in replicate populations of *Drosophila melanogaster* was given by Buri (1956) and discussed in the text by Hartl and Clark (1989). Buri set up 107 replicate populations, each containing 16 heterozygous individuals for an eye color locus. Each subsequent generation was formed from eight male and eight female offspring selected randomly from the previous generation. As time passed, more and more of the replicates had lost one of the original two alleles.

Just as variation between samples from the same population cannot be measured unless there is more than one sample, so variation between replicate populations cannot be measured from data within one such population. This could pose a problem since the replicate populations are often conceptual only. Some idea of the variation can be gained from different loci, and to the extent that loci can be regarded as being independent, their differences in pedigree mimic the differences between populations.

Whether attention is paid to both statistical and genetic sampling depends on the goals of a study. If there is interest in just the present population, it is possible to ignore the genetic sampling that has made this population different from all others that descend from the founder population. If, on the other hand, conclusions are to apply to all populations that may have arisen under similar circumstances, genetic sampling must be incorporated. Similar arguments are made in distinguishing between fixed and random effects in statistical models (e.g., Steel and Torrie 1980, p. 149).

Another situation arises in predicting properties of some future population. The sample size necessary to achieve a certain level of sampling variance in genetic parameter estimates may need to be determined, but the composition of the population, let alone the sample, at that future time is unknown. Both genetic and statistical sampling variation must be anticipated in that situation.

NOTATION AND TERMINOLOGY

The distinction between population size N and sample size n has been made already. It is convenient to speak loosely of a genetic locus, A, either with two alleles A, a, or with a series of alleles A_u. The symbol A will be used whether structural loci, microsatellite markers, restriction sites, or maybe even individual nucleotides, are meant.

An allele A has a population frequency p_A, which will generally mean the average frequency over all replicate populations maintained under the same conditions. If there are no disturbing forces this frequency does not change over time, and it is the frequency of that allele in the reference population. The frequency in one particular replicate does change over time because of genetic sampling, and genetic drift acting alone would cause each replicate finite population eventually to become fixed for one allele.

Frequencies of combinations of alleles at two or more loci will also be written as lowercase p's, as in p_{AB} for the combination of alleles A and B, but uppercase P's will be used for genotypic frequencies. At a single locus, a notation such as P_{AA} is sufficient, but for two loci it may be necessary to indicate which pairs of genes were received from the same parent. Individuals receiving the pair AB from one parent and ab from the other have a frequency written as P_{ab}^{AB}, whereas the other double heterozygotes have frequency P_{aB}^{Ab}. If there is no need to keep track of gametic associations, all alleles will be subscripted, as in P_{AaBb}.

In a sample, counts of genes or genotypes will be written as lowercase n's with appropriate subscripts, so that for a locus A with alleles A and a the sample size is the sum of counts in the three genotypic classes and the allelic counts follow from the genotypic counts:

$$\begin{aligned} n &= n_{AA} + n_{Aa} + n_{aa} \\ n_A &= 2n_{AA} + n_{Aa} \end{aligned}$$

Sample frequencies will be indicated by tildes, and follow from counts as

$$\begin{aligned} \tilde{p}_A &= \frac{1}{2n} n_A \\ \tilde{P}_{AA} &= \frac{1}{n} n_{AA} \end{aligned}$$

whereas estimates will be indicated by carets. Generally, the estimates of gene and genotypic frequencies are just the observed values

$$\hat{p}_A = \tilde{p}_A$$

but this will not be the case for estimates of squares and products of frequencies.

Population frequencies are the basic genetic quantities of interest, and care will be taken to distinguish between such unobservable ***parameters*** and the ***statistics*** used as their estimates. Much effort will be directed to finding estimates of high quality and it is necessary to know their sampling properties. There is a particular interest in the means and variances of estimators, and $\mathcal{E}$ will be used to indicate the mean or average value of a statistic over all possible samples. In statistical language, a mean of a variable is the ***expected value*** of the variable. Unless otherwise stated, these expectations will be taken over all samples from a population, and over all populations. It will be shown that observed gene frequencies are ***unbiased estimates*** of population gene frequencies because

$$\mathcal{E}(\tilde{p}_A) = p_A$$

MENDEL AND FISHER

The work of Mendel forms the basis of genetics, and the analysis of his data provides one of the classic examples in genetic data analysis. Mendel presented data on seven characters in his 1866 paper. Although there is no question about the fundamental model of discrete genes controlling these characters, there has been criticism of the data. It appeared to Fisher (1936a) that the data were so close to the values that Mendel expected under his theory that there must have been some manipulation, or omission, of data. Since Fisher's paper there has been an extensive literature of support for either Fisher or Mendel, but until recently little of substance has been added to the debate. Careful reviews were given by Piegorsch (1983, 1986), and the following discussion is based on Fisher (1936a) and Edwards (1986). The discussion is included here because of its historic interest, and because it introduces the chi-square distribution.

Mendel crossed inbred pea lines that differed in easily scored characters, each of which was controlled by a single locus with two alleles, one dominant to the other. He introduced the notation A for the dominant and a for the recessive alleles. The crossed, or F_1, populations were heterozygous Aa and showed only the dominant character. When such heterozygotes are allowed to self, the offspring are expected to be AA, Aa, and aa in the ratio 1:2:1, or to show dominants $AA + Aa$ and recessives aa in the ratio 3:1. The data in Table 1.1 certainly are consistent with this 3:1 ratio. The characters are labeled A–G in Table 1.1 to maintain consistency with Edwards' paper. The figures in Table 1.1 for the plant characters C–G are repeated in rows 23–27 of Table 1.8.

To illustrate the variability between plants, Mendel reported the counts for the two seed characters in the F_2 generation observed on each of 10 F_1 plants, and these counts are displayed in Table 1.9. For Edwards' analysis, the totals from Table 1.9 are subtracted from the entries in Table 1.1 to avoid using the same data twice. This gives 5138:1749 for seed shape and 5667:1878 for cotyledon color (rows 1 and 2 in Table 1.8). The figures in Table 1.9 are repeated in rows 3 through 22 of Table 1.8.

Individuals in the F_2 generation were selfed to reveal the presence of two genotypes in the dominant class. (The recessive class always gave recessive offspring on selfing.) Under Mendel's model, one-third of the dominant F_2's are homozygous and so should give only dominant F_3 offspring, whereas two-thirds are heterozygous and so should give dominant and recessive F_3's in the ratio 3:1. The expected ratio of heterozygotes to homozygotes among the dominant F_2's is therefore 2:1 and, once again, Mendel's data were consistent with this ratio.

A problem with this second set of experiments is that a heterozygous plant could, by chance, give a series of offspring that did not include any recessives and so could, incorrectly, be included among the dominant homozygous class. For the five plant characters, Mendel cultivated 10 seeds from each of 100 dominant F_2 plants. If all 10 offspring from a single plant were dominant, he concluded that the F_2 parent belonged to the homozygous class. Otherwise it was classified as a heterozygote. With each heterozygous F_2 plant having a 0.75 chance of giving a dominant selfed offspring, there is a $(0.75)^{10} = 0.0563$ chance of 10 offspring all being dominant and the F_2 being misclassified as homozygous. Instead of the probability being 2/3 that an F_2 is heterozygous, the correct probability is 2/3 times the chance of correct classification – i.e., 0.6291. The true expected ratio of homozygotes and heterozygotes among the dominant F_2's is therefore 0.3709:0.6291. Fisher pointed out that Mendel's data were much closer to the 1:2 ratio than to this corrected value, giving rise to the suggestion that some data manipulation had taken place, and that Mendel was demonstrating rather than establishing his theory. The problem does not arise with the two seed characters since they could be scored directly on the F_2 plants, and many F_3 seeds could be observed per plant.

Table 1.8 Normal and chi-square statistics for Mendel's segregations.

	Character	Segregation Expected	Observed		Total	X	X^2
			F_2, seed characters				
1	*A*	3:1	5138	1749	6887	−0.7583	0.58
2	*B*	3:1	5667	1878	7545	+0.2193	0.05
			F_2, plant variability				
3	*A*	3:1	45	12	57	+0.6882	0.47
4	*A*	3:1	27	8	35	+0.2928	0.09
5	*A*	3:1	24	7	31	+0.3111	0.10
6	*A*	3:1	19	10	29	−1.1793	1.39
7	*A*	3:1	32	11	43	−0.0880	0.01
8	*A*	3:1	26	6	32	+0.8165	0.67
9	*A*	3:1	88	24	112	+0.8729	0.76
10	*A*	3:1	22	10	32	−0.8165	0.67
11	*A*	3:1	28	6	34	+0.9901	0.98
12	*A*	3:1	25	7	32	+0.4082	0.17
13	*B*	3:1	25	11	36	−0.7698	0.59
14	*B*	3:1	32	7	39	+1.0170	1.03
15	*B*	3:1	14	5	19	−0.1325	0.02
16	*B*	3:1	70	27	97	−0.6448	0.42
17	*B*	3:1	24	13	37	−1.4237	2.03
18	*B*	3:1	20	6	26	+0.2265	0.05
19	*B*	3:1	32	13	45	−0.6025	0.36
20	*B*	3:1	44	9	53	+1.3482	1.82
21	*B*	3:1	50	14	64	+0.5774	0.33
22	*B*	3:1	44	18	62	−0.7332	0.54
			F_2, plant characters				
23	*C*	3:1	705	224	929	+0.6251	0.39
24	*D*	3:1	882	299	1181	−0.2520	0.06
25	*E*	3:1	428	152	580	−0.6712	0.45
26	*F*	3:1	651	207	858	+0.5913	0.35
27	*G*	3:1	787	277	1064	−0.7788	0.61
			F_3, seed characters				
28	*A*	2:1	372	193	565	−0.4165	0.17
29	*B*	2:1	353	166	519	+0.6518	0.42

(Continued)

Table 1.8 Continued

	Character	Segregation Expected	Observed		Total	X	X^2
		F_3, plant characters					
30	C	0.63:0.37	64	36	100	+0.2252	0.05
31	D	0.63:0.37	71	29	100	+1.6743	2.80
32	E	0.63:0.37	60	40	100	−0.6029	0.36
33	F	0.63:0.37	67	33	100	+0.8462	0.72
34	G	0.63:0.37	72	28	100	+1.8813	3.54
35	E	0.63:0.37	65	35	100	+0.4322	0.19
		F_2, bifactorial experiment					
36	A	3:1	423	133	556	+0.5876	0.35
37	B among 'A'	3:1	315	108	423	−0.2526	0.06
38	B among 'a'	3:1	101	32	133	+0.2503	0.06
		F_3, bifactorial experiment					
39	A among 'AB'	2:1	198	103	301	−0.3261	0.11
40	A among 'Ab'	2:1	67	35	102	−0.2100	0.04
41	B among 'aB'	2:1	68	28	96	+0.8660	0.75
42	B among Aa'B'	2:1	138	60	198	+0.9045	0.82
43	B among AA'B'	2:1	65	38	103	−0.7664	0.59
		F_2, trifactorial, seed characters					
44	A	3:1	480	159	639	+0.0685	0.00
45	B among 'A'	3:1	367	113	480	+0.7379	0.54
46	B among 'a'	3:1	122	37	159	+0.5037	0.25
		F_3, trifactorial, seed characters					
47	A among 'AB'	2:1	245	122	367	+0.0369	0.00
48	A among 'Ab'	2:1	76	37	113	+0.1330	0.02
49	B among 'aB'	2:1	79	43	122	−0.4481	0.20
50	B among Aa'B'	2:1	175	70	245	+1.5811	2.50
51	B among AA'B'	2:1	78	44	122	−0.6402	0.41

(Continued)

Table 1.8 Continued

	Character	Segregation Expected	Segregation Observed		Total	X	X^2
		F_2, trifactorial, plant character					
52	C among *AaBb*	3:1	127	48	175	−0.7419	0.55
53	C among *AaBB*	3:1	52	18	70	−0.1380	0.02
54	C among *AABb*	3:1	60	18	78	+0.3922	0.15
55	C among *AABB*	3:1	30	14	44	−1.0445	1.09
56	C among *Aabb*	3:1	60	16	76	+0.7947	0.63
57	C among *AAbb*	3:1	26	11	37	−0.6644	0.44
58	C among *aaBb*	3:1	55	24	79	−1.1043	1.22
59	C among *aaBB*	3:1	33	10	43	+0.2641	0.07
60	C among *aabb*	3:1	30	7	37	+0.8542	0.73
		F_3, trifactorial, plant character					
61	C among *AaBb*	0.63:0.37	78	49	127	−0.3488	0.12
62	C among *AaBB*	0.63:0.37	38	14	52	+1.5174	2.30
63	C among *AABb*	0.63:0.37	45	15	60	+1.9384	3.76
64	C among *AABB*	0.63:0.37	22	8	30	+1.1816	1.40
65	C among *Aabb*	0.63:0.37	40	20	60	+0.6020	0.36
66	C among *AAbb*	0.63:0.37	17	9	26	+0.2610	0.07
67	C among *aaBb*	0.63:0.37	36	19	55	+0.3903	0.15
68	C among *aaBB*	0.63:0.37	25	8	33	+1.5276	2.33
69	C among *aabb*	0.63:0.37	20	10	30	+0.4257	0.18
		Gametic ratios, first experiment, seed characters					
70	A	1:1	43	47	90	−0.4216	0.18
71	B among *AA*	1:1	20	23	43	−0.4575	0.21
72	B among *Aa*	1:1	25	22	47	+0.4376	0.19
		Gametic ratios, second experiment, seed characters					
73	A	1:1	57	53	110	+0.3814	0.15
74	B among *Aa*	1:1	31	26	57	+0.6623	0.44
75	B among *aa*	1:1	27	26	53	+0.1374	0.02
		Gametic ratios, third experiment, seed characters					
76	A	1:1	44	43	87	+0.1072	0.01
77	B among *AA*	1:1	25	19	44	+0.9045	0.82
78	B among *Aa*	1:1	22	21	43	+0.1525	0.02

(Continued)

Table 1.8 Continued

	Character	Segregation Expected	Observed		Total	X	X^2
		Gametic ratios, fourth experiment, seed characters					
79	A	1:1	49	49	98	0.0000	0.00
80	B among Aa	1:1	24	25	49	−0.1429	0.02
81	B among aa	1:1	22	27	49	−0.7143	0.51
		Gametic ratios, plant characters					
82	G	1:1	87	79	166	+0.6209	0.39
83	C among Gg	1:1	47	40	87	+0.7505	0.56
84	C among gg	1:1	38	41	79	−0.3375	0.11

Source: Edwards (1986).

Mendel's data for the plant characters are shown in rows 30–35 of Table 1.8, and for the two seed characters in rows 28 and 29.

With the single-gene inheritance of dominant characters established, Mendel turned to demonstrating the independent inheritance of two or three characters. In the bifactorial experiments the two seed characters were used. Letting A, a denote the alleles for round and wrinkled seeds and B, b the alleles for yellow and green cotyledons, the cross $AABB \times aabb$ gave all $AaBb$ round yellow seeds, which, when selfed, gave all four possible classes of offspring. Edwards divided the data into three comparisons for the 3:1 expected ratios for each character separately: round versus wrinkled, yellow versus green among round seeds, and yellow versus green among wrinkled seeds (see Table 1.8, rows 36, 37, and 38). Within each dominant class, the expected ratio of 2:1 for heterozygote versus homozygote was looked for by planting the F_2 seeds, and the resulting F_3 data are shown in Table 1.10. When the wrinkled yellow seeds were planted, the F_3's were all wrinkled but were expected to segregate 2:1 for yellow and green versus yellow only (row 41 of Table 1.8). For the offspring of round green seeds there is constant color but 2:1 expected segregation for round and wrinkled versus round only seed shape (row 40). The double dominant class gave all four classes on planting, and Edwards formulated three comparisons with 2:1 expected ratios: round and wrinkled versus round only (row 39), yellow and green versus green only (row 42), and yellow and green versus yellow only among round seeds (row

Table 1.9 Mendel's results for two characters on 10 individual F_1 plants.

	Seed Form		Cotyledon Color	
Plant	Round	Wrinkled	Yellow	Green
1	45	12	25	11
2	27	8	32	7
3	24	7	14	5
4	19	10	70	27
5	32	11	24	13
6	26	6	20	6
7	88	24	32	13
8	22	10	44	9
9	28	6	50	14
10	25	7	44	18

Source: Mendel (1866).

43). All 30 wrinkled green F_2's gave only this kind of offspring.

Seed coat color, with alleles C, c, was added to seed shape and cotyledon color to give a trifactorial experiment. The triple homozygotes *AABBCC* and *aabbcc* were crossed and selfed, leading to the counts set out in Table 1.11 in the format given by Fisher. Edwards detailed the 26 comparisons possible in this data set, as shown in Table 1.8 (rows 44–69). In that table, '*A*' denotes *AA* + *Aa*, and '*a*' denotes *aa*. Taken together, the bifactorial and trifactorial data lend support to independent assortment of characters.

The final set of experiments reported in 1866 by Mendel concerned ga-

Table 1.10 Mendel's F_3 bifactorial data for seed shape and cotyledon color.

		Seed Shape			
		AA	*Aa*	*aa*	Total
Cotyledon Color	*BB*	38	60	28	126
	Bb	65	138	68	271
	bb	35	67	30	132
	Total	138	265	126	529

Table 1.11 Mendel's F_2 trifactorial data for seed shape (A), cotyledon color (B), and seed coat color (C).

	AA	*Aa*	*aa*	Total
		CC		
BB	8	14	8	30
Bb	15	49	19	83
bb	9	20	10	39
		Cc		
BB	22	38	25	85
Bb	45	78	36	159
bb	17	40	20	77
		cc		
BB	14	18	10	42
Bb	18	48	24	90
bb	11	16	7	34
		Total		
BB	44	70	43	157
Bb	78	175	79	332
bb	37	76	37	150
Total	159	321	159	639

metic ratios, designed to demonstrate that egg and pollen cells of F_1 hybrids had alleles *AB*, *Ab*, *aB*, or *ab* at loci *A* and *B*. Five experiments were conducted by crossing two double homozygotes, *AABB* and *aabb*, and then fertilizing the hybrids with the pollen of either double homozygote or fertilizing either double homozygote with pollen from the hybrids. In the first of these four arrangements, *AB* pollen applied to *AaBb* plants is expected to produce equal numbers of *AABB*, *AaBB*, *AABb*, *AaBb* F_2's, all with the dominant appearance for both characters. Selfing this collection, to distinguish the homozygous and heterozygous genotypes within dominant classes as in previous experiments, should reveal this equality of four classes. In Edwards' scheme, the appropriate comparisons are *AA* + *Aa* versus *aa*, *BB* + *Bb* versus *bb* among *AA* + *Aa* plants and *BB* + *Bb* versus *bb* among *aa* plants, and the set of 15 such comparisons is also shown in Table 1.8 (rows 70–84).

The X values shown in Table 1.8 are the square roots of the ***chi-square*** goodness-of-fit statistics. If a category has an observed count of o, but a count e is expected under some hypothesis, then the goodness-of-fit test statistic is formed by summing the quantity $(o-e)^2/e$ over categories:

$$X^2 = \sum_{\text{categories}} \frac{(o-e)^2}{e} \tag{1.1}$$

If the hypothesis is true, it is not likely that the o and e values in any category will be exactly equal, but they should not be too far apart. How far apart they may be and still have the data support the hypothesis is found from properties of the chi-square distribution. In particular, for two categories, there is only a 5% chance that X^2 will exceed 3.84 if the hypothesis is true (see the column headed 0.05 for one degree of freedom in Appendix Table A.2). In the first row of Table 1.8, for example, the chi-square is calculated by comparing the observed values 5138 and 1749 with the values 5165.25 and 1721.75 expected under the 3:1 ratio

$$\begin{aligned} X^2 &= \frac{(5138-5165.25)^2}{5165.25} + \frac{(1749-1721.75)^2}{1721.75} \\ &= 0.5750 \\ X &= -0.7583 \end{aligned}$$

Edwards assigned a negative sign to X when the observed value is less than the expected value in the class with the larger expectation. For cases in which two classes have the same expectation, this rule is applied to that class with the larger number of dominant genes. When the expected ratio is true, the quantity X^2 calculated on two categories has a chi-square distribution with 1 degree of freedom (d.f.)

$$X^2 \sim \chi^2_{(1)}$$

whereas X has a ***standard normal*** distribution

$$X \sim N(0,1)$$

Large values of X^2 cause the hypothesis of the expected ratio to be rejected. In Table 1.8, no X^2 value is greater than 3.84 (which is the same as saying that no X has an absolute value exceeding 1.96). This is surprising since about four such values would be expected just by chance even if all the hypotheses being tested were true.

Some of the criticisms of Mendel's data focus on chi-square values that are unusually small as well as unusually large. The same 5% significance level can be achieved if the hypothesis is rejected for X^2 exceeding 5.02 or being less than 0.000982 (see the columns headed 0.975 and 0.025 for 1 d.f. in Appendix Table A.2). Now the 2.5% closest and the 2.5% furthest away values from the expected ratio cause rejection. If the 84 independent X^2 values in Table 1.8 are added, the resulting sum of 41.15 is unusually low for a chi-square with 84 d.f. Edwards said that such a chi-square may look like a "single lucky shot" but that, in fact, it is a succession of 84 shots showing some kind of pattern requiring investigation.

When the hypothesized ratio is true, and the quantity X has a standard normal distribution, large positive or negative values contribute to large chi-squares, whereas X values close to zero contribute to small chi-squares. Among the 15 values (rows 30–35 and 61–69) that have an expected ratio of 0.6291:0.3709, there are 13 that show a positive X value and only 2 that are negative, suggesting a bias in favor of the dominant class. Fisher's criticism of Mendel for using an expected ratio of 2:1 instead of this corrected value seems borne out by the fact that the data are 57.90 times more likely to have arisen under the incorrect 2:1 ratio than under the correct one. This calculation follows from the likelihood theory taken up in Chapter 2.

Edwards looked at the 69 rows in Table 1.8 that do not have questions over expected ratios. The X values are fairly evenly distributed about zero, with 38 positive, 30 negative, and 1 zero, but there is something unusual about the range of X values. In a sample of 69, the largest three values of the absolute values of X are expected to be 2.6216, 2.2755, and 2.0828 (Edwards 1986), whereas the largest values reported by Mendel are 1.5811, 1.4237, and 1.3482. It appears that Mendel's X values are biased in the sense that there are too few extreme values, positive or negative. It is as though someone had altered figures that appeared to be too extreme.

The 69 X values are plotted in the histogram shown in Figure 1.6, along with the histogram expected for a sample of 69 values from a normal distribution. These expected values can be found from the table of values for the cumulative standard normal distribution (Appendix Table A.1). That table shows the probability of a standard normal variable falling between 0.00 and 0.20 is $0.5000 - 0.4207 = 0.0793$, so that $69 \times 0.0793 = 5.5$ values are expected in a sample of size 69 but 9 were found.

Edwards inclined toward the explanation that some plants were misclassified (perhaps unconsciously) in the direction of conforming to the hypothesized ratios, rather than that some data were discarded. Piegorsch (1986) reviewed other published discussion of Mendel's data, concerning his apparently fortunate choice of seven unlinked characters. Discussion on Mendel's data continues (Novitski 1995).

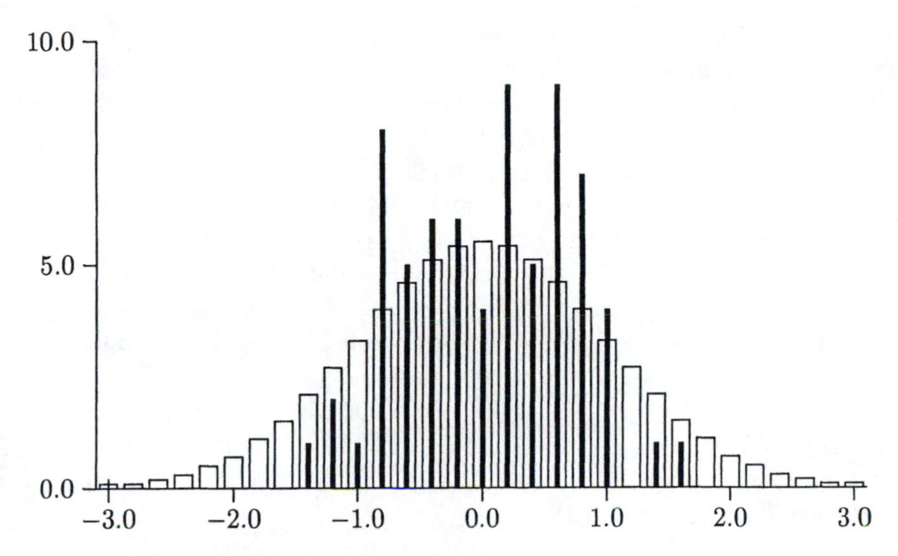

Figure 1.6 Histogram of X values for Mendel's data. Open bars are for a normal distribution, and solid bars are for Mendel's data. Source: Edwards (1986).

SUMMARY

Discrete genetic data may take many forms, ranging from individual nucleotides in a DNA sequence to a qualitative character such as flower color, but underlying all these data is a common genetic model. The data are considered to depend on discrete heritable units that are transmitted from parent to offspring with a random choice of which parent's contribution is passed from offspring to grand-offspring. The resulting genetic variation between populations must be added to the statistical sampling within populations arising from the random sampling of individuals for observation.

Genetic parameters are unobservable and need to be estimated by statistics that are functions of the data. Estimates of parameters are indicated in notation by carets, and observed values of population frequencies are indicated by tildes.

FURTHER READING

A review of recent papers that use statistical analysis for DNA sequences was given by Weir (1993). Of particular interest are those dealing with "African Eve" (Vigilant et al. 1991, Templeton 1992, Hedges et al. 1992) and the Florida dentist (Smith and Waterman 1992). Current thinking on the interpretation of genetic data for identification is contained in the chapters of *Human Identification: The Use of DNA Markers* edited by Weir (1995). A collection of papers on sequence analysis was edited by Doolittle (1990), and a text, *Introduction to Computational Biology*, was written by Waterman (1995). Methods for locating genes by linkage to markers are in the second edition of *Analysis of Human Genetic Linkage* (Ott 1991), and a review of using population associations for disease gene mapping was given by Jorde (1995).

Statisticians who find genetics to be of interest should consult books such *Genes V* (Lewin 1994) for the biological background. Geneticists who wish to enrich their statistical background could consult *Biometry* (Sokal and Rohlf 1995) or *Kendall's Advanced Theory of Statistics* (Stuart and Ord 1987).

EXERCISES

Exercise 1.1

Find a published paper that presents discrete genetic data, and answer the following questions for that paper:

a. Describe the data. Are they clearly genetic?
b. What genetic questions are being addressed with the data?
c. What statistical techniques are being used?
d. Has attention been paid to genetic and statistical sampling?

Exercise 1.2

Draw a sample of 69 values from the standard normal distribution using the method given in Appendix B. Find the upper and lower quartiles and the median of this sample, i.e., find those three values that divide the sample into four (nearly) equal portions. Compare these simulated values with the those for the 69 X values in rows 1–29, 36–60 and 70–84 of Table 1.8.

Exercise 1.3

Perform a chi-square goodness-of-fit test of the hypothesis that round and wrinkled pea seeds occur in the ratio 3:1 for each of the seed-form data sets in Table 1.9. Use a 10% significance level, i.e., reject the hypothesis if the test statistic exceeds 2.71.

Chapter 2

Estimating Frequencies

ESTIMATION

Much of statistics is concerned with estimating quantities of interest. In Chapter 1, the distinction was made between the true but unknown values, or parameters, and the data functions, or statistics, used in their place. Observed allele frequencies are used as estimates of the true allele frequencies. Other genetic parameters, such as inbreeding coefficients, do not have such simple estimates and what is needed is some framework within which to obtain estimates. It is also necessary to establish some criteria with which to judge the quality of estimates. Are the estimates likely to be close to the true values? Do they get closer as the amount of data increases? How do the estimates vary among different samples? Can the same sample be used to provide both an estimate and an indication of the properties of the estimate? Such questions have been answered many times in the statistics literature, and good discussions are contained in Chapter 17 of Kendall and Stuart (1979) and Chapters 4 and 7 of Sokal and Rohlf (1995).

The rule by which a statistic is constructed to provide an estimate is called an ***estimator***. The sample mean is an estimator for the population mean – a specific numerical value of the sample mean is an estimate. Estimators are said to be ***consistent*** if they are increasingly accurate as the sample size increases. An ***unbiased*** estimator has an expected value equal to the parameter for any sample size. Recall that expected value means the average over very many replicate samples. Consistent estimators will always be unbiased for very large sample sizes, and may be unbiased for all sample sizes.

A genetic parameter may have several consistent, or even unbiased, estimators. Other criteria are needed to distinguish among them, and the first is the extent to which they vary between samples. Smaller variation is more desirable, and the amount of variation is quantified with the ***variance*** of the estimator. This is the expected value of the squared deviation of estimates from their mean. An unbiased consistent estimator with a smaller variance will deviate less, on the average, from the true value than one with a larger variance. Later in the chapter, a lower bound on the variance for an unbiased estimator will be given. There is a value below which the variance of an unbiased estimator cannot fall, and this minimum variance bound follows from the likelihood function. Estimators with minimum variance are said to be ***efficient***.

The last general concept for evaluating estimators is that of ***sufficiency***. Sufficient statistics contain all the information there is in a sample about the parameter(s) being estimated. To estimate the frequency of an allele, it is sufficient to know how many copies of that allele appear in a sample, not the order on which they are seen, for example. There is an estimator with the minimum variance bound only if there is a sufficient statistic for that parameter.

MULTINOMIAL GENOTYPIC COUNTS

Most population genetic data sets consist of counts of genotypes, so the statistical properties of such counts need to be studied. The statistical sampling model appropriate for sampling from a single population assumes that every individual in the sample has the same probability of being a particular genotype. The ***multinomial distribution*** is then appropriate. Although this assumption greatly simplifies the theory, it cannot be exactly true in practice because the sampling of one individual reduces the frequency of that type in the population and so reduces the probability of choosing that type for subsequent sample members. Sampling individuals without replacement requires the ***hypergeometric distribution***, but it is generally assumed that the population being sampled is very large so that sampling can be considered to be with replacement. It is worth noting, however, that even at the very outset of this development, an assumption is made that may not always hold. All the statistical analyses in this book are going to rest on assumptions, although they may not always be made explicit.

To set up the multinomial distribution, suppose that the population members fall into a set of k categories indexed by the subscript i. Each individual sampled has probability Q_i of being in the ith category, which is the same as saying that the population frequency of category i is Q_i. When each individual is sampled independently, with the category of each having no effects on the categories of the others, the probability of sampling an individual of type i and then one of type j is Q_iQ_j. Extending this argument to samples of size n leads to the probability of n_i members in category i:

$$\Pr(n_1, n_2, ..., n_k) = \frac{n!}{\prod_{i=1}^{k} n_i!} \prod_{i=1}^{k} Q_i^{n_i} \tag{2.1}$$

where $\prod$ indicates a product, n ***factorial*** ($n!$) is the product of the first n integers, and the coefficient $n!/(\prod_i n_i!)$ is the number of observably different orderings of the n individuals. Equation 2.1 defines the multinomial distribution.

When there are two categories, the multinomial distribution reduces to the ***binomial***. If the probabilities for the two categories are written as Q and $1 - Q$, and the counts in those categories as a and $n - a$, the binomial probability is

$$\Pr(a, n - a) = \frac{n!}{a!(n-a)!} Q^a (1 - Q)^{n-a} \tag{2.2}$$

To illustrate the multinomial distribution for the case of three categories, Table 2.1 shows the 45 possible samples of size 8 from a population in which the three genotypes AA, Aa, and aa have frequencies 0.64, 0.32, and 0.04, respectively. Samples have been ordered according to probability. The cumulative probabilities shown for each sample in Table 2.1 are the probabilities for that sample and for all less probable samples.

When multinomial categories are combined into one category versus the rest, adding the appropriate probabilities gives the binomial distribution for those two new categories. For the example in Table 2.1, the distribution of AA versus $Aa + aa$ genotypes is binomial with probabilities 0.64 and 0.36. The nine possible samples are shown in Table 2.2, with probabilities that can be obtained directly from the binomial formula, Equation 2.2, or from adding terms in Table 2.1.

Multinomial Moments

Although the multinomial distribution is completely characterized by Equation 2.1, most use is made of summary quantities, or ***moments***. These quantities have to do with the expected values of powers of the counts, and the two moments in most use are the mean and variance. The ***mean***, or expected value, of the number in any category is obtained by multiplying each possible value of that number by its probability and adding these products. Since each category can be handled with a binomial distribution it is easiest to

Table 2.1 Multinomial probabilities, shown to four decimal places, for a sample size of 8 and category probabilities of 0.64 (AA), 0.32 (Aa), and 0.04 (aa).

n_{AA}	n_{Aa}	n_{aa}	Prob.	Cum.	n_{AA}	n_{Aa}	n_{aa}	Prob.	Cum.
0	0	8	0.0000	0.0000	1	5	2	0.0006	0.0020
0	1	7	0.0000	0.0000	3	2	3	0.0010	0.0030
1	0	7	0.0000	0.0000	4	1	3	0.0010	0.0039
0	2	6	0.0000	0.0000	1	6	1	0.0015	0.0055
1	1	6	0.0000	0.0000	1	7	0	0.0018	0.0072
2	0	6	0.0000	0.0000	2	4	2	0.0029	0.0101
0	3	5	0.0000	0.0000	6	0	2	0.0031	0.0132
1	2	5	0.0000	0.0000	3	3	2	0.0077	0.0209
3	0	5	0.0000	0.0000	2	5	1	0.0092	0.0301
0	4	4	0.0000	0.0000	5	1	2	0.0092	0.0394
2	1	5	0.0000	0.0000	4	2	2	0.0115	0.0509
0	5	3	0.0000	0.0000	2	6	0	0.0123	0.0632
1	3	4	0.0000	0.0000	7	0	1	0.0141	0.0773
4	0	4	0.0000	0.0001	8	0	0	0.0281	0.1054
2	2	4	0.0000	0.0001	3	4	1	0.0308	0.1362
0	6	2	0.0000	0.0002	3	5	0	0.0493	0.1855
3	1	4	0.0001	0.0002	6	1	1	0.0493	0.2347
0	7	1	0.0001	0.0003	4	3	1	0.0616	0.2963
0	8	0	0.0001	0.0004	5	2	1	0.0739	0.3702
1	4	3	0.0001	0.0006	7	1	0	0.1126	0.4828
5	0	3	0.0004	0.0009	4	4	0	0.1231	0.6059
2	3	3	0.0005	0.0014	6	2	0	0.1970	0.8030
					5	3	0	0.1970	1.0000

derive the mean from the binomial. For category i the expected number in a sample of size n is

$$\begin{aligned} \mathcal{E}(n_i) &= \sum_{r=0}^{n} r \Pr(n_i = r) \\ &= \sum_{r=0}^{n} r \frac{n!}{r!(n-r)!} Q_i^r (1 - Q_i)^{n-r} = nQ_i \end{aligned}$$

The sample proportion for category i is $\tilde{Q}_i = n_i/n$ and, dividing the previous equations by n, the expected value of the sample proportion is found to be the population proportion

$$\mathcal{E}(\tilde{Q}_i) = Q_i \tag{2.3}$$

Table 2.2 Binomial probabilities, shown to four decimal places, for sample size of 8 and category probabilities of 0.36 ($Aa + aa$) and 0.64 (AA).

$n_{Aa} + n_{aa}$	n_{AA}	Prob.	Cum.
0	8	0.0281	0.0281
1	7	0.1267	0.1548
2	6	0.2494	0.4042
3	5	0.2805	0.6847
4	4	0.1972	0.8819
5	3	0.0889	0.9708
6	2	0.0250	0.9958
7	1	0.0040	0.9998
8	0	0.0002	1.0000

In other words, the sample proportion is an ***unbiased estimator*** of the population proportion. Expectation here is referring to all possible samples of size n from the population and so is taking account of the statistical sampling.

Some idea of the variability of the category numbers is found by taking the expected value of the squared difference of the number from its mean, which is the definition of ***variance***. Still using the binomial distribution for a particular category

$$\begin{aligned} \mathrm{Var}(n_i) &= \sum_{r=0}^{n}[r - \mathcal{E}(n_i)]^2 \Pr(n_i = r) \\ &= \sum_{r=0}^{n}(r^2 - 2rnQ_i + n^2Q_i^2)\frac{n!}{r!(n-r)!}Q_i^r(1-Q_i)^{n-r} \\ &= [n(n-1)Q_i^2 + nQ_i] - 2n^2Q_i^2 + n^2Q_i^2 \\ &= nQ_i(1-Q_i) \end{aligned} \tag{2.4}$$

For a sample proportion, division by n occurs inside the square brackets of the first line of Equation 2.4, and this leads to a variance of

$$\begin{aligned} \mathrm{Var}(\tilde{Q}_i) &= \frac{1}{n^2}\mathrm{Var}(n_i) \\ &= \frac{1}{n}Q_i(1-Q_i) \end{aligned}$$

In some later expressions there is a need for the expected values of the squares of counts or proportions, rather than variances, but these are easily found. For any random variable X,

$$\mathcal{E}(X^2) = [\mathcal{E}(X)]^2 + \text{Var}(X)$$

For sample proportions, manipulating the expression for the variance gives

$$\begin{aligned}\mathcal{E}(\tilde{Q}_i^2) &= [\mathcal{E}(\tilde{Q}_i)]^2 + \text{Var}(\tilde{Q}_i) \\ &= Q_i^2 + \frac{1}{n}Q_i(1-Q_i)\end{aligned}$$

showing that $\tilde{Q}_i^2$ is a biased estimate of Q_i^2.

This result is exact. Approximate expressions can be found for $\tilde{Q}_i^3$ and $\tilde{Q}_i^4$ that ignore terms with squares or higher powers of sample size n in the denominator:

$$\begin{aligned}\mathcal{E}(\tilde{Q}_i^3) &= Q_i^3 + \frac{3}{n}Q_i^2(1-Q_i) \\ \mathcal{E}(\tilde{Q}_i^4) &= Q_i^4 + \frac{6}{n}Q_i^3(1-Q_i)\end{aligned}$$

All the results from Equation 2.2 onward were derived for the binomial distribution, and they apply to any one of the categories in a multinomial setting. The multinomial probabilities need to be considered when the joint sampling properties for counts or proportions in two or more categories are sought. To find the expected value of the product of two counts, $n_i n_j$, it is sufficient to work with a trinomial, for categories i, j, and the amalgamation of all other categories. The expected value of the product takes account of all its values:

$$\begin{aligned}\mathcal{E}(n_i n_j) &= \sum_{r=0}^{n}\sum_{s=0}^{n-r} rs\,\Pr(n_i = r, n_j = s) \\ &= \sum_{r=0}^{n}\sum_{s=0}^{n-r} rs\frac{n!}{r!s!(n-r-s)!}Q_i^r Q_j^s(1-Q_i-Q_j)^{n-r-s} \\ &= n(n-1)Q_iQ_j \qquad (2.5)\end{aligned}$$

and

$$\mathcal{E}(\tilde{Q}_i\tilde{Q}_j) = \frac{(n-1)}{n}Q_iQ_j \qquad (2.6)$$

The ***covariance*** of counts is defined as the expected value of the product of the deviations of counts from their means, leading to

$$\begin{aligned}\text{Cov}(n_i, n_j) &= \mathcal{E}\{[n_i - \mathcal{E}(n_i)][n_j - \mathcal{E}(n_j)]\} \\ &= \mathcal{E}(n_i n_j) - \mathcal{E}(n_i)\mathcal{E}(n_j)\end{aligned}$$

From Equations 2.5 and 2.6

$$\begin{aligned}\mathrm{Cov}(n_i, n_j) &= -nQ_iQ_j \\ \mathrm{Cov}(\tilde{Q}_i, \tilde{Q}_j) &= -\frac{1}{n}Q_iQ_j\end{aligned}$$

Since the sum of the counts of two categories cannot be larger than the sample size, increasing one count causes a reduction in possible values for the other count, and this causes the two counts to have a negative covariance. To complete the study of a pair of counts, the ***correlation*** is defined as

$$\begin{aligned}\mathrm{Corr}(n_i, n_j) &= \frac{\mathrm{Cov}(n_i, n_j)}{\sqrt{\mathrm{Var}(n_i)\mathrm{Var}(n_j)}} \\ &= -\frac{Q_iQ_j}{\sqrt{Q_i(1-Q_i)Q_j(1-Q_j)}} \\ &= \mathrm{Corr}(\tilde{Q}_i, \tilde{Q}_j)\end{aligned}$$

Notice that in the binomial case the two probabilities sum to one, $Q_1 = 1 - Q_2$, and the correlation of the two counts or sample proportions is minus one.

Within-Population Variance of Allele Frequencies

Counts of alleles are obtained by adding twice the homozygote counts to the sum of appropriate heterozygote counts. In a sample with n_{uu} homozygotes A_uA_u and n_{uv} heterozygotes A_uA_v, the number n_u of A_u alleles is

$$n_u = 2n_{uu} + \sum_{v \neq u} n_{uv}$$

Throughout this book, a sum such as that in this equation is meant to include every possible heterozygote for allele A_u, but to include it only once. For three alleles, for example, the equation becomes

$$\begin{aligned}n_1 &= 2n_{11} + n_{12} + n_{13} \\ n_2 &= 2n_{22} + n_{12} + n_{23} \\ n_3 &= 2n_{33} + n_{13} + n_{23}\end{aligned}$$

As a convention, the subscripts on heterozygotes will be written in numerical (or alphabetical) order.

Writing population, or expected, frequencies of genotypes as P_{uu} for A_uA_u and P_{uv} for A_uA_v, the expected value of the number of alleles A_u is

$$\begin{aligned}\mathcal{E}(n_u) &= 2nP_{uu} + \sum_{v \neq u} nP_{uv} \\ &= 2np_u\end{aligned}$$

where

$$p_u = P_{uu} + \frac{1}{2}\sum_{v \neq u} P_{uv}$$

is the population frequency of A_u alleles. Dividing by $2n$, the number of alleles sampled, shows that the sample allele frequency $\tilde{p}_u = n_u/2n$ is also unbiased for the population value

$$\mathcal{E}(\tilde{p}_u) = p_u$$

This last relation says that if many samples were taken from the same population the average of all the sample allele frequencies would be the allele frequency in the population. The various samples may differ from each other, and the extent of this difference is indicated by the variance of the sample allele frequency. The size of this variance (of the allele count) can be found by using the result that the variance of a sum of random variables (the genotype counts) is the sum of their variances plus twice the sum of the covariances between all pairs of the variables

$$\begin{aligned}\text{Var}(n_u) &= \text{Var}(2n_{uu}) + \sum_{v \neq u} \text{Var}\,(n_{uv}) + 2\sum_{v \neq u} \text{Cov}\,(2n_{uu}, n_{uv}) \\ &\quad + 2\sum_{v \neq u}\sum_{w \neq u,v} \text{Cov}\,(n_{uv}, n_{uw}) \\ &= 4nP_{uu}(1 - P_{uu}) + \sum_{v \neq u} nP_{uv}(1 - P_{uv}) - 4\sum_{v \neq u} nP_{uu}P_{uv} \\ &\quad - 2\sum_{v \neq u}\sum_{w \neq u,v} nP_{uv}P_{uw} \\ &= 2n\left(2P_{uu} + \frac{1}{2}\sum_{v \neq u} P_{uv}\right) - n\left(2P_{uu} + \sum_{v \neq u} P_{uv}\right)^2 \\ &= 2n(p_u + P_{uu} - 2p_u^2)\end{aligned}$$

so that

$$\text{Var}(\tilde{p}_u) = \frac{1}{2n}(p_u + P_{uu} - 2p_u^2) \qquad (2.7)$$

To make the argument more concrete, it is now repeated in the two-allele case to find the variance of the count for the first allele:

$$\begin{aligned}
\text{Var}(n_1) &= \text{Var}(2n_{11} + n_{12}) \\
&= \text{Var}(2n_{11}) + \text{Var}\,(n_{12}) + 2\text{Cov}\,(2n_{11}, n_{12}) \\
&= 4nP_{11}(1 - P_{11}) + nP_{12}(1 - P_{12}) - 4nP_{11}P_{12} \\
&= 2n\left(2P_{11} + \frac{1}{2}P_{12}\right) - n\,(2P_{11} + P_{12})^2 \\
&= 2n(p_1 + P_{11} - 2p_1^2)
\end{aligned}$$

so that

$$\text{Var}(\tilde{p}_1) = \frac{1}{2n}(p_1 + P_{11} - 2p_1^2)$$

This expression and Equation 2.7 involve the population frequencies p_u and P_{uu}. In practice, data are available only from a single sample. If the sample frequencies $\tilde{p}_u$ and $\tilde{P}_{uu}$ are substituted into Equation 2.7, an estimate $\widehat{\text{Var}}(\tilde{p}_u)$ of the variance of $\tilde{p}_u$ is obtained. This, in turn, gives an idea of the range of values within which the population frequency p_u lies. Providing the sample size is reasonably large ($n \geq 30$, say), there is about a 95% chance that the interval

$$\tilde{p}_u \;\pm\; 2\sqrt{\widehat{\text{Var}}(\tilde{p}_u)}$$

includes the population frequency p_u.

The process of constructing a ***confidence interval*** like the one just given for a genetic parameter ϕ, here the allele frequency p_u, rests on an assumption that the statistic $\hat{\phi}$ used to estimate the parameter has an approximately normal distribution. Equation 2.20 later in the chapter refers to this approximate normality. To construct an interval for which there is $(1 - \alpha) \times 100\%$ confidence that it includes the parameter value, it is necessary to identify the values $z_{\alpha/2}, z_{1-\alpha/2}$ between which lie $(1 - \alpha) \times 100\%$ of the values of the standard normal distribution. Such values are found in Appendix Table A.1. For a 95% confidence interval, that table shows that

$$z_{0.975}, z_{0.025} = \pm 1.96 \approx \pm 2$$

The general expression for the limits of a confidence interval for ϕ is

$$\hat{\phi} \;\pm\; z_{1-\alpha/2}\sqrt{\text{Var}(\hat{\phi})}$$

Equation 2.7 for the variance of a sample allele frequency is important for it demonstrates that although each genotype number can be regarded as being binomially distributed, the same is not true for allele numbers. The variance in Equation 2.7 does not have the form of a binomial variance. There is a special case of great importance, however, and that is when the population sampled is in ***Hardy-Weinberg equilibrium*** (see Chapter 3). This means that alleles as well as genotypes are sampled at random from the population since, at any locus, a genotype is just a random pair of alleles. Genotypic frequencies are equal to the products of corresponding allele frequencies:

$$\begin{aligned} P_{uu} &= p_u^2 \\ P_{uv} &= 2p_u p_v, \text{ for } u \neq v \end{aligned}$$

and Equation 2.7 reduces to

$$\text{Var}(\tilde{p}_u) = \frac{1}{2n} p_u(1 - p_u) \tag{2.8}$$

which is the variance of a binomial distribution with parameters p_u and $2n$. For populations in Hardy-Weinberg equilibrium, the allele numbers as well as genotype numbers are multinomially distributed.

The variance expressions derived in this section refer to repeated samples from one population, and the allele frequencies p in these expressions refer to that population.

Indicator Variables

Many expressions for the expected values of functions of allele or genotype frequencies can be found very easily by using a set of indicator variables. Such variables take the value of 1 when an event is true and 0 when it is false. To find the variance of an allele frequency, for example, an indicator variable is set up for that particular allele. Suppose the two alleles at a locus within an individual are indexed with j, $j = 1, 2$, and the individuals sampled are indexed with i, $i = 1, 2, \ldots, n$. The indicator variable x_{ij} is then defined by

$$x_{ij} = \begin{cases} 1, & \text{if allele } j \text{ in individual } i \text{ is type } A \\ 0, & \text{otherwise} \end{cases}$$

The sample frequency for allele A can then be expressed as

$$\tilde{p}_A = \frac{1}{2n} \sum_{i=1}^{n} \sum_{j=1}^{2} x_{ij}$$

Taking expectations over all possible samples from the population shows that

$$\begin{aligned}\mathcal{E}(x_{ij}) &= 1 \times \Pr(x_{ij} = 1) + 0 \times \Pr(x_{ij} = 0) \\ &= 1 \times p_A + 0 \times (1 - p_A) \\ &= p_A\end{aligned}$$

so that, once again,

$$\mathcal{E}(\tilde{p}_A) = p_A$$

To derive the variance, recognize that the product $x_{ij}x_{ij'}$ $(j \neq j')$ will be nonzero only when both alleles, j and j', in individual i are type A, whereas the square x_{ij}^2 is just the same as x_{ij} itself. In other words,

$$\begin{aligned}\mathcal{E}(x_{ij}^2) &= p_A \\ \mathcal{E}(x_{ij}x_{ij'}) &= P_{AA}\end{aligned}$$

Since individuals are sampled independently from the population, the expectation of the product of x_{ij} and $x_{i'j'}$, for alleles in different individuals, is the product of expectations of these two indicator variables:

$$\begin{aligned}\mathcal{E}(x_{ij}x_{i'j'}) &= \mathcal{E}(x_{ij})\mathcal{E}(x_{i'j'}) \\ &= p_A^2\end{aligned}$$

Expanding the square of the expression for $\tilde{p}_A$, and then taking expectations leads to

$$\begin{aligned}\mathcal{E}(\tilde{p}_A^2) &= \frac{1}{4n^2}\left(\mathcal{E}\sum_i\sum_j x_{ij}^2 + \mathcal{E}\sum_i\sum_j\sum_{j'\neq j} x_{ij}x_{ij'}\right. \\ &\qquad \left. + \mathcal{E}\sum_i\sum_{i'\neq i}\sum_j\sum_{j'} x_{ij}x_{i'j'}\right) \\ &= \frac{1}{4n^2}\left[2np_A + 2nP_{AA} + 4n(n-1)p_A^2\right] \\ &= p_A^2 + \frac{1}{2n}(p_A + P_{AA} - 2p_A^2) \qquad (2.9)\end{aligned}$$

and, when the square of the expected allele frequency is subtracted,

$$\text{Var}(\tilde{p}_A) = \frac{1}{2n}(p_A + P_{AA} - 2p_A^2)$$

as in Equation 2.7. The great power of this indicator variable method will become apparent in more complex cases.

It is often convenient to express genotypic frequencies in terms of allele frequencies and some measure of departure from Hardy-Weinberg equilibrium. One parameterization uses the inbreeding measure f within populations (Cockerham 1969), which is the same as the quantity F_{IS} defined by Wright (1951). For two alleles, A and a,

$$\begin{aligned} P_{AA} &= p_A^2 + p_A p_a f \\ P_{Aa} &= 2p_A p_a(1-f) \\ P_{aa} &= p_a^2 + p_a p_A f \end{aligned}$$

and, putting these into the variance formula Equation 2.7,

$$\text{Var}(\tilde{p}_A) = \frac{1}{2n} p_A(1-p_A)(1+f) \tag{2.10}$$

Once again the binomial variance result holds if there is Hardy-Weinberg equilibrium, $f = 0$. For any particular set of allele frequencies, the variance increases as the amount of inbreeding increases, with an eventual doubling of the noninbred value. A completely inbred population will be homozygous, with alleles being either fixed or lost, and then the within-population variance of a sample allele frequency will be zero since $p_A(1-p_A)$ will be zero. Remember that this variance refers to repeated samples from the same population.

Within-Population Covariance of Allele Frequencies

Indicator variables can also be used to find the covariance between frequencies of two alleles. Let x_{ij} and y_{ij} be the indicator variables for alleles A_1 and A_2, so that the two sample frequencies are

$$\begin{aligned} \tilde{p}_1 &= \frac{1}{2n}\sum_i \sum_j x_{ij} \\ \tilde{p}_2 &= \frac{1}{2n}\sum_i \sum_j y_{ij} \end{aligned}$$

and the expected value of their product is

$$\mathcal{E}(\tilde{p}_1\tilde{p}_2) = \frac{1}{4n^2}\mathcal{E}\left[(\sum_i \sum_j x_{ij})(\sum_i \sum_j y_{ij})\right]$$

$$\begin{aligned} &= \frac{1}{4n^2}\mathcal{E}\left(\sum_i\sum_j x_{ij}y_{ij} + \sum_i\sum_{j\neq j'}\sum_{j'} x_{ij}y_{ij'}\right. \\ &\qquad \left. + \sum_i\sum_{i'\neq i}\sum_j\sum_{j'} x_{ij}y_{i'j'}\right) \\ &= \frac{1}{4n^2}[2n\times 0 + nP_{12} + 4n(n-1)p_1p_2] \\ &= p_1p_2 + \frac{1}{4n}(P_{12} - 4p_1p_2) \end{aligned}$$

The zero results from it not being possible for allele j in individual i to be both type A_1 and type A_2. Subtracting p_1p_2 provides the covariance

$$\text{Cov}(\tilde{p}_1, \tilde{p}_2) = \frac{1}{4n}(P_{12} - 4p_1p_2) \tag{2.11}$$

For noninbred populations, with $P_{12} = 2p_1p_2$, the covariance reduces to the binomial value of

$$\text{Cov}(\tilde{p}_1, \tilde{p}_2) = -\frac{1}{2n}p_1p_2 \tag{2.12}$$

Examples

Calculation of variances and covariances of allele frequencies can be illustrated with data from Chapter 1. For the blood group data shown in Table 1.2, the genotypic counts for the *MN* locus, for mothers and fathers, are given in Table 2.3. By regarding these 93 couples as a random sample of 186 individuals from a population, sample homozygote and allele frequencies for M are calculated as

$$\begin{aligned} \tilde{P}_{MM} &= 53/186 = 0.2849 \\ \tilde{p}_M &= 201/372 = 0.5403 \end{aligned}$$

Substituting these observed values into the right-hand side of Equation 2.7 gives an estimate of the variance of $\tilde{p}_M$:

$$\begin{aligned} \text{Var}(\tilde{p}_M) &\hat{=} [0.5403 + 0.2849 - 2(0.5403)^2]/372 \\ &= 0.00064880 \end{aligned}$$

whereas substituting into Equation 2.8 gives a very similar result:

$$\begin{aligned} \text{Var}(\tilde{p}_M) &\hat{=} 0.5403(1 - 0.5403)/372 \\ &= 0.00066768 \end{aligned}$$

Table 2.3 Genotypic counts for *MN* blood groups among mothers and fathers in Table 1.2.

Genotype	Father	Mother	Total
MM	26	27	53
MN	44	51	95
NN	23	15	38
Total	93	93	186

The symbol $\hat{=}$ means "is estimated by." The similarity in these two numerical results is a reflection of the genotypic frequencies for the *MN* blood group in this sample being very close to Hardy-Weinberg proportions.

A difference in the results from Equations 2.7 and 2.8 does arise for the *Pgm* locus data in Table 1.3. The genotypic and allele counts are shown in Table 2.4, and for allele 1 at that locus,

Table 2.4 Allele and genotype counts for *Pgm* locus in mosquito data of Table 1.3.

Genotype	Count	Allele	Count
11	9	1	19
12	1	2	26
22	5	3	17
13	0	4	18
23	7		
33	0	Total	80
14	0		
24	8		
34	10		
44	0		
Total	40		

$$\begin{aligned}\tilde{P}_{11} &= 9/40 &= 0.2250\\ \tilde{p}_1 &= 19/80 &= 0.2375\end{aligned}$$

so that

$$\mathrm{Var}(\tilde{p}_1) \mathrel{\hat{=}} \begin{cases} 0.00437109 \text{ from Equation 2.7} \\ \\ 0.00226367 \text{ from Equation 2.8} \end{cases}$$

The two variance estimates differ because the data had more homozygotes for allele 1 than expected from the Hardy-Weinberg theorem. Turning to covariances, for alleles 1 and 2, Equation 2.11 provides

$$\mathrm{Cov}(\tilde{p}_1, \tilde{p}_2) \mathrel{\hat{=}} -0.00177344$$

whereas Equation 2.12 gives

$$\mathrm{Cov}(\tilde{p}_1, \tilde{p}_2) \mathrel{\hat{=}} -0.00096484$$

The difference is due to a deficiency of heterozygotes for alleles 1 and 2 compared to Hardy-Weinberg expectations (tests for Hardy-Weinberg are taken up in Chapter 3). The opposite situation holds for alleles 3 and 4, where there is even a difference in sign:

$$\mathrm{Cov}(\tilde{p}_3, \tilde{p}_4) \mathrel{\hat{=}} \begin{cases} 0.00036719 \text{ from Equation 2.11} \\ \\ -0.00059766 \text{ from Equation 2.12} \end{cases}$$

A great deal of attention has been paid here to the difference between results that assume Hardy-Weinberg equilibrium and those that do not. Generally, samples from natural populations of outbreeding species can be treated as though they are indeed from Hardy-Weinberg populations, but there can be occasions when this is not the case. As a final example, reference is made to some unpublished data of R. W. Allard. Genotypic counts at an esterase locus in Composite Cross V of barley, *Hordeum vulgare*, are shown in Table 2.5. The data are described in Allard et al. (1992) and consist of samples from several generations of an experimental population grown in Davis, California. The alleles are labeled by their electrophoretic migration distances. With over 99% selfing, random-mating equilibrium results cannot possibly hold. For allele $B_{1.6}$ in generation 15,

$$\begin{aligned}\tilde{P}_{1.6,1.6} &= 161/2843 &= 0.0566\\ \tilde{p}_{1.6} &= 437/5686 &= 0.0769\end{aligned}$$

Table 2.5 Genotypic counts for *Esterase B* locus in barley data.

Generation	$B_{1.6}B_{1.6}$	$B_{2.7}B_{2.7}$	$B_{3.9}B_{3.9}$	$B_{1.6}B_{2.7}$	$B_{1.6}B_{3.9}$	$B_{2.7}B_{3.9}$	Total
4	58	1132	40	39	0	10	1279
5	68	1356	42	9	1	10	1486
6	91	891	24	0	0	0	1006
14	101	1709	82	30	1	5	1928
15	161	2331	227	115	0	9	2843
16	142	2132	68	27	0	0	2369
17	124	2243	92	2	0	0	2461
24	662	3518	311	68	3	25	4587
25	620	3043	262	23	0	19	3967
26	537	2274	252	10	0	10	3083

Source: R.W. Allard (personal communication).

so that the estimated variances are

$$\text{Var}(\tilde{p}_{1.6}) \;\hat{=}\; \begin{cases} 0.00002140 \text{ from Equation 2.7} \\ \\ 0.00001248 \text{ from Equation 2.8} \end{cases}$$

It is more appropriate to use Equation 2.7 in all situations, unless there is good reason to assume Hardy-Weinberg equilibrium.

Total Variance of Allele Frequencies

The variance of allele frequencies shown in Equations 2.7 or 2.10 referred to variation over repeated samples from the same population. To make statements about allele frequency that are not limited to one particular replicate population, the total variance of a sample allele frequency is used. This total variance must also take account of the genetic sampling that gives rise to variation between replicate populations. It is based on the variation that exists among all possible samples from all possible populations maintained under the same conditions. Obviously the total variance will be greater than the within-population variance, as it contains a contribution for the variation between populations.

The within-population variance for genotypic frequencies was found from properties of the multinomial distribution. There is no simple equivalent distribution that takes into account genetic sampling and describes the variation in frequencies among different populations. Approximations are available that treat frequencies as continuous rather than discrete. For populations obeying the "Wright-Fisher" model for selectively neutral genes with frequencies affected only by drift, Crow and Kimura (1970) gave an approximate distribution and explicit expressions for the first four moments of the distribution. For populations that have reached an equilibrium under the joint effects of drift and mutation or migration, Wright (1945) found that allele frequencies for loci with two alleles had a beta distribution, and for multi-allele loci the distribution was Dirichlet (Wright 1951). The ramifications for estimating frequencies were discussed by Jiang (1987). The treatment in this chapter will not require knowledge of the genetic sampling distribution, but will derive variances in a way similar to that for statistical sampling through the use of indicator variables. The results will be analogous to those given earlier by Wright.

As in the last section, x_{ij} refers to the jth allele in the ith sampled individual, with $x_{ij} = 1$ if the allele is type A. Expectations of x_{ij}, x_{ij}^2, $x_{ij}x_{ij'}$ have the same functional form as before but the product $x_{ij}x_{i'j'}$ must be changed. In the total sampling framework, different individuals can not be regarded as having been sampled independently. Even if there is random mating within populations, there are frequency differences between populations. In statistical language, the component of variance between populations is given by the covariance of individuals within populations. The genetic sampling process now plays a role, and the expectation for alleles from different individuals is written as

$$\mathcal{E}(x_{ij}x_{i'j'}) = P_{A/A}$$

which is the frequency with which two individuals in one population both carry an A allele. Following the same argument in the last section, with this one change, leads to the total variance of

$$\text{Var}(\tilde{p}_A) = (P_{A/A} - p_A^2) + \frac{1}{2n}(p_A + P_{AA} - 2P_{A/A}) \quad (2.13)$$

The first term in this expression will remain even when the sample size n becomes very large, and it represents the between-population contribution to the variance. The frequencies $p_A, P_{AA}, P_{A/A}$ all refer to values expected over replicate populations — another difference from the within-population frequencies that referred to one specific population.

Equation 2.13 can be expressed more conveniently by introducing analogs of the within-population inbreeding coefficient. The quantities F and θ (Cockerham 1969) are the ***total inbreeding coefficient*** and the ***coancestry coefficient***, respectively. They refer to pairs of alleles within and between individuals, and were called F_{IT} and F_{ST} by Wright (1951). They allow the expression of the frequencies P_{AA} and $P_{A/A}$, both of which are expected values over all samples *and* over all replicate populations (see Figure 1.5), as is the allele frequency p_A:

$$\begin{aligned} P_{AA} &= p_A^2 + p_A(1-p_A)F \\ P_{A/A} &= p_A^2 + p_A(1-p_A)\theta \end{aligned}$$

Using these expressions allows Equation 2.13 to be rewritten as

$$\mathrm{Var}(\tilde{p}_A) = p_A(1-p_A)\left(\theta + \frac{F-\theta}{n} + \frac{1-F}{2n}\right) \quad (2.14)$$

with three components that can be identified with the variation between populations, between individuals within populations, and between alleles within individuals within populations, respectively. The term in brackets is the ***group coancestry coefficient***, θ_L. Notice that the between-population component depends on the relation between alleles of different individuals within populations, and it could be written with $(\theta - 0)$, instead of just θ, where 0 indicates the zero relationship between alleles of different populations.

In a random mating population, pairs of alleles have the same relationship whether or not they are located in the same individual, so $F = \theta$ and the total variance of allele frequency becomes

$$\mathrm{Var}(\tilde{p}_A) = p_A(1-p_A)\theta + p_A(1-p_A)\frac{1-\theta}{2n} \quad (2.15)$$

If the sample size is very large, $n \to \infty$, there is no need for the within-population component, and the variance refers to variation, caused by genetic sampling, among populations:

$$\mathrm{Var}(\tilde{p}_A) = p_A(1-p_A)\theta$$

This result was given by Crow and Kimura (1970) for populations affected by drift, in which case θ depends on population size and the number of generations of drift. If equilibrium under drift is reached and each population becomes fixed for one of the alleles at a locus, $\theta = 1$, the variance of allele frequency tends to $p_A(1-p_A)$, reflecting the possibility of the fixation of different alleles in different populations. For populations at equilibrium under the joint effects of drift and recurrent mutation or migration, the same result holds (Jiang 1987) but the value of θ is different. This last result follows from Dirichlet distribution theory and, when the statistical sampling term in Equation 2.15 is included, was given by Mosimann (1962).

Estimation of parameters F and θ will be treated later, but note here that it is not possible to estimate them, or the total variance of an allele frequency, from data from a single population. The three measures of relationship that have been used for pairs of alleles are related as

$$f = \frac{F - \theta}{1 - \theta}$$

With random mating, $F = \theta$ and $f = 0$.

Fisher's Approximate Variance Formula

Two approaches have now been used for finding variances of functions of multinomial frequencies, first by direct combination of the multinomial variances and then by use of indicator variables. Both methods give exact results, but both become unwieldy when complicated functions are considered. Furthermore, estimators of genetic parameters often involve ratios of multinomial frequencies and these do not allow exact expressions for variances to be found. There is an approximate method, often called the ***delta method***, based on Taylor's series expansions of the function whose variance is to be found. Suppose that the variance is wanted for some function T of variables n_i. The delta method gives the variance

$$\mathrm{Var}(T) \approx \sum_i \left(\frac{\partial T}{\partial n_i}\right)^2 \mathrm{Var}(n_i) + \sum_i \sum_{j \neq i} \frac{\partial T}{\partial n_i}\frac{\partial T}{\partial n_j}\mathrm{Cov}(n_i, n_j) \tag{2.16}$$

where each of the derivatives is evaluated with the n_i's replaced by their expected values. If the variables n_i are multinomial counts, the expectations have the form nQ_i for a sample of size n. Since the variances and covariances of multinomial proportions n_i/n are of ***order*** n^{-1} (meaning that they tend to zero as n tends to infinity), this expression is also of order n^{-1}. Terms that involve higher powers of n^{-1} are ignored as being of negligible size.

Because the variances and covariances of multinomial counts n_i's have a very simple form

$$\begin{aligned} \mathrm{Var}(n_i) &= nQ_i(1 - Q_i) \\ \mathrm{Cov}(n_i, n_j) &= -nQ_iQ_j \end{aligned}$$

the variance in Equation 2.16 reduces to

$$\mathrm{Var}(T) \approx n\sum_i \left(\frac{\partial T}{\partial n_i}\right)^2 Q_i - n\left(\sum_i \frac{\partial T}{\partial n_i}Q_i\right)^2$$

When T is a homogeneous function of degree zero in these counts, or in other words is either a function of (n_i/n)'s or is a ratio of functions of the same degree in the n_i's, Fisher (1925) found the further reduction to

$$\text{Var}(T) \approx n \sum_i \left(\frac{\partial T}{\partial n_i}\right)^2 Q_i - n\left(\frac{\partial T}{\partial n}\right)^2 \tag{2.17}$$

In this expression, the derivatives of the function T are evaluated at the expected values of the n_i. The derivative with respect to the total sample size n is needed only when T explicitly involves n. It is important to note that T must be written as a function of *counts* n_i and must be of degree zero. In other words, it must be a ratio of functions of the same order in the counts, or it must have every count divided by the total. Further discussion is given by Bailey (1961, Appendix 2).

As a simple example, consider the variance of the frequency of allele A_1 in a sample of size n for a locus with three codominant alleles. The function $T = \tilde{p}_1$, written in terms of multinomial counts, is

$$T = \frac{1}{2n}(2n_{11} + n_{12} + n_{13})$$

where n_{uv} is the number of A_uA_v genotypes in the sample. The derivatives, evaluated with the n_{uv}'s replaced by their expected values, are

$$\frac{\partial T}{\partial n_{11}} = \frac{1}{n}, \quad \frac{\partial T}{\partial n_{12}} = \frac{1}{2n}$$

$$\frac{\partial T}{\partial n_{13}} = \frac{1}{2n}, \quad \frac{\partial T}{\partial n} = -\frac{p_1}{n}$$

so that the variance becomes

$$\begin{aligned}\text{Var}(\tilde{p}_1) &\approx n\left[\left(\frac{1}{n}\right)^2 P_{11} + \left(\frac{1}{2n}\right)^2 P_{12} + \left(\frac{1}{2n}\right)^2 P_{13}\right] - n\left(\frac{-p_1}{n}\right)^2 \\ &= \frac{1}{2n}(p_1 + P_{11} - 2p_1^2)\end{aligned}$$

as before (Equation 2.7). This expression made use of the relation

$$p_1 = P_{11} + \frac{1}{2}P_{12} + \frac{1}{2}P_{13}$$

Of course this result is exact and the method was not much simpler than the direct methods. For more complex functions, the method is invaluable, however, as can be seen by using it to derive an approximate variance for the quantity $\tilde{p}_A(1 - \tilde{p}_A)$:

$$\text{Var}[\tilde{p}_A(1 - \tilde{p}_A)] \approx \frac{1}{2n}(1 - 2p_A)^2(p_A + P_{AA} - 2p_A^2)$$

The terms involving higher powers of n in the denominator that are ignored by Fisher's formula do not all disappear when $p_A = 0.5$. Fisher's formula gives a variance of zero when $p_A = 0.5$.

It is not possible, however, to use Fisher's formula for a function such as $n_1 = (2n_{11} + n_{12} + n_{13})$ since this is not a homogeneous function of degree zero.

NUMERICAL RESAMPLING

There are occasions when even Fisher's approximate method is not available to provide sampling variances, and then the method of ***numerical resampling*** (Efron 1982, Efron and Tibshirani 1993) is of great benefit. Within-population sampling properties of statistics refer to the variation over repeated samples from a population, and numerical resampling provides these repeated samples by constructing new samples from the original dataset. In other words, numerical resampling mimics the drawing of new samples by resampling the one sample at hand from each population. Two methods are commonly used: ***jackknifing*** and ***bootstrapping***.

The Jackknife

The simplest of the numerical resampling techniques is the jackknife, first discussed by Quenouille (1956). For a parameter ϕ and a set of observations $X_1, X_2, \ldots, X_n$ there is some procedure that leads to an estimate $\hat{\phi}$. The estimate may result from an explicit formula, such as that given below in Equation 2.28 for the inbreeding coefficient f, or it may result from an iterative technique as shown below for maximum likelihood estimation. The jackknife procedure requires that n new estimates $\hat{\phi}_{(i)}$ be formed – the ith new estimate being after removal of the ith observation, $i = 1, 2, \ldots, n$. Each of the new estimates is therefore based on $(n-1)$ observations instead of n, and the average of the new estimates is

$$\hat{\phi}_{(\cdot)} = \frac{1}{n}\sum_i \hat{\phi}_{(i)}$$

Two uses are made of this set of new estimates. First, they provide a new estimator $\hat{\phi}_{\mathrm{J}}$ that should have less bias than did the original $\hat{\phi}$

$$\hat{\phi}_{\mathrm{J}} = n\hat{\phi} - (n-1)\hat{\phi}_{(\cdot)}$$

and, second, they provide an estimate $\mathrm{Var}(\hat{\phi})_{\mathrm{J}}$ of the variance of $\hat{\phi}$

$$\mathrm{Var}(\hat{\phi})_{\mathrm{J}} = \frac{n-1}{n}\sum_i \left[\hat{\phi}_{(i)} - \hat{\phi}_{(\cdot)}\right]^2$$

The justification for the bias correction rests on assuming that the expected value of the original estimator is

$$\mathcal{E}(\hat{\phi}) = \phi + \frac{1}{n}a_1 + \frac{1}{n^2}a_2 + \cdots$$

so that the bias is a series of terms in increasing powers of $(1/n)$. The statistic $\hat{\phi}_J$ will also be biased, but it does not have the first-order term in the bias. Jackknifing removes the first-order bias. To see this, notice that

$$\mathcal{E}(\hat{\phi}_{(i)}) = \phi + \frac{1}{n-1}a_1 + \frac{1}{(n-1)^2}a_2 + \cdots$$

and that, therefore,

$$\mathcal{E}(\hat{\phi}_{(\cdot)}) = \phi + \frac{1}{n-1}a_1 + \frac{1}{(n-1)^2}a_2 + \cdots$$

Substituting these terms into the definition of the jackknife estimator shows that the a_1 term cancels out

$$\mathcal{E}(\hat{\phi}_J) = \phi - \frac{1}{n(n-1)}a_2 + \cdots$$

Tukey (1958) motivated the variance estimate by introducing a set of n "pseudovariables"

$$\hat{\phi} + (n-1)(\hat{\phi} - \hat{\phi}_{(i)})$$

It is these quantities that have mean $\hat{\phi}_J$ and variance estimated by $\mathrm{Var}(\hat{\phi})_J$.

As a simple example, consider the estimation of the mean μ of a variable X when a sample of size n is available. The initial estimator is the sample mean

$$\hat{\mu} = \frac{1}{n}\sum_i X_i = \bar{X}$$

and omitting the ith observation gives

$$\hat{\mu}_{(i)} = \frac{1}{n-1}\sum_{j \neq i} X_j$$

with a mean of

$$\hat{\mu}_{(\cdot)} = \frac{1}{n}\sum_i \left(\frac{1}{n-1}\sum_{j \neq i} X_j\right) = \bar{X}$$

Evidently the jackknife estimator $\hat{\mu}_J$ is just the sample mean $\bar{X}$ again:

$$\begin{aligned}\hat{\mu}_J &= n\hat{\mu} - (n-1)\hat{\mu}_{(\cdot)} \\ &= n\bar{X} - (n-1)\bar{X}\end{aligned}$$

Using the jackknife procedure to estimate the sampling variance of the estimate gives

$$\begin{aligned}\text{Var}(\hat{\mu})_J &= \frac{n-1}{n}\sum_i \left(\hat{\mu}_{(i)} - \hat{\mu}_{(\cdot)}\right)^2 \\ &= \frac{n-1}{n}\sum_i \left[\left(\frac{1}{n-1}\sum_{j\neq i} X_j\right) - \bar{X}\right]^2 \\ &= \frac{1}{n(n-1)}\sum_i (X_i - \bar{X})^2 \\ &= \frac{1}{n}s^2\end{aligned}$$

where $s^2 = \sum_i (X_i - \bar{X})^2/(n-1)$ is the sample variance of the n observations. The jackknife gives the usual variance estimate for the sample mean — the sample variance divided by the sample size. This quantity has an expected value of σ^2/n where σ^2 is the variance of X.

The Bootstrap

The jackknife provides estimates of bias and variance for genetic parameter estimates, but little information for the distribution of the estimates. The bootstrap, on the other hand, is not limited by the number of resampling units, such as individuals or loci, and can provide as many new estimates as needed to give a good approximation to the distribution of the original estimator. Instead of forming new samples by omitting one observation at a time, bootstrapping operates by drawing random samples of the same size as the original sample from that sample. This ***Monte Carlo*** sampling process requires the use of random numbers and does not allow explicit expressions for the new estimates.

If a sample consists of n observations $X_i, i = 1, 2, \ldots, n$, a ***bootstrap*** sample is a set of n observations drawn at random, with replacement, from this set in such a way that every one of the original observations has an equal chance of being chosen at any stage. Some of the original sample elements will not, therefore, appear in any particular bootstrap sample while some may appear many times. The generation of "random" numbers to allow such choices is discussed in Appendix B. Many of these bootstrap samples are drawn, and the parameter of interest estimated for each one. Many estimates become available for determining the sampling properties of the estimator.

To show some of the theoretical consequences of this procedure, the population mean μ is estimated again from a sample of size n. The original estimator is the sample mean, and the mean is also calculated for each bootstrap sample. Properties of the original estimator are determined by calculating the mean and variance of this collection of bootstrap means. Suppose that a bootstrap sample has original observation X_i represented r_i times, which requires $\sum_i r_i = n$. Because each observation is equally likely to be chosen at each draw, r_i is a binomial variable and the set of r_i values is multinomially distributed

$$r_i \sim B(n, 1/n)$$

with moments

$$\begin{aligned}
\mathcal{E}(r_i) &= 1 \\
\mathrm{Var}(r_i) &= \frac{n-1}{n} \\
\mathrm{Cov}(r_i, r_{i'}) &= -\frac{1}{n}, \ i' \neq i
\end{aligned}$$

The sample mean of the bootstrap sample is

$$\bar{X}_{\mathrm{B}} = \frac{1}{n}\sum_i r_i X_i$$

and a double set of expectations is needed to determine its properties. First regard the X_i's as fixed quantities and the r_i's as variables and take expectations $\mathcal{E}_{\mathrm{B}}$ over all possible r_i values

$$\mathcal{E}_{\mathrm{B}}(\bar{X}_{\mathrm{B}}) = \frac{1}{n}\sum_i X_i = \bar{X}$$

which is just the original sample mean again. Still taking into account only the variation of variables r_i, the variance of a bootstrap mean is

$$\begin{aligned}
\mathrm{Var}_{\mathrm{B}}(\bar{X}_{\mathrm{B}}) &= \frac{1}{n^2}\left[\sum_i \frac{n-1}{n} X_i^2 + \sum_i \sum_{i' \neq i} \left(-\frac{1}{n} X_i X_{i'}\right)\right] \\
&= \frac{n-1}{n^2} s^2
\end{aligned}$$

To complete the argument, expectations are taken over all sets of X_i values to find total mean and variance:

$$\begin{aligned}\mathcal{E}(\bar{X}_{\text{B}}) &= \mathcal{E}(\bar{X}) = \mu \\ \text{Var}(\bar{X}_{\text{B}}) &= \mathcal{E}\left(\frac{n-1}{n^2}s^2\right) = \frac{n-1}{n^2}\sigma^2\end{aligned}$$

Whereas jackknifing gave an estimate of the mean with correct mean and variance, bootstrapping gives an estimate with correct mean but slightly biased variance. There is also some judgment called for in applying the bootstrap technique. It is necessary to take a sufficient number of samples so that the mean and variance over samples give good approximations to the theoretical values, but not such a large number that computing time becomes prohibitive. For variances, about 100 bootstrap samples are generally regarded as being sufficient, but for the distribution of the estimator it may be better to use 1000 samples.

When a sample distribution is obtained, confidence intervals for the estimator can be constructed by ordering the estimates. A 95% confidence interval for the parameter being estimated, for example, can be constructed as the interval from the 26th to the 975th of 1000 ordered bootstrap estimates.

Bootstrapping within each of a set of samples can provide confidence intervals for allele frequencies without requiring allele counts to be binomially distributed, or requiring Hardy-Weinberg equilibrium. Two populations can be judged to have different allele frequencies if the estimated (observed) frequencies have nonoverlapping confidence intervals. In other words, numerical resampling provides a convenient way of making inferences when there is not an evolutionary basis for a distribution of allele frequencies over populations (see section on Random Populations in Chapter 5).

MAXIMUM LIKELIHOOD ESTIMATION

Multinomially distributed counts provide estimators of the corresponding population quantities, and functions of such counts can often be manipulated to provide estimators of other parameters. A general procedure of providing estimators that does not require such manipulation will now be considered. Provided the distribution of a random variable (i.e., the probabilities of the variable taking each of its possible values) is known, the method of ***maximum likelihood*** can be used to estimate the parameters of that distribution. For within-population analyses, the multinomial distribution is used as a basis for likelihood estimation.

If the multinomial counts n_i depend on a set of s *known* parameters ϕ_j, so that the expected values Q_i are functions of the ϕ_j's, the probability of observing the n_i's was given in Equation 2.1. Conversely, if the parameter values are *unknown*, but the observations are at hand, then the same expression can be regarded as providing the ***likelihood*** of any particular set of ϕ_j values. The likelihood is written as $L(\phi_1, \ldots, \phi_s)$, so that in the multinomial case

$$L(\phi_1, \ldots, \phi_s) = \frac{n!}{\prod_{i=1}^{k} n_i!} \prod_{i=1}^{k} [Q_i(\phi_1, \ldots, \phi_s)]^{n_i} \tag{2.18}$$

and the maximum likelihood estimates, MLE's, of the ϕ_j's are those values that maximize this likelihood. In this case, as in several others, it is easier and equivalent to work with $\ln L$, the natural logarithm of the likelihood, which is called the ***support***. The MLE's are found by setting to zero the derivatives of $\ln L$ with respect to each parameter ϕ_j, and verifying that this maximizes the likelihood. These derivatives are written as S_j and are called ***scores***:

$$S_j = \frac{\partial \ln L}{\partial \phi_j}$$

To verify that the procedure leads to sensible results it will now be used to estimate allele frequencies at a two-allele locus known to be in Hardy-Weinberg equilibrium. The genotypes AA, Aa, and aa have frequencies that can be expressed in terms of the single parameter p_A

$$\begin{aligned} P_{AA} &= p_A^2 \\ P_{Aa} &= 2p_A(1 - p_A) \\ P_{aa} &= (1 - p_A)^2 \end{aligned}$$

and the likelihood of that parameter is

$$L(p_A) = \frac{n!}{n_{AA}! n_{Aa}! n_{aa}!} (p_A^2)^{n_{AA}} [2p_A(1 - p_A)]^{n_{Aa}} [(1 - p_A)^2]^{n_{aa}}$$

The support is

$$\ln L(p_A) = \text{Constant} + (2n_{AA} + n_{Aa}) \ln(p_A) + (n_{Aa} + 2n_{aa}) \ln(1 - p_A)$$

where the constant term is the logarithm of the ratio of factorials and does not involve the parameter p_A. Differentiating with respect to p_A provides

$$S_{p_A} = \frac{2n_{AA} + n_{Aa}}{p_A} - \frac{n_{Aa} + 2n_{aa}}{1 - p_A}$$

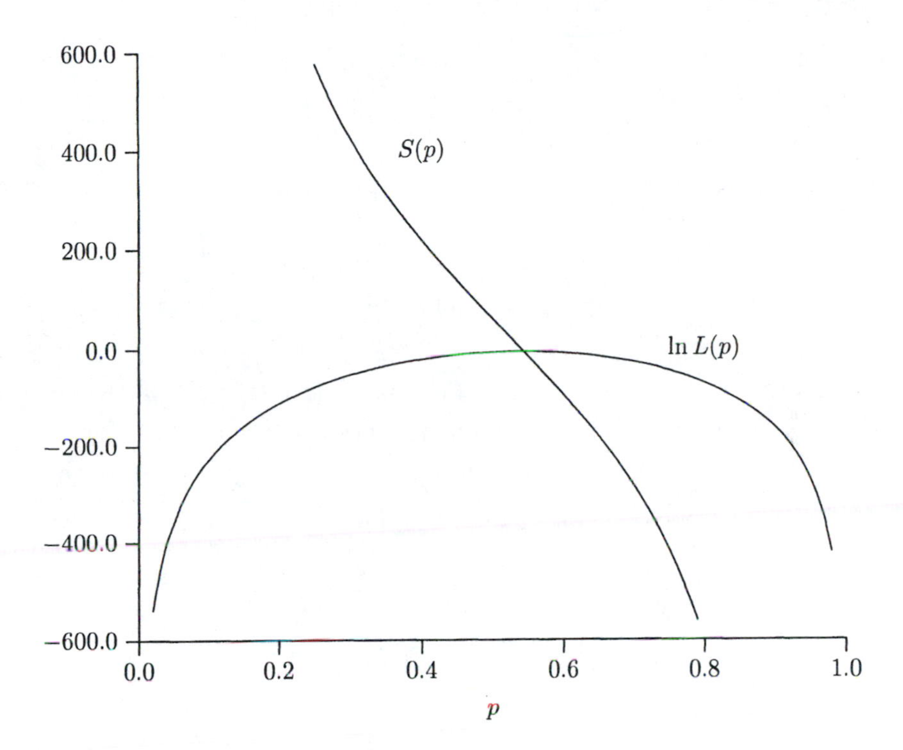

Figure 2.1 Support [$\ln L(p)$] and score [$S(p)$] for the frequency of M allele in the data of Table 2.3.

and setting this score to zero leads to the MLE of

$$\hat{p}_A = \frac{1}{2n}(2n_{AA} + n_{Aa})$$

which is just the observed allele frequency $\tilde{p}_A$. Furthermore, the score S_{p_A} has a derivative with respect to p_A that is everywhere negative, which means that the second derivative of the likelihood is negative when $p_A = \tilde{p}_A$. Therefore, this estimate does indeed maximize the likelihood. As an illustration, the log-likelihood and support curves for p_M for the data in Table 2.3 are shown in Figure 2.1. The shape of these curves makes it clear that there is no ambiguity in estimating an allele frequency for a locus with two codominant alleles. The log-likelihood has its maximum value of -5.38 when $p = 0.54$, and the score is zero at the same p value.

A less trivial case is provided by not assuming Hardy-Weinberg, and using the p_A, f parameterization for a within-population analysis

$$\begin{aligned} P_{AA} &= p_A^2 + p_A(1-p_A)f \\ P_{Aa} &= 2p_A(1-p_A)(1-f) \\ P_{aa} &= (1-p_A)^2 + p_A(1-p_A)f \end{aligned}$$

The likelihood and support for the two parameters are now

$$\begin{aligned} L(p_A, f) &= \frac{n!}{n_{AA}!n_{Aa}!n_{aa}!}\{p_A[p_A + (1-p_A)f]\}^{n_{AA}} \\ &\quad \times [2p_A(1-p_A)(1-f)]^{n_{Aa}}\{(1-p_A)[(1-p_A)+p_Af]\}^{n_{aa}} \end{aligned}$$

$$\begin{aligned} \ln L(p_A, f) &= \text{Constant} + (n_{AA}+n_{Aa})\ln(p_A) + n_{AA}\ln[p_A + (1-p_A)f] \\ &\quad + (n_{Aa}+n_{aa})\ln(1-p_A) + n_{Aa}\ln(1-f) \\ &\quad + n_{aa}\ln[(1-p_A)+p_Af] \end{aligned}$$

There are two scores, one for p_A and one for f:

$$\begin{aligned} S_{p_A} &= \frac{n_{AA}+n_{Aa}}{p_A} - \frac{n_{Aa}+n_{aa}}{1-p_A} + \frac{n_{AA}(1-f)}{p_A+(1-p_A)f} \\ &\quad - \frac{n_{aa}(1-f)}{(1-p_A)+p_Af} \\ S_f &= \frac{n_{AA}(1-p_A)}{p_A+(1-p_A)f} - \frac{n_{Aa}}{1-f} + \frac{n_{aa}p_A}{(1-p_A)+p_Af} \end{aligned} \tag{2.19}$$

It is not a simple task to solve the equations $S_{p_A} = S_f = 0$, although it will be shown below that this is a case in which a procedure known as Bailey's method leads to a simpler set of equations. In general, there is often a need for numerical methods to solve likelihood equations.

Note that there is not an equivalent likelihood for parameters F and θ in a total analysis, within and between populations, because it is not simple to give a sampling distribution analogous to the multinomial distribution within populations. For populations at equilibrium under drift and mutation, Lange(1995) and Roeder et al. (1995) invoke the Dirichlet distribution for allele frequencies and so can construct a likelihood function.

Properties of Maximum Likelihood Estimates

Although the likelihood equations obtained by equating the scores to zero may be difficult to solve in some cases, the resulting estimates have a number of desirable properties, at least for large samples. These properties will be illustrated by continued reference to the multinomial distribution. More details are contained in the book on likelihood by Edwards (1992).

Estimating the multinomial parameters is relatively simple and just requires that care is taken not to attempt to estimate a set of parameters among which there is a functional relationship. Since

$$\sum_{i=1}^{k} Q_i = 1$$

it is necessary to remove the dependency by rewriting one of the Q_i's, say Q_k, as one minus the sum of the others. Then the likelihood in Equation 2.18, with the Q_i's regarded as the parameters, becomes

$$\begin{aligned} L(Q_1, \ldots, Q_{k-1}) &= \frac{n!}{n_1! \ldots n_k!}(Q_1)^{n_1} \cdots (Q_{k-1})^{n_{k-1}} \\ &\quad \times (1 - Q_1 - \cdots - Q_{k-1})^{n_k} \end{aligned}$$

and the score for Q_i is

$$S_{Q_i} = \frac{n_i}{Q_i} - \frac{n_k}{(1 - Q_1 - \cdots - Q_{k-1})}$$

Manipulating the $(k-1)$ equations $S_{Q_i} = 0$ leads to the result that the MLE's of the category probabilities Q_i are just the observed proportions $\tilde{Q}_i$:

$$\hat{Q}_i = \tilde{Q}_i$$

A more elegant way of obtaining this result is by the use of a Lagrange multiplier λ. The constraint, $\sum_i Q_i = 1$ is added to the log-likelihood as

$$\ln L = \text{Constant} + \sum_i n_i \ln(Q_i) + \lambda(1 - \sum_i Q_i)$$

and the score for Q_i is

$$S_{Q_i} = \frac{n_i}{Q_i} - \lambda$$

Setting this equation to zero, and adding over i gives

$$\sum_i n_i = \lambda \sum_i \hat{Q}_i$$

so $\lambda = n$ because $\sum_i n_i = n$ and $\sum_i \hat{Q}_i = 1$. Therefore $\hat{Q}_i = n_i/n = \tilde{Q}_i$. Furthermore, the MLE of a function of parameters is the same function of the MLE's of the separate parameters. As an example, the MLE of a population allele frequency is the observed allele frequency since they are both the same function of population or observed genotypic frequencies. A less obvious result is that the MLE of a squared allele frequency p_i^2 for a Hardy-Weinberg population is the square of the MLE of the allele frequency p_i

$$\widehat{(p_i^2)} = (\tilde{p}_i)^2$$

From Equation 2.9 it is known that the expected value of the squared allele frequency, however, is not the square of the expected allele frequency, and this demonstrates that MLE's may be biased. In other words, the MLE of p_i^2 is $\tilde{p}_i^2$, but if this estimate was calculated in very many samples from the same population the average value of all the estimates would not be the quantity p_i^2 being estimated.

The desirable general properties of MLE's follow, in part, from them being functions of ***sufficient statistics***, which themselves are functions of data that contain all the relevant information about the parameters. In the multinomial case the counts are sufficient, whereas the complete description of the outcomes of multinomial sampling would also give the order in which each of the categories was observed. Provided there is a set of sufficient statistics for the set of parameters being estimated, the MLE's of those parameters will be unique (see Kendall and Stuart 1979 for further discussion in this area).

Although it cannot be said that MLE's are unbiased, something can be said about their values in very large samples. It can be shown that, under very general conditions, MLE's are ***consistent*** estimators, which means that, as the sample size becomes very large, the MLE $\hat{\phi}$ for a parameter ϕ becomes arbitrarily close to ϕ.

Large sample sizes also allow statements to be made about the variances of MLE's, and these statements are phrased in terms of the ***information***. Recall that the score of a parameter is the derivative of the support, or log-likelihood, with respect to that parameter. The second derivative of the support, multiplied by -1, is called the information. For a single parameter ϕ and likelihood $L(\phi)$, the score is

$$S_\phi = \frac{\partial \ln L(\phi)}{\partial \phi}$$

and the information is

$$I(\phi) = -\left(\frac{\partial^2 \ln L(\phi)}{\partial \phi^2}\right)$$

For large samples, the inverse of the expected value of the information provides the variance of the MLE $\hat{\phi}$

$$\mathrm{Var}(\hat{\phi}) = 1/\mathcal{E}[I(\phi)]$$

When the information is a linear function of counts, the expected value is found by replacing each count n_i by its expected value. For any unbiased estimator (not necessarily an MLE), the inverse of the expected information provides a lower bound for the variance of an estimate for any sample size.

There is a straightforward extension to the case when several parameters are estimated. The information then has a matrix form, with (i, j)th element being the second derivative of the log-likelihood with respect to the ith and jth parameters. This information matrix provides the variances and covariances for the estimates. For s independent parameters ϕ_i written as a vector $\boldsymbol{\phi}$ the expected information matrix is

$$\mathcal{E}[I(\boldsymbol{\phi})] = \begin{bmatrix} -\mathcal{E}\left(\frac{\partial^2 \ln L(\phi)}{\partial\phi_1^2}\right) & -\mathcal{E}\left(\frac{\partial^2 \ln L(\phi)}{\partial\phi_1\partial\phi_2}\right) & \cdots & -\mathcal{E}\left(\frac{\partial^2 \ln L(\phi)}{\partial\phi_1\partial\phi_s}\right) \\ -\mathcal{E}\left(\frac{\partial^2 \ln L(\phi)}{\partial\phi_2\partial\phi_1}\right) & -\mathcal{E}\left(\frac{\partial^2 \ln L(\phi)}{\partial\phi_2^2}\right) & \cdots & -\mathcal{E}\left(\frac{\partial^2 \ln L(\phi)}{\partial\phi_2\partial\phi_s}\right) \\ \vdots & \vdots & \vdots & \vdots \\ -\mathcal{E}\left(\frac{\partial^2 \ln L(\phi)}{\partial\phi_s\partial\phi_1}\right) & -\mathcal{E}\left(\frac{\partial^2 \ln L(\phi)}{\partial\phi_s\partial\phi_2}\right) & \cdots & -\mathcal{E}\left(\frac{\partial^2 \ln L(\phi)}{\partial\phi_s^2}\right) \end{bmatrix}$$

Inverting this matrix gives the large-sample variances and covariances

$$\{\mathcal{E}[I(\boldsymbol{\phi})]\}^{-1} = \begin{bmatrix} \mathrm{Var}(\hat{\phi}_1) & \mathrm{Cov}(\hat{\phi}_1, \hat{\phi}_2) & \cdots & \mathrm{Cov}(\hat{\phi}_1, \hat{\phi}_s) \\ \mathrm{Cov}(\hat{\phi}_2, \hat{\phi}_1) & \mathrm{Var}(\hat{\phi}_2) & \cdots & \mathrm{Cov}(\hat{\phi}_2, \hat{\phi}_s) \\ \vdots & \vdots & \vdots & \vdots \\ \mathrm{Cov}(\hat{\phi}_s, \hat{\phi}_1) & \mathrm{Cov}(\hat{\phi}_s, \hat{\phi}_2) & \cdots & \mathrm{Var}(\hat{\phi}_s) \end{bmatrix}$$

As an illustration of these results, consider the estimation of allele frequencies for a locus with three alleles that is known to be in Hardy-Weinberg equilibrium. Although it is known that the estimators are the observed allele frequencies, and that these are multinomially distributed in the Hardy-Weinberg situation, it is instructive to calculate scores, information matrix, and variance-covariance matrix. Label the counts for the six possible genotypes as

A_1A_1	A_1A_2	A_2A_2	A_1A_3	A_2A_3	A_3A_3
n_1	n_2	n_3	n_4	n_5	n_6

and let n be the sum of these counts. Take p_1 and p_2 as the two independent parameters that can be estimated. The likelihood is

$$\begin{aligned} L(p_1, p_2) &\propto (p_1^2)^{n_1}(2p_1p_2)^{n_2}(p_2^2)^{n_3}[2p_1(1-p_1-p_2)]^{n_4} \\ &\quad \times [2p_2(1-p_1-p_2)]^{n_5}[(1-p_1-p_2)^2]^{n_6} \end{aligned}$$

and the two scores are

$$\begin{aligned} S_1 &= \frac{\partial \ln L}{\partial p_1} = \frac{2n_1+n_2+n_4}{p_1} - \frac{2n_6+n_4+n_5}{1-p_1-p_2} \\ S_2 &= \frac{\partial \ln L}{\partial p_2} = \frac{2n_3+n_2+n_5}{p_2} - \frac{2n_6+n_4+n_5}{1-p_1-p_2} \end{aligned}$$

Setting the scores to zero provides the estimates

$$\begin{aligned} \hat{p}_1 &= \tilde{p}_1 = \frac{2n_1+n_2+n_4}{2n} \\ \hat{p}_2 &= \tilde{p}_2 = \frac{2n_3+n_2+n_5}{2n} \end{aligned}$$

and the derivatives of the scores are

$$\frac{\partial S_1}{\partial p_1} = \frac{\partial^2 \ln L}{\partial p_1^2} = -\frac{2n_1+n_2+n_4}{p_1^2} - \frac{2n_6+n_4+n_5}{(1-p_1-p_2)^2}$$

$$\frac{\partial S_2}{\partial p_2} = \frac{\partial^2 \ln L}{\partial p_2^2} = -\frac{2n_3+n_2+n_5}{p_2^2} - \frac{2n_6+n_4+n_5}{(1-p_1-p_2)^2}$$

$$\frac{\partial S_1}{\partial p_2} = \frac{\partial S_2}{\partial p_1} = \frac{\partial^2 \ln L}{\partial p_1 \partial p_2} = -\frac{2n_6+n_4+n_5}{(1-p_1-p_2)^2}$$

Evaluating these at the expected values of the counts gives the expected information matrix

$$\mathcal{E}(I) = 2n \begin{bmatrix} \frac{1}{p_1} + \frac{1}{(1-p_1-p_2)} & \frac{1}{(1-p_1-p_2)} \\ \frac{1}{(1-p_1-p_2)} & \frac{1}{p_2} + \frac{1}{(1-p_1-p_2)} \end{bmatrix}$$

which has inverse

$$[\mathcal{E}(I)]^{-1} = \begin{bmatrix} \frac{p_1(1-p_1)}{2n} & -\frac{p_1p_2}{2n} \\ -\frac{p_1p_2}{2n} & \frac{p_2(1-p_2)}{2n} \end{bmatrix} = \begin{bmatrix} \mathrm{Var}(\hat{p}_1) & \mathrm{Cov}(\hat{p}_1, \hat{p}_2) \\ \mathrm{Cov}(\hat{p}_1, \hat{p}_2) & \mathrm{Var}(\hat{p}_2) \end{bmatrix}$$

The elements of this matrix are just the multinomial variances and covariances.

For large samples, MLE's are unbiased for the parameters they estimate and have variances supplied by the information, and it can further be shown that they are approximately normally distributed. For a single parameter ϕ, the MLE $\hat{\phi}$ has the approximate normal distribution with mean ϕ and variance $1/\mathcal{E}[I(\phi)]$

$$\hat{\phi} \sim N\left(\phi, \{\mathcal{E}[I(\phi)]\}^{-1}\right) \tag{2.20}$$

in large samples. For several parameters, the MLE's are asymptotically multivariate normal. This normality allows tests of hypotheses to be set up very easily (see Chapter 3), although these tests do require large sample sizes.

Bailey's Method for MLE's

If the number of independent parameters to be estimated equals the number of independent pieces of information, or degrees of freedom, in the data, Bailey (1951) showed that MLE's for the parameters may be found by equating observations to their expected values. Doing so often gives equations that are easier to solve than those from setting the scores equal to zero.

Suppose there are s parameters to estimate, and s degrees of freedom in the data. Bailey's rule says to write the expectation of n_i, the observed count in category i, as m_i and then solve the equations

$$m_i = n_i$$

for the s parameters ϕ_j. To see that this does indeed provide MLE's, note that, because

$$\sum_{i=1}^{k} m_i = n$$

where n is the sample size (a constant) and k is the number of categories in the data, the derivative with respect to ϕ_j is zero

$$\sum_{i=1}^{k} \frac{\partial m_i}{\partial \phi_j} = 0 \qquad j = 1, 2, \ldots, s$$

Now the log-likelihood in this multinomial case is

$$\ln L = \text{Constant} + \sum_i n_i \ln m_i$$

which has derivatives

$$\begin{aligned} S_j &= \frac{\partial \ln L}{\partial \phi_j} \\ &= \sum_i n_i \frac{\partial \ln m_i}{\partial \phi_j} \\ &= \sum_i \frac{n_i}{m_i} \frac{\partial m_i}{\partial \phi_j} \end{aligned}$$

and when $m_i = n_i$ this reduces to

$$S_j = \sum_i \frac{\partial m_i}{\partial \phi_j}$$

As the sum on the right-hand side in this last equation has been shown to be zero, putting $m_i = n_i$ has made the score S_j equal to zero and the solutions must be MLE's. For the two sets of equations to have solutions, it is necessary that $s = k - 1$. The number of parameters must equal the number of independent categories.

The power of this procedure is illustrated by returning to the case of estimating allele frequency p_A and inbreeding coefficient f for a two-allele locus. The two independent categories can be taken to be AA homozygotes and Aa heterozygotes, leading to

$$\begin{aligned} n[\hat{p}_A^2 + \hat{p}_A(1-\hat{p}_A)\hat{f}] &= n_{AA} \\ \text{i.e., } m_1 &= n_1 \\ n[2\hat{p}_A(1-\hat{p}_A)(1-\hat{f})] &= n_{Aa} \\ \text{i.e., } m_2 &= n_2 \end{aligned}$$

These equations may be solved very easily:

$$\begin{aligned} \hat{p}_A &= \frac{1}{2n}(2n_{AA} + n_{Aa}) \\ \hat{f} &= 1 - \frac{n_{Aa}}{2n\hat{p}_A(1-\hat{p}_A)} \end{aligned} \tag{2.21}$$

The MLE for allele frequency is still the observed value, and that for the inbreeding coefficient is one minus the observed heterozygosity divided by the heterozygosity expected under Hardy-Weinberg equilibrium. These solutions do make the scores in Equations 2.19 equal to zero. If the parameter f is zero, then it transpires that the expected value of the estimate $\hat{f}$ is $-1/(2n-1)$ so there is a small downward bias. (This result can be shown from the distribution of n_{Aa} conditional on allele counts n_A, n_a introduced in Chapter 3.) For other values of f, the expected value of $\hat{f}$ also depends on the allele frequencies p_A, p_a.

If genotypic frequencies were to be expressed in terms of allele frequencies and a single inbreeding coefficient for the k-allele case $(k > 2)$

$$\begin{aligned} P_{uu} &= p_u^2 + p_u(1-p_u)f, \quad u = 1, 2, \ldots, k \\ P_{uv} &= 2p_u p_v(1-f), \quad u, v = 1, 2, \ldots, k;\ u \neq v \end{aligned} \tag{2.22}$$

there are a total of k independent parameters, f and $k-1$ allele frequencies. This is not the same as the number of independent categories, $k(k+1)/2-1$, and Bailey's method cannot be applied.

Even if Bailey's method provides values for MLE's very easily, there is still the problem of attaching variances to the estimates. Even in the relatively simple case of estimating p_A and f for a two-allele locus, it is not easy to invert the information matrix. For a large-sample variance, Fisher's variance formula (Equation 2.17) can be employed instead, after rewriting $\hat{f}$ as

$$\hat{f} = 1 - 2n\frac{n_{Aa}}{(2n_{AA}+n_{Aa})(2n_{aa}+n_{Aa})}$$

The necessary derivatives are

$$\begin{aligned} \frac{\partial \hat{f}}{\partial n_{AA}} &= 4n\frac{n_{Aa}}{(2n_{AA}+n_{Aa})^2(2n_{aa}+n_{Aa})} \\ \frac{\partial \hat{f}}{\partial n_{Aa}} &= 2n\frac{(n_{Aa}^2 - 4n_{AA}n_{aa})}{(2n_{AA}+n_{Aa})^2(2n_{aa}+n_{Aa})^2} \\ \frac{\partial \hat{f}}{\partial n_{aa}} &= 4n\frac{n_{Aa}}{(2n_{AA}+n_{Aa})(2n_{aa}+n_{Aa})^2} \\ \frac{\partial \hat{f}}{\partial n} &= -\frac{2n_{Aa}}{(2n_{AA}+n_{Aa})(2n_{aa}+n_{Aa})} \end{aligned}$$

Evaluating these at the expected values for the three genotypic counts and simplifying the variance expression

$$\frac{1}{n}\mathrm{Var}(\hat{f}) = \left(\frac{\partial \hat{f}}{\partial n_{AA}}\right)^2 P_{AA} + \left(\frac{\partial \hat{f}}{\partial n_{Aa}}\right)^2 P_{Aa} + \left(\frac{\partial \hat{f}}{\partial n_{aa}}\right)^2 P_{aa} - \left(\frac{\partial \hat{f}}{\partial n}\right)^2$$

gives the approximate variance of

$$\mathrm{Var}(\hat{f}) = \frac{1}{n}(1-f)^2(1-2f) + \frac{f(1-f)(2-f)}{2np_A(1-p_A)}$$

This expression, given by Fyfe and Bailey (1951), is written in terms of the parameters p_A and f, which are generally unknown, but estimated values could be substituted to estimate the variance of $\hat{f}$. The corresponding formula for the variance of the estimated allele frequency is

$$\mathrm{Var}(\hat{p}_A) = \frac{1}{2n}(1+f)p_A(1-p_A)$$

as in Equation 2.7 if P_{uu} is replaced by $p_u^2+p_u(1-p_u)f$. When the parameter f is zero, the variance of the estimated value is $\mathrm{Var}(\hat{f}) = 1/n$, and the binomial variance $p_A(1-p_A)/2n$ obtains for $\mathrm{Var}(\hat{p}_A)$.

Iterative Solutions of Likelihood Equations

There are often situations when the maximum likelihood equations obtained by setting the scores to zero do not yield analytical solutions and when Bailey's method cannot be applied. Numerical methods are needed to find the estimates for any particular set of data. The most direct method is simply to evaluate the likelihood for a range of values of the parameter(s) to be estimated, and choose the value(s) that give the largest likelihood. This ***grid search*** procedure can be made quite sophisticated – the range of parameter values could be divided into tenths, and then the tenth that appears to contain the maximum likelihood estimate itself could be divided into 10 parts. As many rounds of the search could be performed as significant digits are required in the solution. It is helpful to plot the likelihood to show how it is responding to changes in parameter values, and to guard against a local maximum being confused for a global maximum.

An alternative approach is given by ***Newton-Raphson iteration***. Briefly, some initial value is chosen for the estimate and then this value is modified by using the score. The modified value is modified in turn and the process continues until successive iterates differ by less than some specified amount. For a parameter ϕ, write the required MLE as $\hat{\phi}$, and the initial value, or guess, as ϕ'. The score S_ϕ is to be zero at $\hat{\phi}$, and that score can be expanded by Taylor's theorem as

$$S_{\hat{\phi}} = 0 = S_{\phi'} + (\hat{\phi} - \phi')\left[\frac{\partial S_\phi}{\partial \phi}\right]_{\phi=\phi'}$$

where higher order terms in $(\hat{\phi}-\phi')$ are ignored. Rearranging this expression provides an approximate value ϕ'' for $\hat{\phi}$:

$$\begin{aligned}\phi'' &= \phi' - S_{\phi'} / \left[\frac{\partial S_\phi}{\partial \phi}\right]_{\phi=\phi'} \\ &= \phi' + S_{\phi'}/I(\phi')\end{aligned}$$

The initial value ϕ' is modified by adding to it the score divided by the information, both evaluated at the initial value. The new value then serves as an initial value for a further modification

$$\phi''' = \phi'' + S(\phi'')/I(\phi'')$$

and the iteration continues until convergence, i.e., successive values are very close to each other.

For situations with many parameters, the information $I(\phi)$ is a matrix, and the iteration procedure requires matrix inversion:

$$\phi'' = \phi' + I^{-1}(\phi')S(\phi')$$

Obviously the method breaks down if the information is zero, or the information matrix is singular. Whether or not such problems occur, it is always advisable to try several starting values and to compare the likelihoods found after convergence. This guards against problems of nonconvergence, or convergence to solutions other than maximum likelihood.

The EM Algorithm

Another iterative procedure that can lead to MLE's was discussed by Dempster et al. (1977) to handle cases in which the observations can be regarded as being incomplete data. In other words, there are more categories in the data than can be distinguished. The model for which the likelihood is defined is used by the procedure to estimate the frequencies of these hidden categories. Each iteration in this ***EM algorithm*** consists of an Expectation step followed by a Maximization step. For multinomial data the method has also been called ***gene-counting*** (Ceppellini et al. 1955, Smith 1957).

One of the simplest situations in which the procedure may be used is the estimation of allele frequencies at a two-allele locus when one allele, A, is dominant to the other, a. Although there are three genotypic classes, AA, Aa, and aa, only two can be distinguished, AA+Aa and aa. To find the MLE of the frequency p_a of the recessive allele, the first step is to estimate the two unknown genotypic frequencies P_{AA} and P_{Aa}. These two add to $1 - p_a^2$

and are in the proportion $(1-p_a)^2$ to $2p_a(1-p_a)$. They are estimated to be those proportions of the estimate $(n-n_{aa})/n$ of their sample total. If some initial value p'_a is given to p_a and used to construct the unknown genotypic frequencies, the maximization step consists of saying that the MLE of p_a is just the "observed" frequency found using the estimated heterozygote count n^*_{Aa}:

$$\begin{aligned}\hat{p}_a &= \frac{1}{2n}(n^*_{Aa}+2n_{aa}) \\ &= \frac{1}{2n}\left[\frac{2p'_a(1-p'_a)}{1-(p'_a)^2}(n-n_{aa})+2n_{aa}\right] \qquad (2.23)\end{aligned}$$

The value found for the left-hand side of this equation is then regarded as the next iterate and is substituted into the right-hand side for the next iteration. Iteration is continued until successive values are (nearly) equal. As with Newton-Raphson iteration, several starting values should be used.

In this example of estimating the frequency of a recessive allele, an analytical solution can be found by recognizing that convergence means that the value of $\hat{p}_a$ will be unchanged by Equation 2.23:

$$\hat{p}_a = \frac{1}{n}\left[\frac{\hat{p}_a(1-\hat{p}_a)}{1-\hat{p}_a^2}(n-n_{aa})+n_{aa}\right]$$

and this equation can be solved to yield

$$\hat{p}_a = \sqrt{n_{aa}/n} \qquad (2.24)$$

Although this example illustrated the EM algorithm, it can be handled much more directly by noting that there are two observable classes or one degree of freedom in the data and that there is one parameter to be found. Bailey's method leads to the estimate in Equation 2.24 directly. Note that the method rests on an assumption about the missing observations: they were estimated on the assumption of Hardy-Weinberg equilibrium. Without this assumption, it is not possible to estimate allele frequencies at loci that show dominance.

Fisher's approximate variance method, Equation 2.17, applied to the estimate in Equation 2.24 shows that

$$\begin{aligned}\text{Var}(\hat{p}_a) &\approx \frac{1}{4n}(1-P_{aa}) \\ &\hat{=} \frac{n-n_{aa}}{4n^2}\end{aligned}$$

Table 2.6 Frequencies of genotypes and phenotypes for ABO blood groups. Alleles A, B, O have frequencies p, q, r.

Genotype	Phenotype	Count	Expected Frequency	Estimated Count
AA	A	n_A	p^2	$n^*_{AA} = [p/(p+2r)]n_A$
AO			$2pr$	$n^*_{AO} = [2r/(p+2r)]n_A$
BB	B	n_B	q^2	$n^*_{BB} = [q/(q+2r)]n_B$
BO			$2qr$	$n^*_{BO} = [2r/(q+2r)]n_B$
AB	AB	n_{AB}	$2pq$	n_{AB}
OO	O	n_O	r^2	n_O

For Hardy-Weinberg proportions, $P_{aa} = p_a^2$, this variance is larger than the value found for codominant alleles ("complete" data):

$$\text{Var}(\hat{p}_a) = \frac{1}{2n}p_a(1 - p_a)$$

reflecting the fact that less information is available. Smith (1957) related the ratio of these two variances to the limiting rate of convergence of the iterative EM process. Since explicit estimating formulas do not usually result from the EM algorithm, estimation of variances of the estimates is something of a problem. One solution is to substitute the estimates into the information matrix and invert that, as if the estimates had been obtained by solving the likelihood equations. Another solution is to use the numerical resampling methods mentioned earlier.

An interesting example is the classic one of estimating the three allele frequencies for the ABO blood group system. With three alleles; A, B, and O, there are six genotypes but only four phenotypic classes as shown with the conventional notation in Table 2.6. Alleles A, B, O have frequencies p, q, r. Although there are only two independent parameters, EM equations can be set up for all three after estimating the missing counts for genotypes involving the recessive allele O:

$$p'' = \frac{1}{2n}(2n^*_{AA} + n^*_{AO} + n_{AB})$$

Table 2.7 Genotypic counts for Ss blood group in mothers and fathers of Table 1.2.

Genotype	Father	Mother	Total
SS orSs	32	32	64
Ss	20	23	43
ss	41	38	79
Total	93	93	186

$$
\begin{aligned}
&= \frac{p'+r'}{p'+2r'}\frac{n_A}{n} + \frac{n_{AB}}{2n} \\
q'' &= \frac{q'+r'}{q'+2r'}\frac{n_B}{n} + \frac{n_{AB}}{2n} \\
r'' &= \frac{r'}{p'+2r'}\frac{n_A}{n} + \frac{r'}{q'+2r'}\frac{n_B}{n} + \frac{2n_O}{2n}
\end{aligned}
$$

The initial values are p', q', r' and the next iterates are p'', q'', r''. A convenient set of initial values was given by Bernstein (1925):

$$p' = 1 - \sqrt{(n_O + n_B)/n},\ q' = 1 - \sqrt{(n_O + n_A)/n},\ r' = \sqrt{n_O/n}$$

Examples

Turn back to the *MNS* blood group data in Table 1.2. The system can be regarded as being determined by a codominant locus with alleles M, N and a dominant locus with alleles S, s — although other interpretations are possible, such as that of a single locus with alleles *MS, Ms, NS*, and *ns*. The data in Table 1.2 were incomplete in that family data were used to infer all three genotypes at the S locus in some families but not in others. The counts at this locus are given in Table 2.7.

Finding the MLE of the frequency of allele s under the assumption of Hardy-Weinberg equilibrium requires that the 64 dominant phenotypes be partitioned into n^*_{SS} *SS* homozygotes and n^*_{Ss} *Ss* heterozygotes, leading to an iterative equation

$$p'_s = \frac{1}{2n}(2n_{ss} + n_{Ss})$$

$$\begin{aligned} &= \frac{2\times 79 + (43 + n^*_{Ss})}{2\times 186} \\ &= \frac{1}{2\times 186}\left(2\times 79 + 43 + 64\frac{2p_s}{1+p_s}\right) \end{aligned}$$

that can be solved iteratively or explicitly to give $\hat{p}_s = 0.68$.

Data of Morton (1964), discussed by Yasuda and Kimura (1968), will be used to illustrate the EM algorithm for ABO frequencies. The genotypic counts they used were

$$n_A = 725,\ n_B = 258,\ \ n_{AB} = 72,\ \ \ n_O = 1073,\ n = 2128$$

so that the Bernstein estimators are

$$p'_A = 0.2091,\ \ p'_B = 0.0808,\ \ \ p'_O = 0.7101$$

The EM equations given above produce the numerical values displayed in Table 2.8 for both Bernstein's initial values and equally frequent initial values. The convergence is very rapid, and Bernstein's values are very close to the MLE's.

To complete this example, the Newton-Raphson iterative technique is applied to the likelihood equations obtained by setting the scores for frequencies p_A and p_B to zero. The likelihood is

$$\begin{aligned} L \ &\propto\ [p_A(2-p_A-2p_B)]^{n_A}[p_B(2-2p_A-p_B)]^{n_B}[2p_Ap_B]^{n_{AB}} \\ &\quad\times [(1-p_A-p_B)^2]^{n_O} \end{aligned}$$

and the scores are

$$\begin{aligned} S_A &= \frac{n_A+n_{AB}}{p_A} - \frac{n_A}{2-p_A-2p_B} - \frac{2n_B}{2-2p_A-p_B} - \frac{2n_O}{1-p_A-p_B} \\ S_B &= \frac{n_B+n_{AB}}{p_B} - \frac{2n_A}{2-p_A-2p_B} - \frac{n_B}{2-2p_A-p_B} - \frac{2n_O}{1-p_A-p_B} \end{aligned}$$

with derivatives

$$\begin{aligned} -\frac{\partial S_A}{\partial p_A} &= \frac{n_A+n_{AB}}{p_A^2} + \frac{n_A}{(2-p_A-2p_B)^2} \\ &\quad + \frac{4n_B}{(2-2p_A-p_B)^2} + \frac{2n_O}{(1-p_A-p_B)^2} \\ -\frac{\partial S_B}{\partial p_B} &= \frac{n_B+n_{AB}}{p_B^2} + \frac{4n_A}{(2-p_A-2p_B)^2} \\ &\quad + \frac{n_B}{(2-2p_A-p_B)^2} + \frac{2n_O}{(1-p_A-p_B)^2} \end{aligned}$$

Table 2.8 Numerical values of EM iterates for *ABO* data of Morton (1964).

Iterate	$\hat{p}_A$	$\hat{p}_B$	$\hat{p}_O$
	Bernstein's initial values		
0	0.20913343	0.08080208	0.71009107
1	0.20913030	0.08080095	0.71006875
2	0.20913061	0.08080100	0.71006838
3	0.20913065	0.08080101	0.71006834
4	0.20913065	0.08080101	0.71006834
5	0.20913065	0.08080101	0.71006834
	Equal initial values		
0	0.33333333	0.33333333	0.33333333
1	0.24404762	0.09774436	0.65820802
2	0.21390645	0.08172757	0.70436598
3	0.20972135	0.08086164	0.70941701
4	0.20920200	0.08080616	0.70999184
5	0.20913921	0.08080154	0.71005925
6	0.20913168	0.08080107	0.71006725
7	0.20913078	0.08080102	0.71006821
8	0.20913067	0.08080101	0.71006832
9	0.20913066	0.08080101	0.71006834
10	0.20913065	0.08080101	0.71006834

$$\begin{aligned} -\frac{\partial S_A}{\partial p_B} &= -\frac{\partial S_B}{\partial p_A} \\ &= \frac{2n_A}{(2-p_A-2p_B)^2} + \frac{2n_B}{(2-2p_A-p_B)^2} + \frac{2n_O}{(1-p_A-p_B)^2} \end{aligned}$$

Writing the two parameters and two scores as vectors

$$\begin{aligned} p &= \begin{bmatrix} p_A \\ p_B \end{bmatrix} \\ S &= \begin{bmatrix} S_A \\ S_B \end{bmatrix} \end{aligned}$$

and the derivatives in an information matrix

$$I = \begin{bmatrix} -\frac{\partial S_A}{\partial p_A} & -\frac{\partial S_A}{\partial p_B} \\ -\frac{\partial S_B}{\partial p_A} & -\frac{\partial S_B}{\partial p_B} \end{bmatrix}$$

allows a pair of iterative equations to be expressed in matrix form as

$$p' = p + I^{-1}S$$

where p is the initial value and p' is the next iterate.

The information matrix is evaluated by replacing the counts with their expected values using the initial values. Numerical values for this scheme, using two sets of initial values, are shown in Table 2.9, with the variances and covariances being the elements of the inverted information matrix. The possible problems with Newton-Raphson iteration, where the iterates can go outside the permissible range, are illustrated for the case of equally frequent initial values in Table 2.9 . The values in the table were obtained by replacing negative iterates by very small positive quantities.

The advantage of the Newton-Raphson procedure is that estimates of variances and covariances are obtained as part of the process. In addition to the variances shown in Table 2.9, the variance of the estimated frequency $\hat{p}_O$, that is obtained by subtraction, can be found from

$$\begin{aligned} \mathrm{Var}(\hat{p}_O) &= \mathrm{Var}(1 - \hat{p}_A - \hat{p}_B) \\ &= \mathrm{Var}(\hat{p}_A) + \mathrm{Var}(\hat{p}_B) + 2\mathrm{Cov}(\hat{p}_A, \hat{p}_B) \\ &= 0.00004394 + 0.00001821 - 2(0.00000395) \\ &= 0.00005425 \end{aligned}$$

Gametic Frequencies

When pairs of loci are scored, even for codominant loci, it is generally not possible to distinguish between the two double heterozygotes for any two pairs of alleles. This makes the determination of gametic frequencies difficult. Consider loci A and B with alleles A, a and B, b, respectively. The frequency with which the alleles A and B are transmitted from a parent to an offspring is the gametic frequency p_{AB} given by

$$p_{AB} = P_{AB}^{AB} + \frac{1}{2}(P_{Ab}^{AB} + P_{aB}^{AB} + P_{ab}^{AB}) \tag{2.25}$$

but the frequency P_{ab}^{AB} cannot be disentangled from the observed total frequency $P_{ab}^{AB} + P_{aB}^{Ab}$ of double heterozygotes. Provided the population is

Table 2.9 Numerical values of Newton-Raphson iterates for ABO data of Morton (1964).

Iterate	$\hat{p}_A$	$\hat{p}_B$	Var($\hat{p}_A$)	Var($\hat{p}_B$)	Cov($\hat{p}_A, \hat{p}_B$)
			Bernstein's initial values		
0	0.20913343	0.08080208			
1	0.20913065	0.08080101	0.00004394	0.00001821	−0.00000395
2	0.20913065	0.08080101	0.00004394	0.00001821	−0.00000395
3	0.20913065	0.08080101	0.00004394	0.00001821	−0.00000395
4	0.20913065	0.08080101	0.00004394	0.00001821	−0.00000395
5	0.20913065	0.08080101	0.00004394	0.00001821	−0.00000395
			Equal initial values		
0	0.33333333	0.33333333			
1	0.40009792	−0.00377658	0.00009574	0.00010628	−0.00008008
2	0.19331630	0.00019990	0.00008372	0.00000000	0.00000000
3	0.22262186	0.00039935	0.00003961	0.00000000	0.00000000
5	0.22625982	0.00158651	0.00005049	0.00000000	0.00000000
6	0.22592373	0.00314427	0.00005043	0.00000001	0.00000000
8	0.22403217	0.01191091	0.00004999	0.00000012	−0.00000002
10	0.21828522	0.03852170	0.00004854	0.00000148	−0.00000032
12	0.21033711	0.07524251	0.00004552	0.00001014	−0.00000219
14	0.20913096	0.08079959	0.00004396	0.00001806	−0.00000392
15	0.20913065	0.08080101	0.00004394	0.00001821	−0.00000395

assumed to be mating at random, genotypic frequencies are the products of gametic frequencies and the EM algorithm may be employed. The two double heterozygote frequencies are estimated as the products of initial estimates of gametic frequencies, and then used to provide the next values of gametic frequencies. None of this affects the estimation of allele frequencies at each locus since these can be estimated directly as the observed frequencies

$$\tilde{p}_A = \tilde{P}_{AB}^{AB} + \tilde{P}_{Ab}^{AB} + \tilde{P}_{Ab}^{Ab} + \frac{1}{2}(\tilde{P}_{aB}^{AB} + \tilde{P}_{ab}^{AB} + \tilde{P}_{aB}^{Ab} + \tilde{P}_{ab}^{Ab})$$

$$\tilde{p}_B = \tilde{P}_{AB}^{AB} + \tilde{P}_{aB}^{AB} + \tilde{P}_{aB}^{aB} + \frac{1}{2}(\tilde{P}_{Ab}^{AB} + \tilde{P}_{ab}^{AB} + \tilde{P}_{aB}^{Ab} + \tilde{P}_{ab}^{aB})$$

If the initial estimate of the frequency of AB gametes is p'_{AB}, the frequencies of the three other gametes can be found from

$$\begin{aligned} p'_{Ab} &= \tilde{p}_A - p'_{AB} \\ p'_{aB} &= \tilde{p}_B - p'_{AB} \\ p'_{ab} &= 1 - \tilde{p}_A - \tilde{p}_B + p'_{AB} \end{aligned}$$

so that the frequencies of AB/ab and Ab/aB heterozygotes can be estimated as

$$\begin{aligned} P_{ab}^{AB\prime} &= \frac{2p'_{AB}p'_{ab}}{2p'_{AB}p'_{ab} + 2p'_{Ab}p'_{aB}}\tilde{P}_{AaBb} \\ P_{aB}^{Ab\prime} &= \frac{2p'_{Ab}p'_{aB}}{2p'_{AB}p'_{ab} + 2p'_{Ab}p'_{aB}}\tilde{P}_{AaBb} \end{aligned}$$

which are functions of the single unknown quantity p'_{AB}. This is the estimation step of the EM procedure. The maximization (gene-counting) step then provides the new value p''_{AB} from Equation 2.25 of

$$p''_{AB} = \tilde{P}_{AB}^{AB} + \frac{1}{2}\left(\tilde{P}_{Ab}^{AB} + \tilde{P}_{aB}^{AB} + \frac{2p'_{AB}p'_{ab}}{2p'_{AB}p'_{ab} + 2p'_{Ab}p'_{aB}}\tilde{P}_{AaBb}\right) \quad (2.26)$$

This new value then serves as the initial value for another iteration, and the process continues until successive values are sufficiently close. The estimates obtained in this way will be maximum likelihood. There still remain the dangers inherent in all iterative schemes of the process not converging, or perhaps converging to the wrong value, and some examples of these cases are given by Weir and Cockerham (1979) (see Exercise 2.3). It appears that there is a greater chance of problems when either one of the loci does not have frequencies consistent with Hardy-Weinberg equilibrium, suggesting that gametes are not uniting at random as required by the model.

One way to avoid problems in this case is to recognize that Equation 2.26 can be regarded as a cubic equation in $\hat{p}_{AB}$ if the relation $p''_{AB} = p'_{AB} = \hat{p}_{AB}$ is used and the other gametic frequencies are replaced by expressions involving $\hat{p}_{AB}$, $\hat{p}_A$, and $\hat{p}_B$. Numerical methods can be used to solve the cubic, each of the three roots checked to make sure that all gametic frequencies are between zero and appropriate allele frequencies, and then the three likelihoods are calculated. The valid solution that maximizes the likelihood is the required MLE for p_{AB}. As with all multinomial samples, the log-likelihood is the sum over genotypes of the products of the count and the logarithm of the probability of that genotype. Assuming random mating, the probabilities

Table 2.10 Two-locus genotypic counts for *Idh1* and *Mdh* loci in mosquito data of Table 1.3.

		Mdh			
		BB	Bb	bb	
$Idh1$	AA	$n_{AABB} = 19$	$n_{AABb} = 5$	$n_{AAbb} = 0$	$n_{AA} = 24$
	Aa	$n_{AaBB} = 8$	$n_{AaBb} = 8$	$n_{Aabb} = 0$	$n_{Aa} = 16$
	aa	$n_{aaBB} = 0$	$n_{aaBb} = 0$	$n_{aabb} = 0$	$n_{aa} = 0$
		$n_{BB} = 27$	$n_{Bb} = 13$	$n_{bb} = 0$	$n = 40$

are the products of gametic probabilities and the log-likelihood becomes

$$\begin{aligned} \ln L(p_A, p_B, p_{AB}) &= \text{Constant} + (2n_{AABB} + n_{AABb} + n_{AaBB}) \ln(p_{AB}) \\ &\quad + (2n_{AAbb} + n_{AABb} + n_{Aabb}) \ln(p_A - p_{AB}) \\ &\quad + (2n_{aaBB} + n_{AaBB} + n_{aaBb}) \ln(p_B - p_{AB}) \\ &\quad + (2n_{aabb} + n_{Aabb} + n_{aaBb}) \ln(1 - p_A - p_B + p_{AB}) \\ &\quad + n_{AaBb} \ln[p_{AB}(1 - p_A - p_B + p_{AB}) \\ &\quad + (p_A - p_{AB})(p_B - p_{AB})] \end{aligned} \tag{2.27}$$

As an example, consider the data shown in Table 1.3. Two-locus genotypic counts for loci A = *Idh1* and B = *Mdh* are displayed in Table 2.10 in the notation of this section. From Table 2.10 the sample allele frequencies $\tilde{p}_A, \tilde{p}_B$ for alleles number 1 (A and B) at loci *Idh1* and *Mdh* are 0.80 and 0.84. These are the MLEs of the two frequencies, and then the log-likelihood for the frequency p_{AB} of the AB gamete is found from Equation 2.27 as

$$\begin{aligned} \ln L(p_{AB}) &= \text{Constant} + 51 \ln(p_{AB}) + 5 \ln(0.80 - p_{AB}) \\ &\quad + 8 \ln(0.84 - p_{AB}) + 8 \ln[p_{AB}(p_{AB} - 0.64) \\ &\quad + (0.80 - p_{AB})(0.84 - p_{AB})] \end{aligned}$$

This support, and the score, are plotted in Figure 2.2 over the range [0, 0.8] of values for p_{AB}. The score is zero at the three roots 0.46, 0.52, and 0.73, although only the third is a permissible value for p_{AB}. Values less than 0.64 make $p_{ab} = 1 - p_A - p_B + p_{AB}$ negative, and values greater than 0.8 make $p_{Ab} = p_A - p_{AB}$ negative. The first root gives a local maximum for the support, the second root gives a local minimum, and the third maximizes the support. The maximum likelihood solution is therefore $p_{AB} = 0.73$.

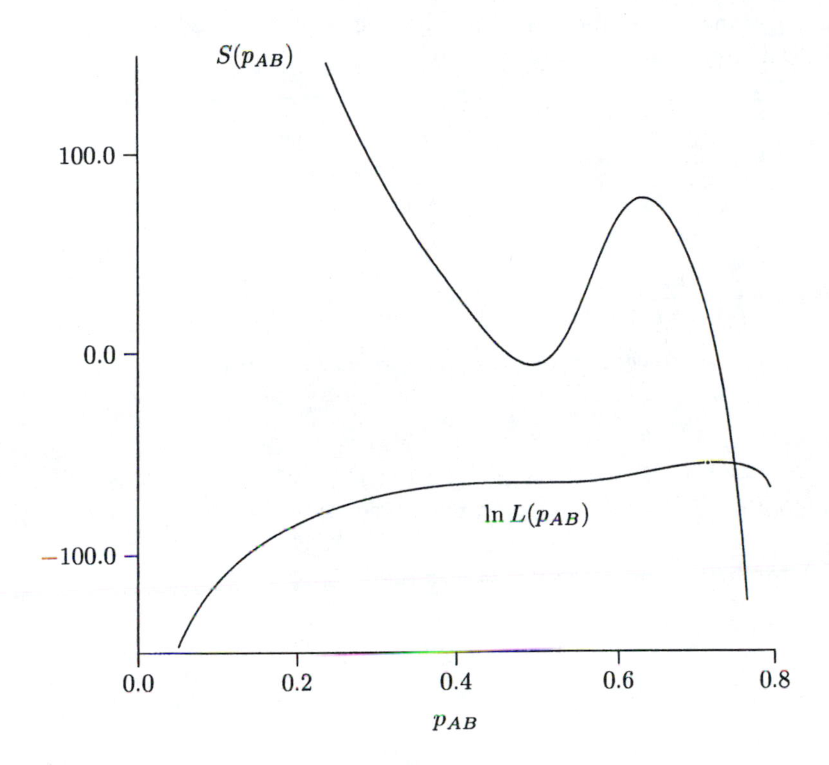

Figure 2.2 Support and score for frequency of *Idh1*-1, *Mdh*-1 gametes for mosquito data of Table 1.3.

Within-population Inbreeding Coefficient

Hill et al. (1995) gave iterative equations for finding the maximum likelihood estimate of the within-population inbreeding coefficient f, for the formulation in Equation 2.22. If homozygotes A_uA_u have sample count n_{uu} in a sample of size n, and allele A_u has sample count n_u, the likelihood

$$L \propto \prod_u [p_u^2 + p_u(1-p_u)f]^{n_{uu}} \prod_{u,v;v<u} [2p_up_v(1-f)]^{n_{uv}}$$

leads to, apart from a constant,

$$\begin{aligned}\ln L &= \sum_u (n_u - n_{uu})\ln(p_u) + \sum_u n_{uu}\ln[p_u + (1-p_u)f] \\ &\quad + (n - \sum_u n_{uu})\ln(1-f) + \lambda(1 - \sum_u p_u)\end{aligned}$$

The Lagrange multiplier term at the end was discussed earlier in the chapter. Differentiating with respect to p_u gives

$$\begin{aligned}\frac{\partial L}{\partial p_u} &= \frac{n_u - n_{uu}}{p_u} + \frac{n_{uu}(1-f)}{p_u + (1-p_u)f} - \lambda \\ &= \frac{n_u - x_u}{p_u} - \lambda\end{aligned}$$

where

$$x_u = \frac{f n_{uu}}{f + p_u(1-f)}$$

Assuming x_u was known, setting $\partial L/\partial p_u$ to zero and summing over u provides

$$p_u = \frac{n_u - x_u}{\sum_u (n_u - x_u)}$$

Differentiating the log-likelihood with respect to f gives

$$\frac{\partial L}{\partial f} = \sum_u \frac{n_{uu}(1-p_u)}{p_u + (1-p_u)f} - \frac{n - \sum_u n_{uu}}{1-f}$$

so that

$$f = \frac{1}{n}\sum_u x_u$$

The procedure begins with assigned values for p_u and f, so that x_u can be regarded as being known and new estimates formed. Convenient starting values are

$$p_u = \frac{n_u}{2n}$$

$$f = \frac{\sum_u n_{uu} - n\sum_u p_u^2}{n(1 - \sum_u p_u^2)}$$

Iteration proceeds by performing each of the following three steps in turn, in each case replacing previously by newly computed values, until convergence is reached (there may be problems if f is negative):

$$x_u = \frac{f n_{uu}}{f + p_u(1-f)}$$

$$p_u = \frac{n_u - x_u}{\sum_u (n_u - x_u)}$$

$$f = \frac{1}{n}\sum_u x_u$$

METHOD OF MOMENTS

Throughout this chapter, estimation has been discussed in the context of maximum likelihood. Such estimates have desirable properties for large samples, although they may not be particularly good for small samples. MLE's may be biased in small samples, for example. There may be situations in which unbiasedness is of great concern, and other cases in which it is not possible to write down a likelihood because the appropriate sampling distribution is unknown. In these circumstances some other means of constructing estimates is needed, and one of the simplest is the ***method of moments***. Briefly, statistics are sought that are unbiased for the parameters of interest, and these are used as estimates. For multinomial data, Fisher's method then provides variance estimates, but for other data it may be necessary to turn to numerical resampling methods, as discussed earlier, to determine the properties of the estimates.

The method of moments is somewhat ad hoc, and depends on choosing statistics whose expectations suggest a functional form for an estimator. As one example of this approach, look again at the case of estimating a single inbreeding coefficient f for a locus with more than two alleles. Although it is possible to find an MLE, it has been shown that this will require numerical solution of a nonlinear likelihood equation. After expressing homozygote frequencies in terms of p_u and f, Equation 2.9 gives

$$\mathcal{E}(\tilde{p}_u^2) = p_u^2 + \frac{1}{2n}p_u(1-p_u)(1+f)$$

so that, adding over alleles,

$$\mathcal{E}(\sum_u \tilde{p}_u^2) = \sum_u p_u^2 + \frac{1}{2n}(1-\sum_u p_u^2)(1+f)$$

while, adding over homozygotes,

$$\mathcal{E}(\sum_u \tilde{P}_{uu}) = \sum_u p_u^2 + (1-\sum_u p_u^2)f$$

These two equations lead to two expressions whose expectations differ only by a factor of f:

$$\begin{aligned}\mathcal{E}\sum_u(\tilde{P}_{uu}-\tilde{p}_u^2) + \mathcal{E}\frac{1}{2n}(1-\sum_u \tilde{P}_{uu}) &= (1-\frac{1}{n})f(1-\sum_u p_u^2)\\ \mathcal{E}(1-\sum_u \tilde{p}_u^2) - \mathcal{E}\frac{1}{2n}(1-\sum_u \tilde{P}_{uu}) &= (1-\frac{1}{n})(1-\sum_u p_u^2)\end{aligned}$$

Taking the ratio of these equations, the quantity $1 - \sum_u p_u^2$ cancels in the right-hand side, which suggests the estimator

$$\hat{f} = \frac{\sum_u(\tilde{P}_{uu} - \tilde{p}_u^2) + \frac{1}{2n}(1 - \sum_u \tilde{P}_{uu})}{(1 - \sum_u \tilde{p}_u^2) - \frac{1}{2n}(1 - \sum_u \tilde{P}_{uu})} \tag{2.28}$$

Although this ratio estimator is not itself unbiased, it is the ratio of terms that are unbiased for $f(1-\sum_u p_u^2)$ and $(1-\sum_u p_u^2)$, respectively. Accordingly, it is expected to have a low bias. For two alleles, the estimate reduces to

$$\hat{f} = 1 - \frac{(n-1)n_{Aa}/n}{2n\tilde{p}_A\tilde{p}_a - n_{Aa}/2n}$$

which is not the same as the maximum likelihood estimate in Equation 2.21. The properties of both estimators were discussed by Jiang (1987). The moment estimate has lower bias, but may have larger variance.

When the sample size n is large, applying Fisher's method allows a variance for the moment estimator to be found as described in Exercise 2.4.

BAYESIAN ESTIMATION

The analyses presented so far have been classical, or ***frequentist***, but an alternative basis for statistical inference is becoming more common in genetics. This ***Bayesian*** method (O'Hagan 1994) considers that parameters have some uncertainty associated with them. There is believed to be some prior information about the parameters, and this is modified on the basis of data. Instead of data being used to provide an estimated value (or confidence interval) for the parameter, as in the frequentist approach, the data are used to update information about the parameter. Specifically, a posterior distribution for the parameter is calculated by making use of Bayes' theorem. This distribution provides probabilities for the parameter taking certain values. Some formal treatment of probability is needed to express this theorem.

Definition of Probability

Frequentist statistical theory takes the probability of an event to be the proportion of times the event occurs in very many of the trials that could lead to the event. If a trial consists of tossing a coin, then the probability of a head refers to the number of heads divided by the number of tosses, provided very many tosses are made. Other definitions of probability are possible. For example, a purely formal mathematical approach can be used, by requiring that the probability $\Pr(E)$ of event E satisfies these three axioms (Gnedenko 1963):

- $\Pr(E) \geq 0$, for any event E.
- $\Pr(\Omega) = 1$, where Ω is the "universe" of all events.
- $\Pr(E_1 + E_2 + \cdots + E_n) = \Pr(E_1) + \Pr(E_2) + \cdots + \Pr(E_n)$, where the events $E_1, E_2, \cdots E_n$ are pairwise mutually exclusive.

From this set of axioms, all the mathematical properties of probability can be derived. For example:

- $0 \leq \Pr(E) \leq 1$ for all events E.
- $\Pr(\bar{E}) = 1 - \Pr(E)$, where $\bar{E}$ means not-E.
- $\Pr(A \cup B) = \Pr(A) + \Pr(B) - \Pr(A \cap B)$ for any two events A, B, where the union $\cup$ means "either" and the intersection $\cap$ means "both."

Joint and Conditional Probabilities

For two separate events A, B, the notation $\Pr(A \cap B)$ or $\Pr(A, B)$ denotes the ***joint probability*** that both occurred. If two coins are tossed, for example, A and B may represent the events that the first and second coins show a head. This leads naturally to the concept of ***conditional probability***. What is the probability of the second coin showing a head if the first coin shows a head, for example? Conditional probabilities will be written as $\Pr(B|A)$ for the probability of event B occurring, given that event A has occurred. This leads to the formal definition of the conditional probability of B given A:

$$\Pr(B|A) = \frac{\Pr(A, B)}{\Pr(A)}$$

If two events are such that the probability of one does not depend on the probability of the other, the events are termed ***independent***. Events A and B are independent if

$$\Pr(B|A) = \Pr(B)$$

which implies $\Pr(A|B) = \Pr(A)$.

Bayes' Theorem

Suppose there is some initial information, or ***prior probability*** $\Pr(B)$, of event B. How is this information changed by observing event A? In other words, what is the ***posterior probability*** $\Pr(B|A)$? From the definition of conditional probability

$$\begin{aligned}\Pr(B|A) &= \frac{\Pr(A,B)}{\Pr(A)} \\ &= \frac{\Pr(A|B)}{\Pr(A)}\Pr(B) \qquad (2.29)\end{aligned}$$

This shows how to modify the prior for B after A is observed, on the basis of the probability of A given B. The probability $\Pr(A|B)$ is also known as the likelihood of B given A (O'Hagan 1994). Equation 2.29 is known as ***Bayes' theorem***, although it is generally expressed in the following alternative form. If there are a series of mutually exclusive and exhaustive events B_r, where $\cup_r B_r = B_1 \cup B_2 \cup \ldots \cup B_r = \Omega$, then

$$\begin{aligned}\Pr(A) &= \Pr(A \cap \Omega) \\ &= \Pr[A \cap (\cup_r B_r)] \\ &= \Pr[\cup_r (A \cap B_r)] \\ &= \sum_r \Pr(A, B_r) \\ &= \sum_r \Pr(A|B_r)\Pr(B_r)\end{aligned}$$

This leads to the standard form of Bayes' theorem

$$\Pr(B_s|A) = \frac{\Pr(A|B_s)\Pr(B_s)}{\sum_r \Pr(A|B_r)\Pr(B_r)} \qquad (2.30)$$

This result has been expressed in terms of probabilities of discrete events, A and B_r, in keeping with the focus of this book on discrete data analysis. A Bayesian treatment of estimation, however, restates the theorem in terms of random variables instead of events. Parameter ϕ will replace event B, and data (counts) $\{n\}$ will replace event A. The data are discrete, but the parameters have continuous distributions. The probabilities $\Pr(B)$ and $\Pr(B|A)$ in Equation 2.30 are replaced by prior and posterior probability density functions $\pi(\phi), \pi(\phi|\{n\})$, and the sum is replaced by an integral:

$$\pi(\phi|\{n\}) = \frac{\Pr(\{n\}|\phi)\pi(\phi)}{\int \Pr(\{n\}|\phi)\pi(\phi)d\phi}$$

The density $\pi(\phi)$ represents prior information about ϕ. After data $\{n\}$ are observed, the posterior density is constructed from Bayes' theorem to be proportional to the prior and to the likelihood.

Allele Frequencies

Suppose the population is in Hardy-Weinberg equilibrium, and that allele A has population frequency p_A and count n_A in a sample of $2n$ alleles (n individuals). The Hardy-Weinberg assumption provides a binomial distribution for n_A *given* p_A:

$$\begin{aligned} n_A|p_A &\sim B(2n, p_A) \\ \Pr(n_A|p_A) &= \frac{(2n)!}{n_A!(2n-n_A)!} p_A^{n_A}(1-p_A)^{2n-n_A} \end{aligned}$$

A Bayesian analysis requires a prior distribution for p_A, and a convenient choice is the ***beta distribution***, which is the continuous analog of the binomial. For a beta distribution with
parameters α, β, the notation and density function are

$$\begin{aligned} p_A &\sim Be(\alpha, \beta) \\ \pi(p_A) &= \frac{\Gamma(\alpha+\beta)}{\Gamma(\alpha)\Gamma(\beta)} p_A^{\alpha-1}(1-p_A)^{\beta-1} \end{aligned}$$

The ***gamma function*** $\Gamma(x)$ generally needs to be evaluated numerically, as described by Abramowitz and Stegun (1970), although if x is an integer $\Gamma(x) = (x-1)!$ and the similarity between binomial and beta distributions is clear. The function integrates to 1 over the range [0,1]

$$\int_0^1 \pi(p_A)dp_A = 1$$

and has mean value $\alpha/(\alpha+\beta)$. The beta distribution can take many shapes, ranging from a unimodal with a peak at $p_A = (\alpha-1)/(\alpha+\beta-2)$ when $\alpha, \beta > 1$, to uniform when $\alpha = \beta = 1$, to U-shaped with most of the density near the boundaries $p_A = 0, 1$ when $\alpha, \beta < 1$.

With this prior, Bayes' theorem provides a posterior distribution of

$$\frac{\frac{(2n)!}{n_A!(2n-n_A)!} p_A^{n_A}(1-p_A)^{2n-n_A} \frac{\Gamma(\alpha+\beta)}{\Gamma(\alpha)\Gamma(\beta)} p_A^{\alpha-1}(1-p_A)^{\beta-1}}{\int_0^1 \frac{(2n)!}{n_A!(2n-n_A)!} p_A^{n_A}(1-p_A)^{2n-n_A} \frac{\Gamma(\alpha+\beta)}{\Gamma(\alpha)\Gamma(\beta)} p_A^{\alpha-1}(1-p_A)^{\beta-1} dp_A}$$

Cancelling the terms not involving p_A

$$\begin{aligned} \pi(p_A|n_A) &= \frac{p_A^{\alpha+n_A-1}(1-p_A)^{\beta+2n-n_A-1}}{\int_0^1 p_A^{\alpha+n_A-1}(1-p_A)^{\beta+2n-n_A-1} dp_A} \\ &= \frac{\Gamma(\alpha+\beta+2n)}{\Gamma(\alpha+n_A)\Gamma(\beta+2n-n_A)} p_A^{\alpha+n_A-1}(1-p_A)^{\beta+2n-n_A-1} \end{aligned}$$

So the posterior distribution is also a beta distribution, but with parameters modified by the data. In other words, the beta is a ***conjugate distribution*** for the binomial. Although the whole posterior distribution is now available for the allele frequency p_A, it may be convenient to take a single feature of this distribution to serve as a Bayesian estimator of p_A. For example, the mean of this distribution is

$$\begin{aligned}\mathcal{E}(p_A|n_A) &= \frac{\alpha+n_A}{\alpha+\beta+2n} \\ &= x\frac{\alpha}{\alpha+\beta}+(1-x)\frac{n_A}{2n} \\ &= x\mathcal{E}(p_A)+(1-x)\hat{p}_A\end{aligned}$$

which is the weighted sum of the prior mean and the maximum likelihood estimate of p_A, where the weight x is $(\alpha+\beta)/(\alpha+\beta+2n)$. Alternatively, the maximum (if $\alpha, \beta > 1$) of the posterior density is at

$$\begin{aligned}\max[\pi(p_A|n_A)] &= \frac{\alpha+n_A-1}{\alpha+\beta+2n-2} \\ &= y\frac{\alpha-1}{\alpha+\beta-2}+(1-y)\frac{n_A}{2n} \\ &= y\max[\pi(p_A)]+(1-y)\hat{p}_A\end{aligned}$$

which is the weighted sum of the prior maximum and the maximum likelihood estimate of p_A. The weight y is $(\alpha+\beta-2)/(\alpha+\beta+2n-2)$. Either the mean or the mode of the posterior distribution can serve as an estimate, but each is merely a summary of the whole posterior distribution.

As an example, Gunel and Wearden (1995) take $Be(61, 44)$ to be the prior distribution $\pi(p_M)$ for allele M in the MN blood group system. They based this on information from previous samples. How would this prior affect the estimate for the data in Table 1.2? From Table 2.3, $n_M = 201$ and $2n = 372$. Assuming Hardy-Weinberg equilibrium, the posterior distribution $\pi(p_M|n_M)$ is $Be(61+201, 44+171) = Be(262, 215)$. These two distributions are plotted in Figure 2.3, along with the likelihood function of p_M given n_M

$$L(p_M) = \Pr(n_M = 201|p_M) = \frac{372!}{201!171!}(p_M)^{201}(1-p_M)^{171}$$

The posterior density $\pi(p_M|n_M)$ is more narrow than the prior $\pi(p_M)$, indicating more precise information about the parameter p_M. This is as expected, because there is information from the data. In this example, the prior mean was quite close to the sample mean but it is the sample mean (the maximum likelihood estimate) that has the greatest influence on the Bayesian estimate (the posterior mean or mode).

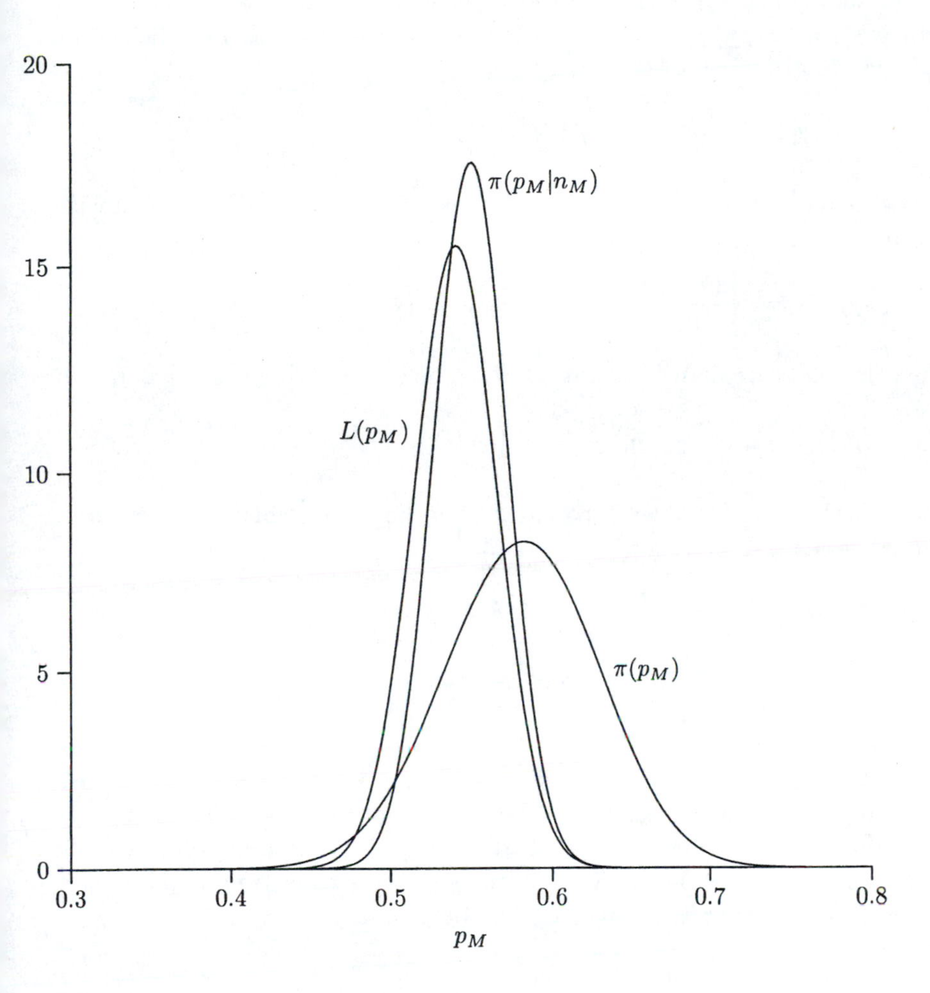

Figure 2.3 Likelihood function, prior and posterior distributions for the frequency of the M allele in the data of Table 2.3.

Multiple Alleles

Just as the multinomial distribution is the extension of the binomial from two to several categories, so is the ***Dirichlet distribution*** the extension of the beta. The Dirichlet distribution was invoked to estimate allele frequencies for a multi-allele locus by Lange (1995) and Gunel and Wearden (1995). The essence of the method is now shown.

Suppose alleles A_i have population frequencies p_i and sample counts n_i, where $\sum_i p_i = 1, \sum_i n_i = 2n$. For a Hardy-Weinberg population, the counts satisfy a multinomial distribution

$$\Pr(\{n_i\}|\{p_i\}) = \frac{(\sum_i n_i)!}{\prod_i (n_i)!} \prod_i (p_i)^{n_i}$$

If the frequencies p_i are supposed to have a Dirichlet prior distribution with parameters γ_i

$$\pi(\{p_i\}) = \frac{\Gamma(\sum_i \gamma_i)}{\prod_i \Gamma(\gamma_i)} \prod_i (p_i)^{\gamma_i - 1}$$

then the posterior distribution is also Dirichlet, but with parameters $\gamma_i + n_i$

$$\pi(\{p_i\}|\{n_i\}) = \frac{\Gamma[\sum_i (\gamma_i + n_i)]}{\prod_i \Gamma(\gamma_i + n_i)} \prod_i (p_i)^{\gamma_i + n_i - 1}$$

showing that the Dirichlet is conjugate for the multinomial. The mean of the posterior distribution is

$$\begin{aligned} \mathcal{E}(p_i|\{n_i\}) &= \frac{\gamma_i + n_i}{\sum_i (\gamma_i + n_i)} \\ &= x \frac{\gamma_i}{\sum_i \gamma_i} + (1 - x) \frac{n_i}{\sum_i n_i} \\ &= x\mathcal{E}(p_i) + (1 - x)\hat{p}_i \end{aligned}$$

where the weight is $x = (\sum_i \gamma_i)/[\sum_i (\gamma_i + n_i)]$.

Lange (1995) proceeded to estimate parameters for the Dirichlet prior from previous data. The likelihood $\Pr(\{n_i\}|\{p_i\})$ is the probability for the sample counts conditional on the parameters. Integrating over all parameter values gives the ***marginal probability*** for the counts

$$\begin{aligned} \Pr(\{n_i\}) &= \int \Pr(\{n_i\}|\{p_i\}) \pi(\{p_i\}) \prod dp_i \\ &= \frac{(\sum n_i)!}{\prod (n_i)!} \frac{\Gamma(\sum \gamma_i)}{\Gamma(\sum \gamma_i + \sum n_i)} \prod_i \frac{\Gamma(\gamma_i + n_i)}{\Gamma(\gamma_i)} \end{aligned}$$

This distribution is called the ***Dirichlet-multinomial***. For the two allele case, with sample counts n_A, n_a, the ***beta-binomial*** distribution is

$$\Pr(n_A, n_a) = \frac{(n_A + n_a)!}{n_A! n_a!} \frac{\Gamma(\gamma_A + \gamma_a)}{\Gamma(\gamma_A + \gamma_a + n_A + n_a)} \frac{\Gamma(\gamma_A + n_A)}{\Gamma(\gamma_A)} \frac{\Gamma(\gamma_a + n_a)}{\Gamma(\gamma_a)}$$

Maximum likelihood estimates of the γ parameters of these distributions require numerical methods (Lange 1995).

Bayesian and Classical Estimation

Debate over the use of Bayesian methods has centered on the assignment of probabilities to parameters. "In Bayesian inference the parameters are random variables, and therefore have both prior and posterior distributions. In classical inference the parameters take unique values, although these values are unknown, and it is not permitted to treat them as random or to give them probabilities." (O'Hagan 1994, page 11).

One of the clearest distinctions between classical and Bayesian methods of inference is provided by confidence intervals. Earlier in this chapter, confidence intervals were constructed by the use of normal-distribution theory or by the use of bootstrapping. Under classical theory, the interval for an allele frequency p_A is itself a random variable and probability statements are made about that interval. For example, 95% of these intervals will contain the parameter value. Under the Bayesian scheme, however, there is a probability statement about p_A. There is 95% probability that p_A lies between the points delimiting the middle 95% of the posterior distribution.

The practical difficulty in applying Bayesian methods is in choosing a prior distribution. There is no universal rule, since an investigator's knowledge about a parameter before seeing a particular set of data depends on the context for the particular problem. Previous estimates will surely be relevant, and estimating priors from previous data is known as ***empirical Bayes*** estimation (Casella 1985). The point is that the prior is crucial to Bayesian inference and it is not employed in classical inference.

For allele frequencies, the posterior mean provides an estimate that is between the prior mean and the maximum likelihood estimate. As Lange (1995) points out, this has the effect of moderating the extreme estimates that can arise with classical estimation. The posterior mean will not be zero even if the sample count n_i is zero, and it will not be one even if all of the sample alleles are A_i ($n_i = 2n$).

SUMMARY

Genotypic counts are generally assumed to be multinomially distributed, and this assumption forms the basis of within-population analyses. If the population is in Hardy-Weinberg equilibrium, but not otherwise, allele counts are also multinomially distributed. Variances of functions of genotypic counts can be determined exactly from multinomial theory, or approximately from Fisher's formula. Variances, including those between populations, can also be found from the use of indicator variables for each allele in the sample.

When inferences are to be extended beyond the one population sampled, it is necessary to use the total variance of a statistic that includes the effects of both statistical and genetic sampling.

When a distribution, such as multinomial, is assumed it is possible to formulate the likelihood of a set of parameters and construct maximum likelihood estimators. Such values have desirable properties in large samples and become normally distributed. Variances of MLE's can be found from the information, which is also derived from the likelihood function. Many cases of interest require numerical determination of MLE's, although when the number of parameters equals the number of degrees of freedom, Bailey's method often gives simpler equations. Incomplete data, such as occurs with dominance, can lead to MLE's when the EM algorithm is used. The problem of distinguishing between different double heterozygotes in calculating gametic frequencies can be overcome with the EM algorithm provided random mating is assumed.

When iterative methods are used to solve likelihood equations, several starting values should be used. To choose among several valid solutions it is necessary to calculate the likelihood for each solution and choose that solution for which the likelihood is maximized.

For cases in which a distribution cannot be assumed for genotypic counts, or when more emphasis is to be placed on unbiased estimates, the method of moments offers an alternative to maximum likelihood estimation. There is no general theory giving properties of moment estimators, and numerical methods may have to be used to find properties such as mean and variance.

Prior information about parameter values can be combined with the data to produce posterior distributions for parameters under the Bayesian approach to statistical inference.

EXERCISES

Exercise 2.1

For generations 5, 15, and 25 of the data in Table 2.5, find:

a. genotype and allele frequencies
b. estimated variances and covariances of allele frequencies
c. a single inbreeding coefficient f

Exercise 2.2

a. Use both Newton-Raphson iteration and the EM algorithm to estimate the frequencies of all four alleles S, M, F, and O if the phenotypic counts are

$$\begin{array}{llll} n_{SS}+n_{SO}=1149 & n_{MM}+n_{MO}=36 & n_{FF}+n_{OF}=17 & n_{OO}=20 \\ n_{SM}=336 & n_{MF}=25 & n_{SF}=203 & n=1786 \end{array}$$

b. Estimate the variances and covariances of the estimates. Assume Hardy-Weinberg equilibrium. (Data of S. Ohba, discussed by Yasuda and Kimura 1968.)

Exercise 2.3

Estimate the frequency of AB gametes for the following three sets of genotypic counts (Weir and Cockerham 1979):

a.	BB	Bb	bb
AA	154	81	8
Aa	37	14	3
aa	1	1	1

b.	BB	Bb	bb
AA	12	3	3
Aa	3	54	3
aa	12	3	3

c.	BB	Bb	bb
AA	12	3	3
Aa	3	51	3
aa	12	6	3

Exercise 2.4

Find an expression for the variance of the estimator of f given in Equation 2.28. To do this write

$$h = 1 - \sum_u p_u^2, \; H = 1 - \sum_u P_{uu}$$

so that

$$\hat{f} = \frac{(\tilde{h} - \tilde{H}) + \frac{1}{2n}\tilde{H}}{\tilde{h} - \frac{1}{2n}\tilde{H}}$$

To apply Fisher's formula, it is necessary to drop the $\frac{1}{2n}\tilde{H}$ term in both numerator and denominator. This provides a function that is of degree zero in the counts, and will not affect the accuracy of the variance since that is of order n^{-1}.

Exercise 2.5

Draw ten bootstrap samples from the sample

1	2	3	4	5	6	7	8	9	10
A_2A_2	A_1A_2	A_1A_1	A_1A_1	A_1A_1	A_1A_2	A_1A_2	A_1A_1	A_1A_2	A_1A_1

Number the original samples from 1 to 10, and then draw 10 random digits. For example, the first block of 10 in Appendix Table B.1 is 30246 86149, and this means that the bootstrap sample is

3	0	2	4	6	8	6	1	4	9
A_1A_1	A_1A_1	A_1A_2	A_1A_1	A_1A_2	A_1A_1	A_1A_2	A_2A_2	A_1A_1	A_1A_2

The first bootstrap proportion for allele A_1 is therefore 0.70. For this exercise, an arbitrary starting point in Table B.1 is needed and it may be convenient to use a birthday: the row corresponding to the month and the column to the day.

Take the lowest 9 of 10 bootstrap proportions to define a 90% one-sided confidence interval, and compare that to the value found from normal theory, using a z value of 1.28.

Chapter 3

Disequilibrium

TESTING HYPOTHESES

This chapter explores the relationships among frequencies for sets of alleles, within or between loci. The general question is whether the frequency of a set of alleles is the same as the product of each of the separate allele frequencies, and differences between joint frequencies and products of individual frequencies will be called disequilibrium coefficients. The question is phrased as a hypothesis to be tested. For two alleles at a single locus, the hypothesis is that of Hardy-Weinberg equilibrium. Is the Hardy-Weinberg disequilibrium coefficient equal to zero? For an allele from each of two loci, the hypothesis is linkage equilibrium or a zero value for the linkage disequilibrium coefficient. Other hypotheses will also be considered.

In Chapter 2, estimation of genetic parameters was considered and methods of that chapter provide estimates of disequilibrium coefficients. This chapter is concerned mainly with determining whether the coefficients are zero. As Kendall and Stuart (1979) explain, this order may be back to front – it may be more important to know if there is *any* disequilibrium than to know how large is the disequilibrium coefficient. This distinction will be made again in Chapter 7, where the existence of linkage between loci may be more important than the magnitude of linkage parameters.

Parametric hypothesis testing proceeds by calculating the probabilities of values of test statistics when some parameters have the hypothesized values. Tests based on standard distribution theory, such as z tests, identify test statistic values known to occur with low probability when the hypothesis is true. For specified parameter values, probability or exact tests calculate probabilities of data explicitly. In either case, test statistics or data sets that have low probability values lead to rejection of the hypothesis. The probability of the test statistic causing rejection of a hypothesis, when that hypothesis is true, is the ***size*** or ***significance level*** of the test. The probability of the event of rejection of a false hypothesis is the ***power*** of the test. Both size and power are of importance in choosing among alternative test procedures for a particular hypothesis. Classical hypothesis testing starts with a specified size, and chooses a test to maximize power. Determining the power of a test requires specification of the alternative(s) to the hypothesis being tested.

Several testing strategies are discussed in the chapter. Traditional chi-square goodness-of-fit tests compare counts in a set of categories with the values expected if the hypothesis being tested is true. If the hypothesis involves a parameter, such as a disequilibrium coefficient, a normal test statistic can be constructed from an estimate of the parameter and (an estimate of) its sampling variance. The probabilities of a dataset under alternative hypotheses can be compared by means of a likelihood ratio test statistic. Rejections of hypotheses follow from large values of chi-square, normal, or likelihood ratio test statistics.

Exact, or probability, tests calculate the probability of a dataset under a hypothesis and cause rejection when the dataset is among the least probable outcomes. These probabilities can be made conditional on some other statistics, to remove the effects of parameters not directly concerned with the hypothesis. A problem with exact tests is in identifying the "least probable" outcomes when the number of outcomes is too large to examine each one. It will be shown that permutation procedures (Good 1994) can then be used. These reduce the computational burden, and Maiste (1993) has shown exact conditional tests to be the most desirable tests in the disequilibrium setting. The traditional chi-square, normal, and likelihood ratio tests are included because they are generally easier to apply and they have historical importance.

No specific evolutionary models are invoked, so that inferences are limited to the particular population sampled.

HARDY-WEINBERG DISEQUILIBRIUM

Once allele and genotype frequencies have been estimated, one of the first analyses performed on population genetic data is that of looking for associations between the two alleles an individual receives at a locus. When there are no disturbing forces such as selection, mutation, or migration that would change allele frequencies over time, and when there is random mating in very large populations, these pairs of alleles are known not to be associated. A consequence of this independence is that genotype frequencies are the products of allele frequencies, as noted in Chapter 2:

$$\begin{aligned} P_{uu} &= p_u^2 && \text{for homozygotes } A_uA_u \\ P_{uv} &= 2p_up_v && \text{for heterozygotes } A_uA_v \end{aligned}$$

Departures from these ***Hardy-Weinberg proportions*** can be characterized in several ways, including the use of the within-population inbreeding coefficient f introduced in Chapter 2. With a single coefficient f, the genotypic frequencies can be written as

$$\begin{aligned} P_{uu} &= p_u^2 + p_u(1-p_u)f \\ P_{uv} &= 2p_up_v(1-f),\ v \neq u \end{aligned}$$

and these frequencies are correctly bounded by zero and by allele frequencies

$$\begin{aligned} 0 &\leq P_{uu} \leq p_u \\ 0 &\leq P_{uv} \leq \min(2p_u, 2p_v) \end{aligned}$$

provided

$$-p_u/(1-p_u) \leq f \leq 1, \qquad \text{for all } u$$

By introducing indicator variables x_j for the jth allele of a random individual:

$$x_j = \begin{cases} 1 & \text{if allele is } A \\ 0 & \text{otherwise} \end{cases}$$

it emerges that f can also be regarded as the correlation of x_j and $x_{j'}, j \neq j'$. To see why this is so, note that

$$\begin{aligned} \text{Var}(x_j) &= p_A(1-p_A) \\ \text{Cov}(x_j, x_{j'}) &= P_{AA} - p_A^2 \\ &= fp_A(1-p_A) \end{aligned}$$

When there are no disturbing forces it is quite appropriate to work with just a single coefficient f, the correlation of any pair of alleles within individuals, but a more general formulation would allow as many parameters as there are degrees of freedom. For a locus with k alleles, there are k allele frequencies and $k(k-1)/2$ heterozygotes, suggesting that the $[k(k+1)/2]$ genotypic frequencies be expressed in terms of the p_u's and a set of $k(k-1)/2$ ***fixation indices*** f_{uv}, one for each heterozygote. Coefficients f_{uu} can also be defined for the homozygotes. These fixation indices are defined by the following relations

$$\begin{aligned} P_{uu} &= p_u^2 + p_u(1-p_u)f_{uu} \\ P_{uv} &= 2p_u p_v(1-f_{uv}),\ v \neq u \end{aligned}$$

but the relation

$$p_u = P_{uu} + \frac{1}{2}\sum_{v\neq u} P_{uv}$$

shows that the f_{uu}'s are just functions of the f_{uv}'s:

$$f_{uu} = \sum_{v\neq u} \frac{p_v}{1-p_u} f_{uv}$$

Although the use of inbreeding or fixation coefficients to describe departures from Hardy-Weinberg equilibrium has some merit, it has the disadvantage that these parameters are estimated as ratios of genotypic frequencies, as shown in Chapter 2. It is difficult to determine the statistical properties of ratios.

There are advantages instead in working with a composite kind of quantity that will be termed a ***disequilibrium coefficient***. This is simply the difference between a frequency and its value expected when there are no associations between alleles. Disequilibria will be denoted by D's and for one-locus genotypic frequencies they are defined by the relations

$$\begin{aligned} P_{uu} &= p_u^2 + D_{uu} \\ P_{uv} &= 2p_u p_v - 2D_{uv},\ v \neq u \end{aligned}$$

There is still a dependency among the coefficients caused by the genotypic frequencies summing to allele frequencies:

$$D_{uu} = \sum_{v\neq u} D_{uv}$$

This dependency implies that there are as many independent D's as there are heterozygote types.

In the two-allele case the notation can be modified a little to use a single disequilibrium coefficient D_A:

$$\begin{aligned} P_{AA} &= p_A^2 + D_A \\ P_{Aa} &= 2p_A p_a - 2D_A \\ P_{aa} &= p_a^2 + D_A \end{aligned}$$

Taking account of the range of possible genotypic frequencies, bounds on D_A are found

$$\max[-p_A^2, -p_a^2] \leq D_A \leq p_A p_a$$

Estimating Disequilibrium D_A

In the two-allele case there are two parameters, the same as the number of degrees of freedom. Bailey's rule (Chapter 2) for maximum likelihood estimates (MLE's) can therefore be applied, and the MLE for the single D_A is

$$\hat{D}_A = \tilde{P}_{AA} - \tilde{p}_A^2 \tag{3.1}$$

From Equations 2.3 and 2.9, the expected value of this estimate over replicate samples from the same population is

$$\begin{aligned}\mathcal{E}(\hat{D}_A) &= D_A - \frac{1}{2n}(p_A + P_{AA} - 2p_A^2) \\ &= D_A - \frac{1}{2n}[p_A(1-p_A) + D_A]\end{aligned} \tag{3.2}$$

showing that the estimate is biased, although the bias decreases as the sample size increases. From Fisher's variance approximation, Equation 2.17, the variance of the MLE, ignoring terms in higher powers of $(1/n)$, is

$$\text{Var}(\hat{D}_A) = \frac{1}{n}[p_A^2(1-p_A)^2 + (1-2p_A)^2 D_A - D_A^2] \tag{3.3}$$

The same approximation can be found with the use of indicator variables.

Testing for Hardy-Weinberg with D_A

When a population has Hardy-Weinberg proportions, the disequilibrium coefficient D_A is zero, suggesting that a test of the hypothesis $H_0 : D_A = 0$ is equivalent to testing for Hardy-Weinberg equilibrium (HWE). A word should be said here about the use of the word "equilibrium." Strictly, an equilibrium state is one in which properties of the population are not changing over successive generations. In the HWE case, this implies the continued absence of disturbing forces such as selection, migration, and mutation as well as the continuation of random mating. The tests presented in this chapter, on the other hand, refer to checking for consistency of sample genotypic frequencies with those expected from the Hardy-Weinberg law. Nothing is necessarily implied about the conditions faced by the population or about future generations. In spite of this distinction, it is usual to speak of testing for HWE, and this convention will be continued here.

For large samples, the MLE $\hat{D}_A$ is approximately normally distributed

$$\hat{D}_A \sim N[\mathcal{E}(\hat{D}_A), \text{Var}(\hat{D}_A)]$$

so that a standard normal variate, z, can be constructed

$$z = \frac{\hat{D}_A - \mathcal{E}(\hat{D}_A)}{\sqrt{\mathrm{Var}(\hat{D}_A)}} \tag{3.4}$$

and compared to tabulated critical values (Appendix Table A.1). Departures from HWE will cause $\hat{D}_A$, and hence z, to be large, and the hypothesis of HWE can be rejected if the calculated z is sufficiently large that it is unlikely to have arisen by chance. In particular, departures from HWE caused by an excess or a deficiency of homozygotes, leading to positive or negative disequilibrium values, would cause the hypothesis to be rejected at the 5% significance level when z exceeds 1.96 or is less than -1.96. Of course, if only the possibility of a deficiency *or* an excess was of interest then a one-tailed test would be used. For example, if it was suspected that there were fewer heterozygotes in a population than predicted by HWE, the null hypothesis $H_0 : D_A = 0$ would be tested against the alternative hypothesis $H_1 : D_A > 0$. The null would be rejected for a large positive D_A value, or a large positive z value. From Appendix Table A.1, a 5% significance level is attained when the hypothesis is rejected for $z > 1.64$.

An equivalent procedure depends on the result that the square of a standard normal variable is distributed as chi-square with 1 d.f., and rejects the hypothesis when

$$z^2 = X^2 \sim \chi^2 \tag{3.5}$$

exceeds the 5% critical value of 3.84 (Appendix Table A.2). Combining Equations 3.1 to 3.5, after setting $D_A = \mathcal{E}\hat{D}_A = 0$, gives the test statistic

$$X_A^2 = \frac{n\hat{D}_A^2}{\tilde{p}_A^2(1-\tilde{p}_A)^2} = n\hat{f}^2 \tag{3.6}$$

Large values of z, whether positive or negative, cause large values of X^2, so the chi-square test is also two-sided in that it leads to rejection of the HWE hypothesis for an excess or a deficiency of heterozygotes. Only values in the upper tail of the chi-square distribution give rejection, but one-sided alternative hypotheses cannot be accommodated as directly as they were with the z statistic.

By going through the derivation of this expression the assumptions on which it is based have been revealed. Under the hypothesis $H_0 : D_A = 0$, the expectation of the MLE $\hat{D}_A$ is zero provided the bias term $p_A(1-p_A)/2n$ is ignored. Similarly, under the hypothesis, the variance of the MLE can be approximated by the sample value $\tilde{p}_A^2(1-\tilde{p}_A)^2/n$ provided the sample is large.

Equation 3.6 can be approached from another direction, that of goodness-of-fit chi-square tests. For samples large enough that bias terms can again be ignored, the following set of expected values hold for the three genotypic *counts* when H_0 is true:

Genotype	AA	$A\bar{A}$	$\bar{A}\bar{A}$
Observed number	n_{AA}	$n_{A\bar{A}}$	$n_{\bar{A}\bar{A}}$
Expected number	$n\tilde{p}_A^2$	$2n\tilde{p}_A(1-\tilde{p}_A)$	$n(1-\tilde{p}_A)^2$
Observed−Expected	$n\hat{D}_A$	$-2n\hat{D}_A$	$n\hat{D}_A$

The notation $\bar{A}$ for "not-A" has been employed to emphasize that there is a focus on allele A but that it is not necessary for the locus to have only two alleles. The goodness-of-fit chi-square statistic is

$$X_A^2 = \sum_{\text{genotypes}} \frac{(\text{Observed} - \text{Expected})^2}{\text{Expected}}$$

$$= \frac{(n\hat{D}_A)^2}{n\tilde{p}_A^2} + \frac{(-2n\hat{D}_A)^2}{2n\tilde{p}_A(1-\tilde{p}_A)} + \frac{(n\hat{D}_A)^2}{n(1-\tilde{p}_A)^2}$$

and this leads back to Equation 3.6.

Coming from this direction to Equation 3.6 indicates that the test procedure has the usual problems associated with chi-square tests, the most common one being that the tests are sensitive to small expected values. Since the expected values occur in the denominator of X^2 they can greatly inflate the statistic when they are small, and it has been suggested that expected values less than some specified level, say five or one, not be used. Other problems, such as using a continuous distribution (normal or chi-square) to test hypotheses with discrete genotypic counts, may be overcome by using a continuity correction of 0.5 in the numerator of chi-square (Yates 1934),

$$X_A^2 = \sum_{\text{genotypes}} \frac{(|\,\text{Observed} - \text{Expected}\,| - 0.5)^2}{\text{Expected}}$$

The absolute value of the difference between observed and expected counts is reduced by one-half before it is squared. All problems with the test stem from using large-sample results for small samples. A procedure that makes no such requirement is discussed in the next section.

Mention should be made here of differing philosophies behind the testing of hypotheses. The discussion just given is for a procedure that results in either rejecting, or not rejecting, an hypothesis. A ***significance level***, α, needs to be specified, where α is the chance of rejecting a true hypothesis. The hypothesis is rejected when the test statistic exceeds the ***critical value***, the value which is exceeded by chance with probability α when the hypothesis is true. It is more informative to give the value of the statistic and the probability, or "p-value," of obtaining such a value, or a more extreme one, when the hypothesis is true. The extent to which the data fail to reject the hypothesis is measured by p. Values of p less than α could be used to reject at the α significance level.

Exact Tests for HWE

The use of "exact" tests dates back to work of Fisher (1935), who noted that an observed sample could be used to reject an hypothesis if the total probability under the hypothesis of that sample, or a less probable one, is small. A simple way to perform such a test would be to determine the probabilities of all possible samples of the same size as the sample at hand assuming the hypothesis is true. Once the samples are ordered according to their probabilities, it is a simple matter to add the probability of the observed sample to the sum of the probabilities of all less probable samples and then to reject the hypothesis if that total probability is less than α. Exact tests are generally used for small sample sizes, when there is the greatest chance of having small expected numbers in the chi-square test formulation. However, if there are rare alleles at a locus, expected numbers can be small even in moderately large samples, and exact tests are desirable.

Testing for HWE is equivalent to looking at whether the observed genotypic frequencies are close enough to products of the observed allele frequencies that there can be confidence in the same relationship holding for the population frequencies. The ***exact test***, or ***probability test***, proceeds by looking at all possible sets of genotypic frequencies for the particular observed set of allele frequencies and rejecting the hypothesis of HWE if the observed genotypic frequencies turn out to be very unusual under HWE.

The test is derived here for a locus with two alleles, A and a. Under the HWE hypothesis, the probability of the observed set of genotypic counts n_{AA}, n_{Aa}, and n_{aa} in a sample of size n is

$$\Pr(n_{AA}, n_{Aa}, n_{aa}) = \frac{n!}{n_{AA}!n_{Aa}!n_{aa}!}(p_A^2)^{n_{AA}}(2p_Ap_a)^{n_{Aa}}(p_a^2)^{n_{aa}} \qquad (3.7)$$

whereas the allele counts n_A and n_a are binomially distributed if HWE holds:

$$\Pr(n_A, n_a) = \frac{(2n)!}{n_A!n_a!}(p_A)^{n_A}(p_a)^{n_a}$$

Putting these expressions together provides the probability of the observed genotypic frequencies, assuming HWE, conditional on the observed allele frequencies (which are sufficient statistics under HWE):

$$\begin{aligned}
\Pr(n_{AA}, n_{Aa}, n_{aa} \mid n_A, n_a) &= \frac{\Pr(n_{AA}, n_{Aa}, n_{aa} \text{ and } n_A, n_a)}{\Pr(n_A, n_a)} \\
&= \frac{\Pr(n_{AA}, n_{Aa}, n_{aa})}{\Pr(n_A, n_a)} \\
&= \frac{n!n_A!n_a!2^{n_{Aa}}}{n_{AA}!n_{Aa}!n_{aa}!(2n)!}
\end{aligned}$$

in which the unknown allele frequencies p_A, p_a have cancelled out. Inferences about Hardy-Weinberg are therefore unaffected by the population allele frequencies, except insofar as these affect the sample allele counts. Probability tests were applied to HWE testing by Haldane (1954), who pointed out that the probabilities could be expressed in terms of one of the two allele numbers and the number, $x = n_{Aa}$, of heterozygotes. The conditional probability can therefore be called $\Pr(x \mid n_A)$ and rewritten as

$$\Pr(x \mid n_A) = \frac{n!n_A!(2n - n_A)!2^x}{[(n_A - x)/2]!x![n - (n_A + x)/2]!(2n)!}$$

These probabilities are evaluated numerically at all valid values of x for a particular pair of n, n_A and then x values are ordered according to these probabilities. The least probable outcomes with a total probability of α form a rejection region of size α. An attempt is made to get α close to a conventional value such as 0.05, but this is unlikely to be met exactly.

As an example, return to the mosquito *Pgm* data displayed in Table 2.4 after collapsing to alleles 1 and not-1:

$$n_{11} = 9,\ n_{1\bar{1}} = 1,\ n_{\bar{1}\bar{1}} = 30; \quad n_1 = 19,\ n_{\bar{1}} = 61$$

The possible numbers of heterozygotes, x, when there are $n_1 = 19$ alleles of type 1 in a sample of size $n = 40$, are the odd numbers from 1 to 19 and the probabilities $\Pr(x|19)$ are shown in Table 3.1. It happens that the observed set of genotypic frequencies has the smallest probability of all possible samples with these allele numbers, and the HWE hypothesis would be rejected with a significance level that is extremely small. Notice that both large and small numbers of heterozygotes can lead to small probabilities,

Table 3.1 Exact test for HWE at *Pgm* locus for mosquito data of Table 1.3.

Possible samples				Cumulative	Disequi–	
11	$1\bar{1}$	$\bar{1}\bar{1}$	Probability	Probability	librium	Chi – square
9	1	30*	0.0000	0.0000†	0.1686	34.67†
8	3	29	0.0000	0.0000†	0.1436	25.15†
7	5	28	0.0001	0.0001†	0.1186	17.16†
6	7	27	0.0023	0.0024†	0.0936	10.69†
5	9	26	0.0205	0.0229†	0.0686	5.74†
0	19	21	0.0594	0.0823	−0.0564	3.88†
4	11	25	0.0970	0.1793	0.0436	2.32
1	17	22	0.2308	0.4101	−0.0314	1.20
3	13	24	0.2488	0.6589	0.0186	0.42
2	15	23	0.3411	1.0000	−0.0064	0.05

*Observed sample.
†Causes rejection of HWE at 5% significance level.

or large chi-square values, and rejection of the hypothesis. The ordering of probabilities is not the same as the ordering on the numbers of heterozygotes. The exact test is two-sided. In this particular example, though, the exact test rejection region consists only of small numbers of heterozygotes.

By adding the probabilities for 1, 3, 5, 7, and 9 heterozygotes a rejection region of size 0.0229 is found. In other words, there is a probability of 2.29% of falsely rejecting the hypothesis of HWE when it is rejected with 9 or fewer heterozygotes. This probability is the significance level or probability of a ***type I error***. Adding the next largest probability, for 19 heterozygotes, would give a test of size 8.23%, which would generally be regarded as being too high. Chi-square test statistics are also shown in Table 3.1 and demonstrate that the two procedures differ even for samples as large as 40. The chi-square test rejects for 19 heterozygotes whereas the exact test does not reject. Applying Yates' continuity correction would bring X^2 down to 2.62 for $x = 19$, below the critical value of 3.84, and the two tests would then agree.

This procedure of adding probabilities for all deviations (of the numbers of heterozygotes) from the value expected under the hypothesis regardless of sign has been criticized by Yates (1984). He advocates keeping track of sign. In the present example, the expected number of heterozygotes is $40 \times 2(19/80)(61/80) = 14.5$, and the observed data have fewer heterozygotes than expected, so that the observed disequilibrium is positive. If the rows in Table 3.1 were ordered the same as the numbers of heterozygotes, attention would be focused only on the tail with small numbers. There is a probability of 0.0229 of observing 9 or fewer heterozygotes, so that the hypothesis would be rejected with a one-tailed probability of 0.0229. In other words, the hypothesis of no disequilibrium is rejected in favor of the hypothesis of positive disequilibrium at that significance level. If a two-tailed test is wanted, meaning that either positive or negative disequilibria can cause rejection, Yates says that the one-tail significance level should be doubled, to 0.0458. The case for approaching exact testing from a one-sided viewpoint does have some appeal. As E.A. Thompson (personal communication) points out, however, this assumes that the alternative hypothesis is relevant. Fisher would not have made that assumption. Which approach is taken should be clearly stated when results are presented. Note that there is no debate about normal distribution-based tests because of symmetry. Positive or negative values of z of the same magnitude are equally likely, whereas exact tests have different probabilities for deviations equal in size but opposite in sign.

Application of the exact test for HWE in the two-allele case is simplified by using tables published by Vithayasai (1973). He gave the rejection regions for significance levels of 0.10, 0.05, and 0.01 for samples of size $n = 20$ to 100 in steps of 5 and $n = 100$ to 200 in steps of 25. The tables show the critical numbers of heterozygotes for each possible number of the least frequent allele. For the present example, he gave as critical values those heterozygote numbers lying outside the ranges (10,18), (10,20), and (9,21) for the three significance levels 0.10, 0.05, and 0.01. With only 19 of the least frequent allele, the number of heterozygotes cannot exceed 19 of course, but the tables give a quick approximation to the test.

Likelihood Ratio Test for HWE

Just as the likelihood function provided a general means of finding estimates by the maximum likelihood method, so too does it provide a general framework for testing hypotheses. Test statistics can be found as ***likelihood ratios***. To test whether a parameter ϕ has some specific value ϕ_0, such as zero, a comparison is made of two likelihoods: that with $\phi = \phi_0$ (or maximized with $\phi = \phi_0$ when there are parameters in addition to ϕ), and that maximized when ϕ is not constrained. These two maximum likelihoods are written as L_0 for the hypothesized parameter value, and L_1 for the unconstrained value. If the hypothesis is true, L_0 and L_1 should be equal in value. The less well the data support the hypothesis, the smaller L_0 will be than L_1, suggesting that the ratio of the two be used as a test statistic. The likelihood ratio is defined as

$$\lambda = \frac{L_0}{L_1}$$

Remember that the L's refer to maximum likelihoods, and they are calculated by using the MLE's of the parameters in the two cases. A convenient chi-square approximation to the distribution of λ is available. When the hypothesis is true, and only a single parameter is being specified by the hypothesis,

$$-2\ln\lambda = -2(\ln L_0 - \ln L_1) \sim \chi^2_{(1)}$$

If L_0 and L_1 differ because of values assigned to s parameters, $-2\ln\lambda$ has a chi-square distribution with s degrees of freedom. Likelihood ratio tests for multinomial proportions have also been called G tests (e.g., Sokal and Rohlf 1995), with the test statistic G^2 being defined as

$$G^2 = -2\ln\lambda = -2\ln\left(\frac{L_0}{L_1}\right)$$

To test for HWE, the hypothesis is that $D_A = 0$. Under the unconstrained model, since there are two degrees of freedom and two parameters, Bailey's method shows that the MLE's of the genotypic frequencies are just the observed genotypic frequencies. Writing observed frequencies in terms of counts, and using the relation

$$n = n_{AA} + n_{Aa} + n_{aa}$$

the maximum likelihood L_1 is

$$L_1 = \frac{n!}{n_{AA}!n_{Aa}!n_{aa}!}\,\frac{(n_{AA})^{n_{AA}}(n_{Aa})^{n_{Aa}}(n_{aa})^{n_{aa}}}{n^n}$$

Under the HWE hypothesis, the MLE's of the genotypic frequencies are the appropriate products of allele frequencies, such as

$$\begin{aligned}\hat{P}_{AA} &= (\tilde{p}_A)^2 \\ &= (n_A/2n)^2\end{aligned}$$

where

$$\begin{aligned}n_A &= 2n_{AA} + n_{Aa} \\ n_a &= 2n_{aa} + n_{Aa}\end{aligned}$$

This allows L_0 to be written as

$$L_0 = \frac{n!}{n_{AA}!n_{Aa}!n_{aa}!}\,\frac{(n_A)^{n_A}(n_a)^{n_a}2^{n_{Aa}}}{(2n)^{2n}}$$

Testing for HWE can therefore be carried out with the test statistic

$$-2\ln\lambda = -2\ln\left[\frac{(n)^n(n_A)^{n_A}(n_a)^{n_a}2^{n_{Aa}}}{(2n)^{2n}(n_{AA})^{n_{AA}}(n_{Aa})^{n_{Aa}}(n_{aa})^{n_{aa}}}\right]$$

Note that, as for the exact test, this test statistic also does not make explicit use of the population allele frequencies p_A, p_a.

Log-Linear Models

A quite different approach to testing for HWE starts with a multiplicative instead of an additive model. Instead of representing genotypic frequencies as the sum of allele frequency terms and coefficients of disequilibrium, they may be written as products of terms, some of which are for alleles and some of which are for disequilibrium. In other words, departures from HWE can be accommodated with a set of multiplicative coefficients. Such a representation for genotypic frequencies is

$$\begin{aligned} P_{AA} &= MM_A^2M_{AA} \\ P_{Aa} &= 2MM_AM_aM_{Aa} \\ P_{aa} &= MM_a^2M_{aa} \end{aligned}$$

Here M_A, M_a represent the allele frequency contributions and M_{AA}, M_{Aa}, M_{aa} represent the associations between allele frequencies. The term M is a mean effect. Taking logarithms of these equations to give

$$\begin{aligned} \ln P_{AA} &= \ln M + 2\ln M_A + \ln M_{AA} \\ \ln P_{Aa} &= \ln 2 + \ln M + \ln M_A + \ln M_a + \ln M_{Aa} \\ \ln P_{aa} &= \ln M + 2\ln M_a + \ln M_{aa} \end{aligned}$$

shows why such models are referred to as log-linear. It is the logarithms of frequencies that are represented as linear combinations of terms for a mean, allele, and genotypic effects. As there are four parameters for three genotypic frequencies, this formulation is overparameterized, and several ways of reducing the number of parameters can be used. One way is to set the terms involving allele a to one, leaving genotypic frequencies as

$$\begin{aligned} P_{AA} &= MM_A^2M_{AA} \\ P_{Aa} &= 2MM_A \\ P_{aa} &= M \end{aligned}$$

which shows the need for the mean term M. Since the frequencies must sum to 1,

$$M = (1 + 2M_A + M_A^2M_{AA})^{-1}$$

This results in two independent parameters, M_{AA} and M_A, for the two degrees of freedom, and Bailey's method gives estimates

$$\begin{aligned}\hat{M} &= \tilde{P}_{aa} \\ \hat{M}_A &= \frac{\tilde{P}_{Aa}}{2\tilde{P}_{aa}} \\ \hat{M}_{AA} &= \frac{4\tilde{P}_{AA}\tilde{P}_{aa}}{\tilde{P}^2_{Aa}}\end{aligned}$$

leading to

$$\begin{aligned}\hat{P}_{AA} &= \tilde{P}_{AA} \\ \hat{P}_{Aa} &= \tilde{P}_{Aa} \\ \hat{P}_{aa} &= \tilde{P}_{aa}\end{aligned}$$

for the unconstrained estimates.

HWE in this system means that the genotypic term M_{AA} is 1, and MLE's constrained by the hypothesis are

$$\begin{aligned}\hat{M} &= \tilde{p}^2_a \\ \hat{M}_A &= \frac{\tilde{p}_A}{\tilde{p}_a}\end{aligned}$$

Therefore

$$\begin{aligned}\hat{P}_{AA} &= \tilde{p}^2_A \\ \hat{P}_{Aa} &= 2\tilde{p}_A\tilde{p}_a \\ \hat{P}_{aa} &= \tilde{p}^2_a\end{aligned}$$

for the estimates constrained by the hypothesis.

Testing uses the likelihood ratio procedure. A comparison is made of likelihoods maximized under models that either include M_{AA} or do not. Under the unconstrained model (or model 1), with the genotypic term M_{AA}, the maximum log-likelihood is

$$\ln L_1 = \text{Constant} + n_{AA}\ln(\tilde{P}_{AA}) + n_{Aa}\ln(\tilde{P}_{Aa}) + n_{aa}\ln(\tilde{P}_{aa})$$

Under the constrained, HWE, model (or model 0) without M_{AA}, the maximum log-likelihood is

$$\ln L_0 = \text{Constant} + n_A\ln(\tilde{p}_A) + n_a\ln(\tilde{p}_a)$$

The ratio of likelihoods measures the relative extents to which the two models fit the data. If HWE does not hold, the full model is expected to provide a higher likelihood and the hypothesis will be rejected for small values of

$$\lambda = \frac{L_0}{L_1}$$

or for large values of

$$-2\ln\lambda = -2(\ln L_0 - \ln L_1)$$

For large samples, the quantity $-2\ln\lambda$ has an approximate chi-square distribution. In the language of log-linear models, this quantity is also called the ***deviance*** of the model 0. The chi-square distribution simplifies the determination of whether to reject the hypothesis, whereas the deviance concept allows the treatment of a series of models. The difference of two deviances, for nested submodels, also has an approximate chi-square distribution. The deviance for the full model is zero.

Note that the likelihood ratio test statistic for the log-linear model, in this case, is the same as was previously found for the additive model, even though the two sets of parameters are different. Notice also that different restrictions could have been put on the $\hat{M}$'s in the log-linear model so that the estimates would have been different, but the final test statistic would have been the same.

Multiple Alleles

The likelihood ratio framework offers a systematic way of testing for HWE when there are more than two alleles at a locus. Each of the genotypes can differ from the Hardy-Weinberg proportions, and it may be of interest to test each of these departures separately. Specifically, each of the disequilibrium coefficients D_{uv}, for alleles A_u and A_v, can be tested. When there are k codominant alleles, the $k(k+1)/2$ genotypic frequencies provide $[k(k+1)/2] - 1$ degrees of freedom and allow $k-1$ allele frequencies to be estimated and $k(k-1)/2$ disequilibrium coefficients to be estimated and tested for departures from zero. The full, or completely unconstrained, model has MLE's of

$$\begin{aligned}
\hat{p}_u &= \tilde{p}_u \\
\hat{D}_{uv} &= \tilde{p}_u\tilde{p}_v - \frac{1}{2}\tilde{P}_{uv}
\end{aligned}$$

with a log-likelihood of

$$\ln L_1 = \text{Constant} + \sum_u n_{uu}\ln\left(\frac{n_{uu}}{n}\right) + \sum_u\sum_{v\neq u} n_{uv}\ln\left(\frac{n_{uv}}{n}\right)$$

The first sum is over all homozygotes, and the second over all heterozygotes. As is the convention in this book, the second sum involves each heterozygous class only once. For convenience, the subscripts for heterozygotes are written in numerical or alphabetical order. When the model is completely constrained by having Hardy-Weinberg proportions for all genotypes, the disequilibria are all zero while the MLE's for allele frequencies are still the observed frequencies. The log-likelihood reduces to

$$\begin{aligned} \ln L_0 &= \text{Constant} + \sum_u n_{uu} \ln\left(\frac{n_u}{2n}\right)^2 + \sum_u \sum_{v \neq u} n_{uv} \ln\left(2\frac{n_u}{2n}\frac{n_v}{2n}\right) \\ &= \text{Constant} + \sum_u n_u \ln\left(\frac{n_u}{2n}\right) + \sum_u \sum_{v \neq u} n_{uv} \ln 2 \end{aligned}$$

using the allele counts of

$$n_u = 2n_{uu} + \sum_{v \neq u} n_{uv}$$

An overall test for HWE is given by the likelihood ratio

$$G_T^2 = -2 \ln \lambda$$

with

$$\begin{aligned} \ln \lambda &= \ln\left(\frac{L_0}{L_1}\right) \\ &= \ln L_0 - \ln L_1 \end{aligned}$$

Under HWE, this quantity has an approximate chi-square distribution with $k(k-1)/2$ degrees of freedom.

The same distribution is found by a goodness-of-fit test on all the genotypic classes:

$$X_T^2 = \sum_u \frac{(n_{uu} - n\tilde{p}_u^2)^2}{n\tilde{p}_u^2} + \sum_u \sum_{v \neq u} \frac{(n_{uv} - 2n\tilde{p}_u\tilde{p}_v)^2}{2n\tilde{p}_u\tilde{p}_v}$$

This chi-square statistic also has $k(k-1)/2$ degrees of freedom. Care must be taken not to use genotypic classes with small expected numbers. Although an absolute rule cannot be given, there should be no problems if all expected numbers are above five. Cochran (1954) showed that there may even be two expectations as low as one with little effect on tests performed at the 5% significance level.

Note that X_T^2 can be expressed in terms of the estimated disequilibrium coefficients:

$$X_T^2 = \sum_u \frac{n\hat{D}_{uu}^2}{\tilde{p}_u^2} + \sum_u \sum_{v \neq u} \frac{2n\hat{D}_{uv}^2}{\tilde{p}_u \tilde{p}_v}$$

where $\hat{D}_{uu}$ is minus the sum of the disequilibria for all heterozygotes involving the uth allele.

To test each of the D_{uv} individually, the likelihood ratio approach requires the comparison of two likelihoods: one with D_{uv} included in the model and one with D_{uv} set to zero. If D_{12} is put to zero in the three-allele case, for example, the maximum likelihood can be written as

$$L_{13,23} = \max\{L(p_1, p_2, p_3, D_{13}, D_{23})\}$$

while for the unconstrained model

$$L_1 = \max\{L(p_1, p_2, p_3, D_{12}, D_{13}, D_{23})\}$$

and the test statistic, G_{12}^2, is:

$$G_{12}^2 = -2(\ln L_{13,23} - \ln L_1)$$

This statistic has an approximate chi-square distribution with 1 d.f. when the hypothesis $H_0 : D_{12} = 0$ is true. The difficulty with the procedure is in maximizing the likelihood under the hypothesis. As there are four parameters to estimate, two allele frequencies and two disequilibria, from five independent genotypic classes, Bailey's method cannot be used. Numerical iterations are required to solve simultaneous equations.

An alternative procedure was described by Hernández and Weir (1989). They used the asymptotic normality of MLE estimates of disequilibria. Fisher's approximate variance formula, Equation 2.17, leads to an expression for the variance of $\hat{D}_{uv}$:

$$\begin{aligned} 2n\text{Var}(\hat{D}_{uv}) = \; & p_u p_v[(1-p_u)(1-p_v) + p_u p_v] \\ & - [(1-p_u-p_v)^2 - 2(p_u-p_v)^2]D_{uv} \\ & + \sum_{w \neq u,v} (p_u^2 D_{vw} + p_v^2 D_{uw}) - D_{uv}^2 \end{aligned}$$

To test that D_{uv} is zero, this condition is used in the variance formula, and the test statistic becomes

$$\begin{aligned} X_{uv}^2 &= \frac{\hat{D}_{uv}^2}{\text{Var}(\hat{D}_{uv})} \\ &= \frac{2n\hat{D}_{uv}^2}{\tilde{p}_u\tilde{p}_v[(1-\tilde{p}_u)(1-\tilde{p}_v) + \tilde{p}_u\tilde{p}_v] + \sum_{w \neq u,v}(\tilde{p}_u^2\hat{D}_{vw} + \tilde{p}_v^2\hat{D}_{uw})} \end{aligned}$$

Table 3.2 Possible genotypic arrays of size 5.

	A_1	A_2	A_3	A_1	A_2	A_3	A_1	A_2	A_3	A_1	A_2	A_3
A_1	3			2			2			1		
A_2	0	1		1	1		2	0		3	0	
A_3	0	1	0	1	0	0	0	1	0	1	0	0
Allele counts	6	3	1	6	3	1	6	3	1	6	3	1
Probability		0.048			0.286			0.286			0.380	
X^2		5.56			1.67			2.78			2.22	

where observed and estimated values have been used for allele frequencies and disequilibria. In the three-allele case, for example,

$$
\begin{aligned}
X_{12}^2 &= \frac{2n\hat{D}_{12}^2}{\tilde{p}_1\tilde{p}_2[(1-\tilde{p}_1)(1-\tilde{p}_2)+\tilde{p}_1\tilde{p}_2]+(\tilde{p}_1^2\hat{D}_{23}+\tilde{p}_2^2\hat{D}_{13})} \\
X_{13}^2 &= \frac{2n\hat{D}_{13}^2}{\tilde{p}_1\tilde{p}_3[(1-\tilde{p}_1)(1-\tilde{p}_3)+\tilde{p}_1\tilde{p}_3]+(\tilde{p}_1^2\hat{D}_{23}+\tilde{p}_3^2\hat{D}_{12})} \\
X_{23}^2 &= \frac{2n\hat{D}_{23}^2}{\tilde{p}_2\tilde{p}_3[(1-\tilde{p}_2)(1-\tilde{p}_3)+\tilde{p}_2\tilde{p}_3]+(\tilde{p}_2^2\hat{D}_{13}+\tilde{p}_3^2\hat{D}_{12})}
\end{aligned}
$$

These test statistics provide single degree-of-freedom chi-squares without the complicated numerical methods needed for likelihood ratio tests.

An advantage of the likelihood methods, however, is that more complex hypotheses can be tested. The hypothesis $H_0 : D_{12} = D_{13} = 0$ can be tested by comparing the likelihoods L_{23} and L_1. The former quantity is found by maximizing the likelihood with respect to the allele frequencies and D_{23} when the other two disequilibria are set equal to zero.

Exact Tests with Multiple Alleles

To illustrate the exact test procedure with more than two alleles, all possible sets of genotype counts in samples of size 5, when the three allele counts are $n_1 = 6, n_2 = 3, n_3 = 1$ are shown in Table 3.2.

The probability of an array with genotypic counts $\{n_{uv}\}$, conditional on the allele counts $\{n_u\}$, is

$$\Pr(\{n_{uv}\}|\{n_u\}) = \frac{n!2^H \prod_u (n_u)!}{(2n!) \prod_{u,v}(n_{uv})!} \tag{3.8}$$

where $H = \sum_u \sum_{v \neq u} n_{uv}$ is the number of heterozygotes in the sample. Note that there are no p's in this equation. For three alleles the equation becomes

$$\Pr(\{n_{ij}\}|\{n_i\}) = \frac{n!2^{(n_{12}+n_{13}+n_{23})}(n_1)!(n_2)!(n_3)!}{(2n)!(n_{11})!(n_{22})!(n_{33})!(n_{12})!(n_{13})!(n_{23})!}$$

The probabilities shown in Table 3.2 come from applying this formula. In order, they are 1/21, 6/21, 6/21, and 8/21 for the four possible genotypic arrays.

If the first genotypic array in Table 3.2 was observed, the Hardy-Weinberg hypothesis would be rejected at the 5% level, because that outcome (or a less probable one) is expected to occur by chance less than 5% of the time. The same conclusion would be reached with a chi-square goodness-of-fit test. The values of the test statistic X^2 in Table 3.2 show that the first array would lead to rejection at the 5% level because $X^2 = 5.56 > 3.84$. However, the chi-square test statistic does not rank the four possible arrays according to their probabilities under the null hypothesis. Goodness-of-fit is measuring something different than these probabilities. Note also that the chi-square test is not conditional on the allele counts, and it uses estimates of the population allele frequencies. Based on some extensive simulations to examine significance level and power, Maiste (1993) concluded that the probability test was to be preferred over the chi-square test. Of course, the chi-square test is very much simpler to apply.

Permutation Version of Exact Test for HWE

The difficulty in applying the exact test is that there are too many possible genotypic arrays to consider, even on a computer, for large samples and large numbers of alleles. A way of circumventing the computing task has been given by Guo and Thompson (1992). For large sample sizes and many alleles, the probability in Equation 3.8 will be very small, but the relevant quantity is the aggregate probability of all the sample genotypic arrays, with the same allele frequencies, that are as probable or less probable than the sample array. This aggregate probability p is the significance level with which the Hardy-Weinberg hypothesis would be rejected. Guo and Thompson gave the following solution to the problem of applying this conditional test to astronomical numbers of genotypic arrays.

In a sample of n individuals, there are n_i copies of allele A_i and the probability of the sample under the Hardy-Weinberg hypothesis can be calculated from Equation 3.8. Suppose a deck of $2n$ cards is constructed, with the first n_1 cards marked to represent allele A_1, the next n_2 marked for allele A_2, and so on. The deck is then shuffled, and successive pairs of cards taken to represent n genotypes. Under the HWE hypothesis, alleles are distributed independently into genotypes, so the genotypic array found by permutation (or card shuffling) corresponds to one of the arrays possible under HWE. It has the same allele frequencies as the original sample, and its probability can be calculated with Equation 3.8. The process is repeated, and if m of S shuffled arrays are as probable or less probable than the sample, the significance level p is estimated as $\hat{p} = m/S$. If it is desired to be 95% confident of being within 0.01 of the true value of p, normal theory provides

$$\begin{aligned} 0.01 &\geq 1.96\sqrt{p(1-p)/S} \\ S &\geq (196)^2 p(1-p) \end{aligned}$$

The largest value of this last expression is when $p = 0.5$, so $S \geq 10,000$.

Although the permutation test is based on the conditional probability, it is necessary to calculate only the portion of Equation 3.8 that varies among the permuted arrays. This portion is $2^H / \prod_{u,v}(n_{uv})!$, and it is computed for the original sample and all the permuted samples.

If the hypothesis to be tested allows for some of the disequilibrium coefficients to be nonzero, the probability of a sample, conditional on the observed allele frequencies, depends on the true allele frequencies *and* on the nonzero disequilibria. These parameters do not cancel out as they did for the overall test, and exact tests are not available for individual disequilibria in the multiple allele case.

Power of Tests for HWE

Statistical tests of hypotheses are subject to two kinds of error: a true hypothesis may be rejected or a false hypothesis may not be rejected. The significance level measures the probability of the first kind of error, while power is one minus the probability of the second kind. In general, power is the probability of rejecting an hypothesis, and the particular case when the hypothesis is true gives a power value equal to the significance level of the test. For a given significance level, those tests with the greatest power (probability of rejecting a false hypothesis) are preferred. Methods for calculating power are now considered.

If a significance value is specified in advance, such as $\alpha = 0.05$, this determines the rejection region. Before any data are collected, it is known what values of the test statistic will cause the hypothesis to be rejected. Powers are calculated for each alternative value of the parameter being tested. As an example, for a given value of α in a test of $H_0 : D_{12} = 0$, the power of the test for detecting a disequilibrium value of 0.10 may be needed. In other words, what is the probability of rejecting H_0 when $D_{12} = 0.10$? The answer depends on the probabilities with which the test statistic takes its various values.

Consider first the chi-square test for HWE, with the test statistic written as in Equation 3.6. Theory provides that, in large samples, X^2 is distributed as approximately chi-square with 1 d.f. when the hypothesis is true, and that it is distributed as a ***noncentral chi-square*** when the hypothesis is false (provided it is not false by too large an amount):

$$\begin{aligned} X^2 &\sim \chi^2_{(1)} \text{ when } H_0 \text{ true} \\ X^2 &\sim \chi^2_{(1,\nu)} \text{ when } H_0 \text{ false} \end{aligned}$$

The degrees of freedom are still 1, but there is now a noncentrality parameter ν given by replacing statistics $\tilde{p}_A$, $\hat{D}_A$ with parametric values p_A, D_A in X^2:

$$\nu = \frac{nD_A^2}{p_A^2(1-p_A)^2} = nf^2$$

The noncentral chi-square distribution holds when $D_A^2/p_A^2(1-p_A)^2$ is of order n^{-1}. When H_0 is true, $D_A = 0$ and so $\nu = 0$ and the distribution is central chi-square. The larger is D_A, the larger is ν and the more likely it is that the hypothesis will be rejected. Recall that an X^2 value of 3.84 or greater causes rejection of H_0 for a significance level of 0.05. There is a probability of 0.05 that X^2 will exceed 3.84 when the hypothesis is true. Tables of the noncentral chi-square (Haynam et al. 1970, Appendix Table A.3) show that there is a probability of 0.90 that X^2 will exceed 3.84 when the hypothesis is false if $\nu = 10.5$. This value of the noncentrality parameter depends on the critical value of 3.84 and hence on the significance level. Power and significance level are related to each other.

It is useful to turn this last argument around to see what value of the disequilibrium coefficient D_A will be detected with 90% probability when a 5% level test is used. Rearranging the expression for ν shows that

$$D_A = p_A(1-p_A)\sqrt{\frac{10.5}{n}}$$

confirming that larger samples will allow the detection of smaller levels of disequilibrium. Alternatively, it can be seen that the number of individuals n must be

$$n = 10.5\frac{p_A^2(1-p_A)^2}{D_A^2} = \frac{10.5}{f^2}$$

for a 5% test to have 90% power.

This use of the noncentral chi-square distribution should be limited to cases in which the alternative values of the parameter being tested are close to the hypothesized values. In other words, it is appropriate only for local departures from the hypothesized values. Hernández and Weir (1989) showed that, otherwise, misleadingly large power values could result from using this theory.

For exact test power calculations, an alteration is needed in Equation 3.7 for calculating the probability of each of the possible samples given the observed allele frequencies. If the true disequilibrium is D_A instead of zero,

$$\begin{aligned}\Pr(n_{AA}, n_{Aa}, n_{aa}) &= \frac{n!}{n_{AA}!n_{Aa}!n_{aa}!}(p_A^2 + D_A)^{n_{AA}} \\ &\quad \times (2p_Ap_a - 2D_A)^{n_{Aa}}(p_a^2 + D_A)^{n_{aa}}\end{aligned}$$

and $\Pr(n_A, n_a)$ needs to be calculated from appropriate sums of these probabilities since allele counts are not binomially distributed when D_A is not zero. The calculations would need to be built into the computer program used to generate the possible samples.

LINKAGE DISEQUILIBRIUM

The logical next step in a study of associations between alleles is to look at the frequencies for alleles at different loci. Any associations found will be referred to generally as ***linkage disequilibrium*** even though they may have nothing to do with linkage. Alleles at different loci may have frequencies that show association whether or not those loci are linked. Between-locus associations are needed in several contexts. Associations involving pairs of loci find use in expressions for the variances of quantities estimated as averages over single loci, as shown in Chapter 4. As discussed in Chapter 7, linkage disequilibrium between a marker locus and a trait value can play a role in the search for genes affecting the trait.

Gametic Disequilibrium at Two Loci

In the first situation to be considered, data are available on gametes and there is no need to worry about genotypic associations. This will be the situation in which single chromosomes have been sampled from natural populations (e.g., Langley et al. 1974), or when family data are available to allow haplotypes to be inferred from genotypes (e.g., Weir and Brooks 1986). For a pair of alleles at two loci, the procedures for defining, estimating, and testing disequilibria are entirely analogous to those for pairs of alleles at a single locus.

The disequilibrium coefficient for alleles A and B at two loci compares the gametic frequency with the product of allele frequencies

$$D_{AB} = p_{AB} - p_A p_B$$

and inferences are based on the assumed multinomial distribution of gametes. As in the one-locus case, the MLE of D_{AB} is found directly from the observed frequencies

$$\hat{D}_{AB} = \tilde{p}_{AB} - \tilde{p}_A \tilde{p}_B$$

Taking expectations over all samples of size $2n$ gametes from the same population shows that the MLE is biased

$$\mathcal{E}(\hat{D}_{AB}) = \frac{2n-1}{2n} D_{AB}$$

and that the large-sample variance is

$$\begin{aligned}\text{Var}(\hat{D}_{AB}) &= \frac{1}{2n}\Big[p_A(1-p_A)p_B(1-p_B) \\ &\quad +(1-2p_A)(1-2p_B)D_{AB} - D_{AB}^2\Big] \qquad (3.9)\end{aligned}$$

with an obvious similarity to the variance of $\hat{D}_A$ shown in Equation 3.3.

A chi-square statistic for the hypothesis of no disequilibrium, $H_0 : D_{AB} = 0$, can be constructed by squaring the asymptotically normal variable z:

$$z = \frac{\hat{D}_{AB} - \mathcal{E}(\hat{D}_{AB})}{\sqrt{\text{Var}(\hat{D}_{AB})}}$$

after setting D_{AB} to zero in both the mean and the variance. The test statistic is written X_{AB}^2

$$\begin{aligned}X_{AB}^2 &= z^2 \\ &= \frac{2n\hat{D}_{AB}^2}{\tilde{p}_A(1-\tilde{p}_A)\tilde{p}_B(1-\tilde{p}_B)} \qquad (3.10)\end{aligned}$$

Table 3.3 Contingency table for counts of four gametic types at two loci.

Counts		Locus B B	$\bar{B}$	Totals
Locus A	A	n_{AB}	$n_{A\bar{B}}$	n_A
	$\bar{A}$	$n_{\bar{A}B}$	$n_{\bar{A}\bar{B}}$	$n_{\bar{A}}$
Totals		n_B	$n_{\bar{B}}$	$2n$

The same chi-square test statistic follows from a goodness-of-fit on the four gametic classes

Gamete	AB	$A\bar{B}$	$\bar{A}B$	$\bar{A}\bar{B}$	Total
Observed number	n_{AB}	$n_{A\bar{B}}$	$n_{\bar{A}B}$	$n_{\bar{A}\bar{B}}$	$2n$
Expected number	$2n\tilde{p}_A\tilde{p}_B$	$2n\tilde{p}_A\tilde{p}_{\bar{B}}$	$2n\tilde{p}_{\bar{A}}\tilde{p}_B$	$2n\tilde{p}_{\bar{A}}\tilde{p}_{\bar{B}}$	$2n$

As before, $\bar{A}, \bar{B}$ mean not-A, not-B. It is more common to arrange the four gamete counts in a 2×2 contingency table, as shown in Table 3.3. The test for gametic linkage disequilibrium can be regarded as a test for independence in this table. The count in each of the four cells is compared to the products of the row and column totals divided by the overall total. If Yates' continuity correction was to be applied, it would be on the differences between observed and expected counts. Each corrected difference has the same value of $(|n_{AB} - 2n\tilde{p}_A\tilde{p}_B| - 0.5)$.

Exact Test for Gametic Disequilibrium

An exact test for gametic linkage disequilibrium depends on the probabilities of all possible samples of gametic numbers for the observed allele numbers. Since gametic numbers are being assumed to be multinomially distributed, the allele numbers are binomially distributed (with the same sample size). The probability $\Pr(n_{AB})$ of the gametic array under the hypothesis of no linkage disequilibrium is

$$\Pr(n_{AB}) = \Pr(n_{AB}, n_{A\bar{B}}, n_{\bar{A}B}, n_{\bar{A}\bar{B}})$$

$$= \frac{(2n)!(p_A p_B)^{n_{AB}}(p_A p_{\bar{B}})^{n_{A\bar{B}}}(p_{\bar{A}} p_B)^{n_{\bar{A}B}}(p_{\bar{A}} p_{\bar{B}})^{n_{\bar{A}\bar{B}}}}{n_{AB}!n_{A\bar{B}}!n_{\bar{A}B}!n_{\bar{A}\bar{B}}!}$$

and the probabilities of the two allele arrays are

$$\begin{aligned} \Pr(n_A, n_{\bar{A}}) &= \frac{(2n)!}{n_A!n_{\bar{A}}!}(p_A)^{n_A}(p_{\bar{A}})^{n_{\bar{A}}} \\ \Pr(n_B, n_{\bar{B}}) &= \frac{(2n)!}{n_B!n_{\bar{B}}!}(p_B)^{n_B}(p_{\bar{B}})^{n_{\bar{B}}} \end{aligned}$$

Taking the ratio of these quantities gives the probability of the gametic numbers conditional on the allele numbers

$$\Pr(n_{AB}, n_{A\bar{B}}, n_{\bar{A}B}, n_{\bar{A}\bar{B}} \mid n_A, n_B) = \frac{n_A!n_{\bar{A}}!n_B!n_{\bar{B}}!}{n_{AB}!n_{A\bar{B}}!n_{\bar{A}B}!n_{\bar{A}\bar{B}}!(2n)!}$$

The unknown allele frequencies have cancelled out in this probability, just as they did in testing for HWE. Notice that the probabilities may just as well have been written as $\Pr(n_{AB} \mid n_A, n_B)$. Tables of significant values of n_{AB} for given values of n, n_A, and n_B have been published (Finney et al. 1963) for $2n$ up to 80.

To illustrate the exact test for gametic data, turn to the *Drosophila* restriction site data shown in Table 1.6. The presence or absence of a restriction site can be regarded as alternative alleles at a locus, and for the sites recognized by *Bam*HI and *Xho*I:

Counts		*Xho*I +	*Xho*I −	Total
*Bam*HI	+	5	6	11
	−	6	0	6
Total		11	6	17

All possible samples of size 17 with the same marginal totals as in the observed data are displayed in Table 3.4, along with the exact probabilities and chi-square statistics. The observed set of frequencies are the third least probable and lead to rejection of the hypothesis of linkage equilibrium at a significance level of 4.27%. There is a probability of 0.0427 of observing a sample as extreme, or more extreme, than 5,6,6,0. The chi-square test has a larger 5% rejection region when calculated as in Equation 3.10 since it would

Table 3.4 Linkage disequilibria for restriction sites in *Drosophila* data of Table 1.6.

Gamete*					Cumulative		
+ +	+ −	− +	− −	Probability	Probability	D_{AB}	Chi-square
11	0	0	6	0.0001	0.0001†	0.2284	17.00†
10	1	1	5	0.0053	0.0054†	0.1696	9.37†
5	6	6	0‡	0.0373	0.0427†	−0.1246	5.06†
9	2	2	4	0.0667	0.1094	0.1107	4.00†
6	5	5	1	0.2240	0.3334	−0.0657	1.41
8	3	3	3	0.2666	0.6000	0.0519	0.88
7	4	4	2	0.4000	1.0000	−0.0069	0.02

* First symbol is for *Bam*HI, second is for *Xho*I.
† Causes rejection of hypothesis of no disequilibrium at 5% level.
‡ Observed data.

reject the hypothesis of linkage equilibrium for the sample 9,2,2,4 as well. However, when Yates' continuity correction is applied, the chi-square test statistic has nonsignificant values (at the 5% level) for the samples 9,2,2,4 ($X^2 = 2.15$) *and* 5,6,6,0 ($X^2 = 2.95$). The correction reduces the absolute value of the observed and expected gametic *counts*. Remember that the exact test gives an accurate accounting of the chance of each possible outcome when the hypothesis is true.

For the exact test of HWE, either large positive or large negative values of the one-locus disequilibrium coefficient D_A can lead to data points of low probability under the null hypothesis. In the linkage disequilibrium case, the same situation holds. Table 3.4 illustrates that small probabilities for a sample under the hypothesis of no disequilibrium can be obtained for either positive or negative values of the coefficient D_{AB}. An alternative approach was advocated by Yates (1984). If the rows in Table 3.4 are ordered by disequilibrium values, the observed value of −0.1246 is the most extreme in the negative tail of the distribution of all possible values for these allele frequencies. The hypothesis of no disequilibrium would be rejected in favor of the one-tailed alternative of negative disequilibrium at a significance level of 0.0373. If a two-tailed alternative is needed, Yates' rule of doubling the one-tailed probability gives a significance level of 0.0746. In presenting results of exact tests, care is needed to state the procedure used.

Power calculations for the exact test are relatively easy since the allele numbers are binomially distributed whether or not there is disequilibrium in this case of samples of gametes (unlike the situation for samples of genotypes). The probabilities will involve both the allele frequencies and the disequilibrium, however, so that power cannot be found without specifying both sets of parameters. Fu and Arnold (1990) give tables of sample sizes needed to achieve power levels of 50% or 90% when a 5% significance level is used with exact tests.

Gametic Disequilibrium with Multiple Alleles

There is no conceptual difficulty in extending the study of gametic disequilibrium at two loci to allow for several alleles at each locus. A separate coefficient D_{uv} is defined for each pair of alleles A_u and B_v

$$D_{uv} = p_{uv} - p_u p_v$$

These coefficients may be tested separately with the chi-square statistics

$$X^2_{uv} = \frac{(2n)\hat{D}^2_{uv}}{\tilde{p}_u(1-\tilde{p}_u)\tilde{p}_v(1-\tilde{p}_v)} \qquad (3.11)$$

while the overall hypothesis that none of the D_{uv}'s is different from zero can be tested with the statistic

$$X^2_T = \sum_{u=1}^{k}\sum_{v=1}^{\ell} \frac{(2n)\hat{D}^2_{uv}}{\tilde{p}_u \tilde{p}_v} \qquad (3.12)$$

which has $(k-1)(\ell-1)$ degrees of freedom when the two loci have k and ℓ alleles, respectively. The sample size is still $2n$ gametes.

Power calculations for the chi-squares in Equations 3.11 and 3.12 follow from the noncentral chi-square distribution provided the departures from the null hypotheses are not too large.

Exact tests with multiple alleles are also straightforward. The probability of a set of counts $\{n_{uv}\}$ for A_uB_v gametes, conditional on the allele counts $\{n_u\}$ and $\{n_v\}$, is

$$\Pr(\{n_{uv}\}|\{n_u\},\{n_v\}) = \frac{\prod_u n_u! \prod_v n_v!}{(2n)! \prod_{u,v} n_{uv}!}$$

and the significance level is obtained by permuting the alleles at one of the loci.

Variances and Covariances of Gametic Linkage Disequilibria

Primary interest here will be in associations between alleles at two loci, but most data sets contain frequencies at more than two loci and many higher-order disequilibria can be defined. Even concentration on pairs of loci, however, requires a consideration of associations for three or four loci when relationships between the pairwise measures are considered. If gametic linkage disequilibria D_{AB} and D_{BC} are estimated for alleles A, B, and C at three loci, they are expected to be related in value because of the common dependence on allele B. It is possible to modify Fisher's approximate variance formula, Equation 2.17, to give approximate covariances of functions S and T of multinomial counts n_i (these counts have expected values nQ_i):

$$\frac{1}{n}\text{Cov}(S,T) = \sum_i \frac{\partial S}{\partial n_i}\frac{\partial T}{\partial n_i}Q_i - \frac{\partial S}{\partial n}\frac{\partial T}{\partial n}$$

Applying this result allows covariances of sample disequilibria to be found:

$$\begin{aligned}\text{Cov}(\hat{D}_{AB},\hat{D}_{BC}) &= \frac{1}{n}[p_B(1-p_B)D_{AC} + (1-2p_B)D_{ABC} - D_{AB}D_{BC}] \\ \text{Cov}(\hat{D}_{AB},\hat{D}_{CD}) &= \frac{1}{n}[D_{AC}D_{BD} + D_{AD}D_{BC} + D_{ABCD}]\end{aligned}$$

Note that the three- and four-locus disequilibria (defined in Equations 3.13 and 3.14 below) are involved, and that the second covariance is likely to be quite small because it involves products of two-locus disequilibria and the four-locus disequilibrium.

The relationships between disequilibria may not be particularly meaningful because of the arbitrary way in which the coefficients are generally defined. There is usually no particular reason for labeling one allele A and the others $\bar{A}$, for example, so that D_{AB} may as well have had the same value but the opposite sign

$$D_{AB} = -D_{A\bar{B}} = -D_{\bar{A}B} = D_{\bar{A}\bar{B}}$$

There is also the feature that, in the absence of disturbing forces, disequilibrium is expected to decay to zero over time. It is common to work with squared disequilibria. This at least eliminates problems with sign, and population genetic theory shows that the ratio of $\hat{D}^2_{AB}$ to $\tilde{p}_A(1-\tilde{p}_A)\tilde{p}_B(1-\tilde{p}_B)$ remains nonzero even as populations drift to fixation for one allele at each locus. It appears then that there is a need also for variances and covariances of squared disequilibria to provide a complete picture of the relationships among two-locus linkage disequilibria (Hill and Weir 1988).

Gametic Disequilibrium at Three or Four Loci

Whether there is interest in three- and four-allele disequilibria in their own right, or merely as components of the covariances between two-locus coefficients, it is useful to define and make inferences about them. There have been alternative parameterizations suggested in the past to describe disequilibrium among alleles at more than two loci. The problem is to remove the effects of two-locus disequilibrium when describing three-locus associations, or two- and three-locus disequilibria when describing four-locus associations. Here the additive formulation of Bennett (1954), which subtracts terms for lower-order disequilibria, will be used. For alleles A, B, and C at three loci:

$$D_{ABC} = p_{ABC} - p_A D_{BC} - p_B D_{AC} - p_C D_{AB} - p_A p_B p_C \quad (3.13)$$

and for alleles A, B, C, and D at four loci

$$\begin{aligned} D_{ABCD} = \; & p_{ABCD} - p_A D_{BCD} - p_B D_{ACD} - p_C D_{ABD} - p_D D_{ABC} \\ & - p_A p_B D_{CD} - p_A p_C D_{BD} - p_A p_D D_{BC} \\ & - p_B p_C D_{AD} - p_B p_D D_{AC} - p_C p_D D_{AB} \\ & - D_{AB} D_{CD} - D_{AC} D_{BD} - D_{AD} D_{BC} - p_A p_B p_C p_D \quad (3.14) \end{aligned}$$

Maximum likelihood estimates follow by replacing frequencies with their observed values, and there is a slight bias in these estimates. Variances of the estimates are found with Fisher's formula, although for four loci the result requires a computerized algebra treatment (B.S. Weir, G.B. Golding, and P.O. Lewis, unpublished results), and are expressed most succinctly with the quantities π_A and τ_A for allele A:

$$\pi_A = p_A(1 - p_A), \quad \tau_A = (1 - 2p_A)$$

The results are

$$\mathrm{Var}(\hat{D}_{AB}) = \frac{1}{n}[\pi_A \pi_B + \tau_A \tau_B D_{AB} - D_{AB}^2]$$

$$\begin{aligned} \mathrm{Var}(\hat{D}_{ABC}) = \; & \frac{1}{n}[\pi_A \pi_B \pi_C + 6 D_{AB} D_{BC} D_{AC} + \pi_A(\tau_B \tau_C D_{BC} - D_{BC}^2) \\ & + \pi_B(\tau_A \tau_C D_{AC} - D_{AC}^2) + \pi_C(\tau_A \tau_B D_{AB} - D_{AB}^2) \\ & + D_{ABC}(\tau_A \tau_B \tau_C - 2\tau_A D_{BC} - 2\tau_B D_{AC} \\ & - 2\tau_C D_{AB} - D_{ABC})] \end{aligned}$$

The coefficient D_{ABC} is set to zero, and all other terms replaced by their observed values, when using the three-locus variance to test the hypothesis

$H_0 : D_{ABC} = 0$ with the chi-square statistic

$$X^2_{ABC} = \frac{\hat{D}^2_{ABC}}{\mathrm{Var}(\hat{D}_{ABC})}$$

A similar procedure applies to the four-locus case.

Normalized Gametic Disequilibria

Additive disequilibrium coefficients are constrained in the values they may take since gametic frequencies cannot be negative or greater than corresponding allele frequencies. Focusing on alleles A and B at loci A and B, the frequencies of the gametes are constrained by

$$\begin{array}{ll} 0 \le p_{AB} = p_A p_B + D_{AB} \le p_A, p_B \quad , & 0 \le p_{A\bar{B}} = p_A p_{\bar{B}} - D_{AB} \le p_A, p_{\bar{B}} \\ 0 \le p_{\bar{A}B} = p_{\bar{A}} p_B - D_{AB} \le p_{\bar{A}}, p_B \quad , & 0 \le p_{\bar{A}\bar{B}} = p_{\bar{A}} p_{\bar{B}} + D_{AB} \le p_{\bar{A}}, p_{\bar{B}} \end{array}$$

so that the linkage disequilibrium coefficient D_{AB} is constrained by

$$-p_A p_B, -p_{\bar{A}} p_{\bar{B}} \le D_{AB} \le p_A - p_A p_B, p_B - p_A p_B$$

or

$$\max(-p_A p_B, -p_{\bar{A}} p_{\bar{B}}) \le D_{AB} \le \min(p_{\bar{A}} p_B, p_A p_{\bar{B}})$$

with $\bar{A}, \bar{B}$ meaning not-A, not-B.

Another view of the amount of disequilibrium is provided by dividing the coefficient by its maximum numerical value, remembering that the maximum values depend on sign. This normalized value D'_{AB} is therefore defined as

$$D'_{AB} = \begin{cases} \dfrac{D_{AB}}{\max(-p_A p_B, -p_{\bar{A}} p_{\bar{B}})}, & D_{AB} < 0 \\ \\ \dfrac{D_{AB}}{\min(p_{\bar{A}} p_B, p_A p_{\bar{B}})}, & D_{AB} > 0 \end{cases}$$

Although D'_{AB} has a range of values from -1 to $+1$ it would be mistaken to regard it as being independent of allele frequencies. The frequencies are very much part of the definition of the normalized coefficient, as has been pointed out by Lewontin (1988).

Bounds on three-locus disequilibria were given by Thomson and Baur (1984).

Table 3.5 Partitioning of the 9 d.f. available from the 10 genotypes at 2 loci.

Description	d.f.
Allele frequencies p_A, p_B	2
One-locus disequilibria D_A, D_B	2
Gametic disequilibrium D_{AB}	1
Nongametic disequilibrium $D_{A/B}$	1
Trigenic disequilibria D_{AAB}, D_{ABB}	2
Quadrigenic disequilibrium D_{AABB}	1
Total	9

Genotypic Disequilibrium at Two Loci

When population genetic data are collected on genotypes, it is possible to check for associations between alleles other than those at one locus or those on one gamete. Recall that HWE testing looks at two alleles at the same locus, but on different gametes. Testing for linkage disequilibrium looks at two alleles on the same gamete, but at different loci. There is a third alternative: two alleles on different gametes and at different loci. Whatever forces are responsible for causing associations of pairs of alleles in the first two cases might also cause an association in the third. Whether or not an immediate biological mechanism is suggested, it is recommended here that at least the possibility for this additional type of disequilibrium be investigated. Continuing this argument, and referring back to the search for associations between frequencies of three or four alleles on a single gamete, a case can be made for looking for associations between the frequencies of three or four of the four alleles present at two loci in a diploid individual. Although such associations will generally be quite small, they complete the disequilibrium analysis at two loci.

When focusing on alleles A and B, there are a total of 10 possible genotypes with 9 d.f. available for estimation. The frequencies of the 10 genotypes can be used to estimate the 9 quantities listed in Table 3.5 and illustrated in Figure 3.1.

Allele frequencies and one-locus coefficients are estimated as described above and the gametic disequilibrium can also be estimated directly from observations, since it is assumed here that gametic frequencies can be recovered from the genotypic frequencies, as in

$$p_{AB} = P^{AB}_{AB} + \frac{1}{2}\left(P^{AB}_{A\bar{B}} + P^{AB}_{\bar{A}B} + P^{AB}_{\bar{A}\bar{B}}\right)$$

The new digenic disequilibrium, $D_{A/B}$, refers to alleles at different loci and on different gametes within individuals, and is defined as

$$D_{A/B} = p_{A/B} - p_A p_B$$

with the nongametic frequency

$$p_{A/B} = P^{AB}_{AB} + \frac{1}{2}\left(P^{AB}_{A\bar{B}} + P^{AB}_{\bar{A}B} + P^{\bar{A}B}_{A\bar{B}}\right)$$

So that there is no confusion over notation, an example of 10 genotypic frequencies is shown in Table 3.6, arranged to show both the one-locus genotypic frequencies as marginal totals, and the two-locus gametic and nongametic marginal totals.

For the trigenic coefficients, frequencies of triples of alleles are compared with the products of allele frequencies, after removing any digenic disequilibria. The two trigenic frequencies are

$$p_{AAB} = P^{AB}_{AB} + \frac{1}{2}P^{AB}_{A\bar{B}}$$

$$p_{ABB} = P^{AB}_{AB} + \frac{1}{2}P^{AB}_{\bar{A}B}$$

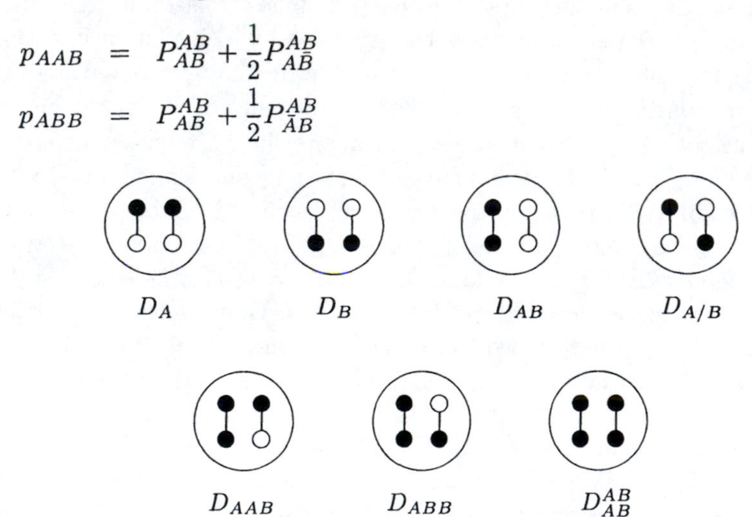

Figure 3.1 Pictorial representation of complete set of two-locus disequilibria. Vertical lines represent two gametes. Upper circles are the alleles at the A locus and lower circles are the alleles at the B locus. Filled circles are those for which the disequilibrium coefficient is defined.

Table 3.6 Example of genotypic frequency notation for two alleles at each of two loci.

$P^{AB}_{AB} = 0.20$	$P^{AB}_{A\bar{B}} = 0.18$	$P^{A\bar{B}}_{A\bar{B}} = 0.02$	$P^{A}_{A} = 0.40$
$P^{AB}_{\bar{A}B} = 0.26$	$P^{AB}_{\bar{A}\bar{B}} = 0.08$ $P^{A\bar{B}}_{\bar{A}B} = 0.04$	$P^{A\bar{B}}_{\bar{A}\bar{B}} = 0.02$	$P^{A}_{\bar{A}} = 0.40$
$P^{\bar{A}B}_{\bar{A}B} = 0.04$	$P^{\bar{A}B}_{\bar{A}\bar{B}} = 0.10$	$P^{\bar{A}\bar{B}}_{\bar{A}\bar{B}} = 0.06$	$P^{\bar{A}}_{\bar{A}} = 0.20$
$P^{B}_{B} = 0.50$	$P^{B}_{\bar{B}} = 0.40$	$P^{\bar{B}}_{\bar{B}} = 0.10$	1.00

$P^{AB}_{AB} = 0.20$	$\frac{1}{2}P^{AB}_{A\bar{B}} = 0.09$	$\frac{1}{2}P^{AB}_{\bar{A}B} = 0.13$	$\frac{1}{2}P^{AB}_{\bar{A}\bar{B}} = 0.04$	$P_{AB} = 0.46$
$\frac{1}{2}P^{A\bar{B}}_{AB} = 0.09$	$P^{A\bar{B}}_{A\bar{B}} = 0.02$	$\frac{1}{2}P^{A\bar{B}}_{\bar{A}B} = 0.02$	$\frac{1}{2}P^{A\bar{B}}_{\bar{A}\bar{B}} = 0.01$	$P_{A\bar{B}} = 0.14$
$\frac{1}{2}P^{\bar{A}B}_{AB} = 0.13$	$\frac{1}{2}P^{AB}_{\bar{A}\bar{B}} = 0.04$	$P^{\bar{A}B}_{\bar{A}B} = 0.04$	$\frac{1}{2}P^{\bar{A}B}_{\bar{A}\bar{B}} = 0.05$	$P_{\bar{A}/B} = 0.26$
$\frac{1}{2}P^{A\bar{B}}_{\bar{A}B} = 0.02$	$\frac{1}{2}P^{\bar{A}\bar{B}}_{A\bar{B}} = 0.01$	$\frac{1}{2}P^{\bar{A}\bar{B}}_{\bar{A}B} = 0.05$	$P^{\bar{A}\bar{B}}_{\bar{A}\bar{B}} = 0.06$	$P_{\bar{A}/\bar{B}} = 0.14$
$P_{A/B} = 0.44$	$P_{A/\bar{B}} = 0.16$	$P_{\bar{A}B} = 0.24$	$P_{\bar{A}\bar{B}} = 0.16$	1.00

and the disequilibria are defined as

$$\begin{aligned} D_{AAB} &= p_{AAB} - p_A D_{AB} - p_A D_{A/B} - p_B D_A - p_A^2 p_B \\ D_{ABB} &= p_{ABB} - p_B D_{AB} - p_B D_{A/B} - p_A D_B - p_A p_B^2 \end{aligned}$$

Finally, the quadrigenic coefficient measures any remaining disequilibrium after the removal of all digenic and trigenic disequilibria

$$\begin{aligned} D^{AB}_{AB} &= P^{AB}_{AB} - 2p_A D_{ABB} - 2p_B D_{AAB} - 2p_A p_B D_{AB} - 2p_A p_B D_{A/B} \\ &\quad - p_A^2 D_B - p_B^2 D_A - D_{AB}^2 - D_{A/B}^2 - D_A D_B - p_A^2 p_B^2 \end{aligned}$$

The MLE's of all seven disequilibria follow by substituting observed values into these expressions. It is now genotypic, rather than gametic, frequencies that are multinomially distributed, and this must be reflected in the sampling variances. The variances for $\hat{D}_A$ and $\hat{D}_B$ remain as in Equation 3.3, but that for $\hat{D}_{AB}$ is changed from Equation 3.9 to

$$\begin{aligned}\text{Var}(\hat{D}_{AB}) &= \frac{1}{2n}\Big(\pi_A\pi_B + \tau_A\tau_B D_{AB} - D_{AB}^2 \\ &\quad + D_A D_B + D_{A/B}^2 + D_{AB}^{AB}\Big)\end{aligned}$$

for an estimate based on n genotypes (Equation 3.9 was for $2n$ gametes). This result is from Fisher's approximate variance formula, as is

$$\begin{aligned}\text{Var}(\hat{D}_{A/B}) &= \frac{1}{2n}\Big(\pi_A\pi_B + \tau_A\tau_B D_{A/B} - D_{A/B}^2 \\ &\quad + D_A D_B + D_{AB}^2 + D_{AB}^{AB}\Big)\end{aligned}$$

Exact results were given by Weir (1979). Further application of Fisher's formula leads to variances for the trigenic and quadrigenic coefficients, but the complexity makes it desirable that the algebraic manipulations be done on a computer. The results were derived by Lisa D. Brooks (personal communication). For the trigenic coefficient $\hat{D}_{AAB}$

$$\begin{aligned}2n\text{Var}(\hat{D}_{AAB}) &= (\pi_A^2 + \tau_A^2 D_A - D_A^2)(\pi_B + D_B) \\ &\quad + \pi_A\tau_A\tau_B(D_{AB} + D_{A/B}) - 2D_{AAB}^2 \\ &\quad + (1 - 5\pi_A + D_A)(D_{AB} + D_{A/B})^2 + 2\pi_A\tau_A D_{ABB} \\ &\quad + [\tau_A^2\tau_B - 2D_A\tau_B - 4\tau_A(D_{AB} + D_{A/B})]D_{AAB} \\ &\quad + (\tau_A^2 - 2D_A)(D_{AB}^{AB} - 2D_{AB}D_{A/B}) \qquad (3.15)\end{aligned}$$

There is a similar expression for $\hat{D}_{ABB}$ and for the quadrigenic coefficient $\hat{D}_{AABB}$

$$\begin{aligned}2nVar(\hat{D}_{AB}^{AB}) &= (\pi_A^2 + \tau_A^2 D_A - D_A^2)(\pi_B^2 + \tau_B^2 D_B - D_B^2) \\ &\quad + \cdots - (D_{AB}^{AB})^2\end{aligned}$$

The complete expression for this variance was given by Weir and Cockerham (1989).

These trigenic and quadrigenic variance expressions are extremely cumbersome and illustrate how quickly the additive model for disequilibrium becomes unmanageable. Extensions to three loci appear to be prohibitively complex. The results given here for two loci can be incorporated into computer programs for a complete analysis of two-locus disequilibrium, and a numerical justification of the testing procedure was given by Weir and Cockerham (1989).

The most noteworthy feature of the variance formulas just displayed is that the variances for the two digenic disequilibria calculated from genotypic data involve the higher-order disequilibrium coefficients. This suggests that an appropriate testing strategy is to begin with these higher-order disequilibria, and first test the hypothesis $H_0 : D_{AB}^{AB} = 0$. This is performed with the chi-square statistic

$$X_{AABB}^2 = \frac{(\hat{D}_{AB}^{AB})^2}{\text{Var}(\hat{D}_{AB}^{AB})}$$

with D_{AB}^{AB} set to zero in the denominator and all other terms set to the observed or estimated values. If the hypothesis is not rejected, the quadrigenic coefficient can be set to zero in the test for trigenic and digenic disequilibria. Otherwise, higher order terms must be included in test statistics for lower-order disequilibria.

When quadrigenic disequilibrium can be ignored, the digenic test statistics reduce to

$$X_{AB}^2 = \frac{2n\hat{D}_{AB}^2}{\tilde{\pi}_A\tilde{\pi}_B + \tilde{D}_A\tilde{D}_B + \tilde{D}_{A/B}^2}$$

$$X_{A/B}^2 = \frac{2n\hat{D}_{A/B}^2}{\tilde{\pi}_A\tilde{\pi}_B + \tilde{D}_A\tilde{D}_B + \tilde{D}_{AB}^2}$$

showing that each test depends on the other coefficient of disequilibrium, which is somewhat unsatisfactory. A reasonable procedure would be to include each coefficient in the test for the other and then to repeat the test for a coefficient without including the other if the other is found to be not significantly different from zero. It is important to note that both tests include a term for departure from HWE at both loci. Linkage disequilibrium may be tested for whether or not there is Hardy-Weinberg equilibrium at each of the two loci.

Test statistics use estimated variances in the denominators, but these estimates incorporate the hypotheses being tested, and maybe the results of prior tests. Note that it would not be appropriate to use variances estimated by numerical resampling since these ignore the hypotheses.

Composite Genotypic Disequilibria

When genotypes are scored, it is often not possible to distinguish between the two double heterozygotes $AB/\bar{A}\bar{B}$ and $A\bar{B}/\bar{A}B$, so that gametic frequencies cannot be inferred. Under the assumption of random mating, in which genotypic frequencies are assumed to be the products of gametic frequencies, it is possible to estimate gametic frequencies with the EM algorithm as discussed in Chapter 2, and so proceed with the full suite of genotypic disequilibria discussed in the previous section. To avoid making the random-mating assumption, however, it is possible to work with a set of composite disequilibrium coefficients.

Although the separate digenic frequencies p_{AB} and $p_{A/B}$ cannot be observed, their sum can be since

$$p_{AB} + p_{A/B} = 2P^{AB}_{AB} + P^{AB}_{A\bar{B}} + P^{AB}_{\bar{A}B} + \frac{1}{2}\left(P^{AB}_{\bar{A}\bar{B}} + P^{A\bar{B}}_{\bar{A}B}\right)$$

This can be verified in Table 3.6, where the sum is seen to depend on double heterozygotes only by their total of $P^{AB}_{\bar{A}\bar{B}} + P^{A\bar{B}}_{\bar{A}B}$. The digenic disequilibrium is measured with a composite measure Δ_{AB} defined as

$$\begin{aligned}\Delta_{AB} &= p_{AB} + p_{A/B} - 2p_A p_B \\ &= D_{AB} + D_{A/B}\end{aligned}$$

which is the sum of the gametic and nongametic coefficients.

Suppose the nine genotypic classes are numbered as

	BB	$B\bar{B}$	$\bar{B}\bar{B}$
AA	1	2	3
$A\bar{A}$	4	5	6
$\bar{A}\bar{A}$	7	8	9

and that the observed count for class i is n_i, with a total sample size of n. The digenic count for $AB + A/B$ is

$$n_{AB} = 2n_1 + n_2 + n_4 + \frac{1}{2}n_5$$

and the MLE for the composite linkage disequilibrium is

$$\hat{\Delta}_{AB} = \frac{1}{n}n_{AB} - 2\tilde{p}_A\tilde{p}_B$$

Note that n is the number of individuals in the sample.

There is no problem with trigenic disequilibria when double heterozygotes cannot be distinguished, but the definitions may be simplified a little to

$$\begin{aligned}D_{AAB} &= p_{AAB} - p_A\Delta_{AB} - p_B D_A - p_A^2 p_B \\ D_{ABB} &= p_{ABB} - p_B\Delta_{AB} - p_A D_B - p_A p_B^2\end{aligned}$$

A modification of the quadrigenic coefficient is needed to account for lack of knowledge of D_{AB} and $D_{A/B}$. A composite coefficient Δ_{AABB} is defined

$$\begin{aligned}\Delta_{AABB} &= D_{AB}^{AB} - 2D_{AB}D_{A/B} \\ &= P_{AB}^{AB} - 2p_A D_{ABB} - 2p_B D_{AAB} - 2p_A p_B \Delta_{AB} - \Delta_{AB}^2 \\ &\quad - p_A^2 D_B - p_B^2 D_A - D_A D_B - p_A^2 p_B^2\end{aligned}$$

Substituting observed frequencies into the definition equations provides MLE's, and Fisher's formula gives approximate variances

$$\begin{aligned}n\mathrm{Var}(\hat{\Delta}_{AB}) &= (\pi_A + D_A)(\pi_B + D_B) + \frac{1}{2}\tau_A\tau_B\Delta_{AB} \\ &\quad + \tau_A D_{ABB} + \tau_B D_{AAB} + \Delta_{AABB}\end{aligned}$$

where n still refers to n individuals. The variances for the trigenic coefficients are unchanged from Equation 3.15, although they can be simplified slightly by using the two composite coefficients Δ_{AB} and Δ_{AABB}. Finally, the quadrigenic variance (Lisa D. Brooks, personal communication) is

$$\begin{aligned}nVar(\hat{\Delta}_{AABB}) &= (\pi_A^2 + \tau_A^2 D_A - D_A^2)(\pi_B^2 + \tau_B^2 D_B - D_B^2) \\ &\quad + \cdots - \Delta_{AABB}^2\end{aligned}$$

Details are in Weir and Cockerham (1989).

Testing employs chi-square statistics. The first test is for the composite quadrigenic coefficient. If quadrigenic and trigenic coefficients can be ignored, the test statistic for composite digenic linkage disequilibrium is

$$X_{AB}^2 = \frac{n\hat{\Delta}_{AB}^2}{(\tilde{\pi}_A + \hat{D}_A)(\tilde{\pi}_B + \hat{D}_B)}$$

Note the explicit way in which departures from Hardy-Weinberg are included in this expression. Linkage disequilibrium can be tested for, even in non-random mating populations.

Exact Tests

Other composite tests can be performed on genotypic data, by comparing the observed two-locus genotypic counts with the values expected under various hypotheses (Zaykin et al. 1995). It might be hypothesized that a two-locus frequency is equal to the product of corresponding allele frequencies, or of corresponding one-locus genotypic frequencies. The probability of the two-locus genotypic array conditional on the allelic or one-locus genotypic arrays is examined to see if it lies in the tail of the distribution generated by permutation.

If the counts of $A_rA_sB_uB_v$ genotypes are n_{rsuv}, of A_rA_s genotypes are n_{rs} and of B_uB_v genotypes are n_{uv}, then the conditional probability required for testing whether two-locus genotype frequencies are the products of one-locus frequencies is

$$\Pr(\{n_{rsuv}\}|\{n_{rs}\},\{n_{uv}\}) = \frac{\prod_{r,s} n_{rs}! \prod_{u,v} n_{uv}!}{n! \prod_{r,s,u,v} n_{rsuv}!}$$

To compare two-locus genotypic frequencies with products of allele frequencies, the conditional probability is

$$\Pr(\{n_{rsuv}\}|\{n_r\},\{n_u\}) = \frac{n! 2^{H_A} 2^{H_B} \prod_r n_r! \prod_u n_u!}{(2n)!(2n)! \prod_{r,s,u,v} n_{rsuv}!}$$

where n_r, n_u are the numbers of A_r, B_u alleles and H_A, H_B are the numbers of heterozygotes at loci *A, B* in the sample. Zaykin et al. (1995) extended this to any number of loci.

Log-Linear Tests for Linkage Disequilibrium

To make inferences about linkage disequilibrium with a log-linear approach, a series of multiplicative models is constructed that includes different numbers of disequilibrium terms. The difference in deviances for two models that differ only by whether a particular coefficient is included provides a chi-square test statistic for that coefficient. Consider the gametic disequilibrium case first.

For alleles A and B at two loci, the multiplicative model for the four gamete frequencies, retaining only the necessary terms, can be expressed as

$$\begin{aligned} P_{AB} &= MM_AM_BM_{AB} \\ P_{A\bar{B}} &= MM_A \\ P_{\bar{A}B} &= MM_B \\ P_{\bar{A}\bar{B}} &= M \end{aligned}$$

with MLE's of the terms

$$\begin{aligned} \hat{M} &= \tilde{p}_{\bar{A}\bar{B}} \\ \hat{M}_A &= \frac{\tilde{p}_{A\bar{B}}}{\tilde{p}_{\bar{A}\bar{B}}} \\ \hat{M}_B &= \frac{\tilde{p}_{\bar{A}B}}{\tilde{p}_{\bar{A}\bar{B}}} \\ \hat{M}_{AB} &= \frac{\tilde{p}_{AB}\tilde{p}_{\bar{A}\bar{B}}}{\tilde{p}_{A\bar{B}}\tilde{p}_{\bar{A}B}} \end{aligned}$$

and maximum log-likelihood

$$\ln L_1 = n_{AB} \ln \tilde{p}_{AB} + n_{A\bar{B}} \ln \tilde{p}_{A\bar{B}} + n_{\bar{A}B} \ln \tilde{p}_{\bar{A}B} + n_{\bar{A}\bar{B}} \ln \tilde{p}_{\bar{A}\bar{B}}$$

The equation for M_{AB}, or its estimate, can be expressed in terms of the additive coefficient D_{AB} and the allele frequencies

$$M_{AB} = \frac{(p_A p_B + D_{AB})(p_{\bar{A}} p_{\bar{B}} + D_{AB})}{(p_A p_{\bar{B}} - D_{AB})(p_{\bar{A}} p_B - D_{AB})}$$

This shows that there is a one-to-one relationship between the additive and multiplicative coefficients. As D_{AB} changes from its minimum of max$(-p_A p_B,$ $-p_{\bar{A}} p_{\bar{B}}$ to its maximum of min$(p_{\bar{A}} p_B, p_A p_{\bar{B}})$, the coefficient M_{AB} changes from 0 to ∞. Bounds on M_{AB} do not depend on allele frequencies, and the coefficient is also known as the ***odds ratio*** for the two-by-two array of gametic frequencies.

Under the hypothesis of linkage equilibrium, the model does not include the M_{AB} term and the MLE's become

$$\hat{M} = \tilde{p}_{\bar{A}} \tilde{p}_{\bar{B}}$$

$$\hat{M}_A = \frac{\tilde{p}_A}{\tilde{p}_{\bar{A}}}$$

$$\hat{M}_B = \frac{\tilde{p}_B}{\tilde{p}_{\bar{B}}}$$

with maximum log-likelihood

$$\ln L_0 = n_A \ln \tilde{p}_A + n_{\bar{A}} \ln \tilde{p}_{\bar{A}} + n_B \ln \tilde{p}_B + n_{\bar{B}} \ln \tilde{p}_{\bar{B}}$$

Minus twice the difference of the log-likelihoods provides a test statistic for linkage disequilibrium.

$$X^2_{AB} = -2(\ln L_0 - \ln L_1)$$

With more loci, explicit expressions cannot be given for the MLE's for all terms under the various models, and there is a need to use numerical methods for estimates and maximum likelihoods.

For genotypic disequilibria, details for the two-allele, two-locus case were given by Weir and Wilson (1986). Changing their conditions on the terms, and setting all terms for alleles $\bar{A}$ or $\bar{B}$ equal to 1, the 10 genotypic frequencies can be expressed as

$$P^{AB}_{AB} = M M^2_A M^2_B M_{AA} M_{BB} M^2_{AB} M^2_{A/B} M^2_{AAB} M^2_{ABB} M^{AB}_{AB}$$

$$P^{AB}_{A\bar{B}} = 2M M^2_A M_B M_{AA} M_{AB} M_{A/B} M_{AAB}$$

$$
\begin{aligned}
P^{A\bar{B}}_{A\bar{B}} &= MM_A^2M_{AA} \\
P^{AB}_{\bar{A}B} &= 2MM_AM_B^2M_{BB}M_{AB}M_{A/B}M_{ABB} \\
P^{AB}_{\bar{A}\bar{B}} &= 2MM_AM_BM_{AB} \\
P^{A\bar{B}}_{\bar{A}B} &= 2MM_AM_BM_{A/B} \\
P^{A\bar{B}}_{\bar{A}\bar{B}} &= 2MM_A \\
P^{\bar{A}B}_{\bar{A}B} &= MM_B^2M_{BB} \\
P^{\bar{A}\bar{B}}_{\bar{A}B} &= 2MM_B \\
P^{\bar{A}\bar{B}}_{\bar{A}\bar{B}} &= M
\end{aligned}
$$

When only nine genotypic classes can be distinguished, it is sufficient to use

$$M, M_A, M_B, M_{AA}, M_{BB}, S_{AB}, Q_{AB}, M_{AAB}, M_{ABB}, M^{AB}_{AB}$$

where

$$S_{AB} = \frac{1}{2}(M_{AB} + M_{A/B}) \quad , \quad Q_{AB} = M_{AB}M_{A/B}$$

are the sum and product of digenic disequilibria. To reduce the number of parameters to nine, the absence of quadrigenic disequilibrium could be assumed, $M^{AB}_{AB} = 1$, which is similar to assuming $\Delta_{AABB} = 0$ in the additive model to allow a test of $\Delta_{AB} = 0$. A computer program is needed (Weir and Wilson 1986) that considers a series of models of increasing complexity and calculates the deviance for each model. The simplest model, model 0, includes only the mean term, M, and this has MLE of $\hat{M} = 1/16$ and maximum log-likelihood

$$
\begin{aligned}
\ln L_0 &= \text{Constant} \\
&\quad + n_1 \ln(\hat{M}) + n_2 \ln(2\hat{M}) + n_3 \ln(\hat{M}) \\
&\quad + n_4 \ln(2\hat{M}) + n_5 \ln(4\hat{M}) + n_6 \ln(2\hat{M}) \\
&\quad + n_7 \ln(\hat{M}) + n_8 \ln(2\hat{M}) + n_9 \ln(\hat{M})
\end{aligned}
$$

where the constant term is the logarithm of the multinomial coefficient and the nine genotypic classes are numbered as they were in the previous section.

The full model, with all nine parameters, equates expected and observed genotypic frequencies and has a maximum log-likelihood of

$$
\begin{aligned}
\ln L_8 &= \text{Constant} \\
&\quad + n_1 \ln(\tilde{P}_1) + n_2 \ln(\tilde{P}_2) + n_3 \ln(\tilde{P}_3) \\
&\quad + n_4 \ln(\tilde{P}_4) + n_5 \ln(\tilde{P}_5) + n_6 \ln(\tilde{P}_6) \\
&\quad + n_7 \ln(\tilde{P}_7) + n_8 \ln(\tilde{P}_8) + n_9 \ln(\tilde{P}_9)
\end{aligned}
$$

The deviance of the model with mean only is

$$\text{Dev}_0 = -2(\ln L_0 - \ln L_8)$$

The quantity Dev_0 is approximately distributed as chi-square with 8 d.f. when model 0 is true. It is likely to be very large, indicating that the genotypic frequencies cannot be fitted just by a constant average frequency. The deviance for the full model, Dev_8, is zero since this model fits the data perfectly. The next pair of models, 1a and 1b, includes the mean and a term for either allele A or B. With M and M_A in the model, since the frequencies sum to 1,

$$M = \frac{1}{[2(M_A + 1)]^2}$$

and the MLE of M_A can be written as

$$\hat{M}_A = \frac{\tilde{p}_A}{\tilde{p}_{\bar{A}}}$$

so that the maximum log-likelihood is

$$\begin{aligned}\ln L_{1a} = \; & \text{Constant} \\ & + n_1 \ln(\hat{M}\hat{M}_A^2) + n_2 \ln(2\hat{M}\hat{M}_A^2) + n_3 \ln(\hat{M}\hat{M}_A^2) \\ & + n_4 \ln(2\hat{M}\hat{M}_A) + n_5 \ln(4\hat{M}\hat{M}_A) + n_6 \ln(2\hat{M}\hat{M}_A) \\ & + n_7 \ln(\hat{M}) + n_8 \ln(2\hat{M}) + n_9 \ln(\hat{M})\end{aligned}$$

and the deviance

$$\text{Dev}_{1a} = -2(\ln L_{1a} - \ln L_8)$$

provides a 7-d.f. chi-square statistic for testing how well a model with only A allele effects fits the data.

The first model likely to be of interest, Model 2, has terms for allele frequencies at each locus, but does not allow for any disequilibria. This test has 6 d.f. and should be performed prior to the single-degree-of-freedom tests for individual disequilibria. The first model not to allow explicit MLE's for the terms is Model 5, with terms M, M_A, M_{AA}, M_B, M_{BB}, and S_{AB} as the two loci cannot now be treated separately. Numerical methods are needed to solve the likelihood equations.

Example

To illustrate the higher-order disequilibria, some *MNS* blood group data of Mourant et al. (1976) are examined. The data, in the present notation, are

		Locus B			
		SS	Ss	ss	Total
	MM	$n_1 = 91$	$n_2 = 147$	$n_3 = 85$	323
Locus A	MN	$n_4 = 32$	$n_5 = 78$	$n_6 = 75$	185
	NN	$n_7 = 5$	$n_8 = 17$	$n_9 = 7$	29
Total		128	242	167	537

Estimates for the set of additive disequilibria are shown in Table 3.7, and the log-linear disequilibria in Table 3.8. Successive deviances in Table 3.8 can be subtracted to provide chi-square statistics for testing the significance of the term by which the two models differ. Subtracting Dev_{3a} from Dev_2 gives a 1-d.f. chi-square for testing for HWE at locus A (i.e., locus M here). This statistic is 0.14, as listed in Table 3.7, and it could also have been found from Equation 3.6. For the other locus, the chi-square is 4.74 (Dev_{3a} subtracted from Dev_2), indicating that blood group S is not in HWE for this population. Similarly, evidence is seen for linkage disequilibrium, and trigenic disequilibrium for two M alleles and one S allele. All these log-linear tests rest on the assumed absence of a quadrigenic effect. Table 3.7 shows general agreement with the log-linear analysis, although the disequilibria between loci are handled differently.

For the log-linear approach, the estimates of a term such as M_A differ according to which model is being fitted, whereas for the additive model the allele frequency p_A is estimated by the observed frequency, regardless of how many disequilibrium coefficients are also being fitted.

Table 3.7 shows the estimated disequilibrium coefficients. For the B locus, the chi-square is 4.74, indicating that blood group S is not in HWE for this population. There is linkage disequilibrium, and trigenic disequilibrium for two M alleles and one S allele. It is not known in Table 3.7 whether the nonsignificance of the composite quadrigenic component implies an absence of a four-allele association or whether the product $2D_{AB}D_{A/B}$ is nearly equal to but of opposite sign to the four-allele term D_{AB}^{AB}.

Table 3.7 Two-locus disequilibrium coefficient analysis of *MNS* data of Table 1.2.

Estimate	SD(Est.)	Test Statistic
$\tilde{p}_A = 0.7737$		
$\tilde{p}_B = 0.4637$		
$\hat{D}_A = 0.0028$	0.0077	$X_A^2 = 0.14$
$\hat{D}_B = 0.0234$	0.0107	$X_B^2 = 4.74^*$
$\hat{\Delta}_{AB} = 0.0273$	0.0090	$X_{AB}^2 = 8.21^*$
$\hat{D}_{ABB} = 0.0063$	0.0026	$X_{ABB}^2 = 5.00^*$
$\hat{D}_{AAB} = 0.0022$	0.0032	$X_{AAB}^2 = 0.46$
$\hat{\Delta}_{AABB} = -0.0034$	0.0020	$X_{AABB}^2 = 3.00$

*Estimate significantly different from zero at the 5% level.

MULTIPLE TESTS

When testing for the presence of associations within or between loci, a problem arises when the data set contains several loci. If there is interest in whether HWE holds at each of these particular loci, then the tests as presented are appropriate. If, however, there is interest in whether HWE holds

Table 3.8 Two-locus log-linear analysis of *MNS* data of Table 1.2.

Model	Effects in Model	Deviance	d.f.
0	M	366.10	8
1*a*	M, M_A	26.49	7
1*b*	M, M_B	361.14	7
2	M, M_A, M_B	20.82	6
3*a*	M, M_A, M_B, M_{AA}	20.69	5
3*b*	M, M_A, M_B, M_{BB}	16.08	5
4	$M, M_A, M_B, M_{AA}, M_{BB}$	15.95	4
5	$M, M_A, M_B, M_{AA}, M_{BB}, S_{AB}$	14.98	3
6	$M, M_A, M_B, M_{AA}, M_{BB}, S_{AB}, Q_{AB}$	6.88	2
7*a*	$M, M_A, M_B, M_{AA}, M_{BB}, S_{AB}, Q_{AB}, M_{ABB}$	3.62	1
7*b*	$M, M_A, M_B, M_{AA}, M_{BB}, S_{AB}, Q_{AB}, M_{AAB}$	1.91	1
8	$M, M_A, M_B, M_{AA}, M_{BB}, S_{AB}, Q_{AB}, M_{ABB}, M_{AAB}$	0	0

at every locus, the loci at hand serve to give multiple tests of the same hypothesis. Among a set of L tests the largest chi-square will exceed 3.84 more than 5% of the time simply because it is the largest. The significance level of the set of tests is the probability that one or more of them causes rejection of the hypothesis when it is true. This ***experimentwise*** error rate α' is

$$\begin{aligned}\alpha' &= \Pr(\text{at least one test causes rejection}|H_0 \text{ true}) \\ &= 1 - \Pr(\text{all tests do not cause rejection}|H_0 \text{ true}) \\ &= 1 - [\Pr(\text{one test does not cause rejection}|H_0 \text{ true})]^L \\ &= 1 - (1-\alpha)^L \qquad (3.16) \\ &\approx L\alpha\end{aligned}$$

where α is the significance level for an individual test. This argument, called the ***Bonferroni procedure***, assumes that all the tests are independent, which is not true, but the error in making this assumption may not be major. With a 5% level used for 10 tests, the actual significance level for the set of tests is

$$\begin{aligned}\alpha' &= 1 - 0.95^{10} \\ &= 0.40\end{aligned}$$

which is substantially larger. To avoid spurious rejections of the hypothesis, each individual test needs to be more stringent. For an overall $\alpha' = 0.05$, it is necessary that individual values are $\alpha = 0.005$. This individual α value is obtained from

$$\begin{aligned}\alpha &= 1 - (1-\alpha')^{1/L} \\ &\approx \alpha'/L\end{aligned}$$

For single degree-of-freedom chi-squares this translates into a rejection value of 7.84 instead of 3.84. Tables can be constructed to provide these values (Rohlf and Sokal 1981). Equation 3.16 is known as ***Sidak's multiplicative inequality***.

The argument given here is for multiple applications of the same test. It is often the case that a test, say for Hardy-Weinberg, is applied once to multiple loci in the same set of data. Whether the Bonferroni correction is applied depends on the inferences to be drawn. If Hardy-Weinberg is of interest at each locus separately, then there is no need for the correction since each test is applied once. If the question concerns Hardy-Weinberg over the whole genome, then the correction should be applied. In this case, there are questions about the dependence of tests, even at different loci, conducted from the same set of individuals. If Hardy-Weinberg disequilibrium at one locus is likely to increase the chance of Hardy-Weinberg disequilibrium at another locus, the tests are not independent but the α value for the second test is increased so setting all the individual α's to the same value is conservative.

TESTS FOR HOMOGENEITY

Use has been made in this chapter of chi-square goodness-of-fit tests. Often data will be available from several samples, and it is generally desirable to combine such data to perform a goodness-of-fit test on all the information available. Before doing so, it is necessary to verify that the samples are homogeneous and can indeed be combined. A test for homogeneity is essentially a test for independence of rows and columns in a contingency table.

For Mendel's data on seed shape from 10 F_2 plants, given in Table 1.8, a 2×10 contingency table can be constructed. The 10 rows are for the samples, and the two columns for seed shape. If the 10 samples are homogeneous, each has the same proportion of round seeds, and the common proportion is estimated by the overall proportion of 336/437. Under the hypothesis of homogeneity, the expected count of round seeds for plant 1 is $57 \times 336/437 = 43.83$. These calculations are given in Table 3.9, and the chi-square for homogeneity found by adding [(Observed − Expected)^2/Expected] values over all 20 cells in the contingency table. This quantity has the value 4.71, and is from a chi-square distribution with 9 d.f. under the homogeneity hypothesis. The hypothesis is not rejected, and information from all 10 plants can be pooled. In the pooled data the ratio of round to wrinkled is observed to be 336:101 and expected to be 327.75:109.25 under the 3:1 hypothesis. The single degree-of-freedom chi-square test statistic for this hypothesis is

$$\begin{aligned} X^2 &= \frac{(336 - 327.75)^2}{327.75} + \frac{(101 - 109.25)^2}{109.25} \\ &= 0.83 \end{aligned}$$

The data fail to reject the hypothesis of a 3:1 ratio.

There are also occasions where there are genetic reasons to hypothesize a specified ratio for all samples. This is the case for Mendel's data on seed shape. Under Mendel's theory, each sample should exhibit a 3:1 ratio of round versus wrinkled seed shape. In Table 3.9 the observed counts are compared with those expected under a 3:1 ratio. If each sample is tested

Table 3.9 Testing for homogeneity of Mendel's data on seed shape (R:round, W:Wrinkled) in 10 F_2 plants from Table 1.8.

	Observed		Expected*				Expected†			
Plant	R	W	R	W	X^2	G	R	W	X^2	G
1	45	12	42.75	14.25	0.47	0.49	43.83	13.17	0.14	0.14
2	27	8	26.25	8.75	0.09	0.09	26.92	8.08	0.00	0.00
3	24	7	23.25	7.75	0.10	0.10	23.84	7.16	0.00	0.00
4	19	10	21.75	7.25	1.39	1.30	22.30	6.70	2.11	1.92
5	32	11	32.25	10.75	0.01	0.01	33.06	9.94	0.15	0.14
6	26	6	24.00	8.00	0.67	0.71	24.60	7.40	0.34	0.36
7	88	24	84.00	28.00	0.76	0.79	86.11	25.89	0.18	0.18
8	22	10	24.00	8.00	0.67	0.63	24.60	7.40	1.19	1.11
9	28	6	25.50	8.50	0.98	1.06	26.14	7.86	0.57	0.61
10	25	7	24.00	8.00	0.17	0.17	24.60	7.40	0.03	0.03
Total	336	101	327.75	109.25	5.31	5.34	336	101	4.71	4.49

* Under the hypothesis of a 3:1 ratio in each sample.
† Under the hypothesis of unspecified equal proportions in each sample.

separately for the 3:1 hypothesis, none of the chi-square values, with 1 d.f., is significant. Their sum of 5.31, which is a chi-square value with 10 d.f., is also nonsignificant. Note that the various chi-squares are not additive (5.31 $\neq$ 4.71 + 0.83), whereas the G-test statistics of Sokal and Rohlf (1981) are additive, as indicated by the G values shown in Table 3.9.

The G statistics are likelihood ratio values. For the first row in Table 3.9 the values are

$$\begin{aligned} G^2 &= 2[45\ln(45/42.75) + 12\ln(12/14.25)] = 0.49 \quad (3:1 \text{ hypothesis}) \\ G^2 &= 2[45\ln(45/43.83) + 12\ln(12/13.17)] = 0.14 \quad (\text{equal proportions}) \end{aligned}$$

and for the pooled data

$$\begin{aligned} G^2 &= 2[336\ln(336/327.5) + 101\ln(101/109.25)] \\ &= 0.85 \end{aligned}$$

Additivity of G-test statistics is shown by 5.34 = 4.49 + 0.85.

Another approach to testing for homogeneity can be used for the case of testing linkage disequilibrium. It requires that the disequilibrium coefficients be transformed to correlation coefficients by Fisher's z transformation

(Fisher 1925). If D_{AB} is the gametic linkage disequilibrium, then the ratio

$$r_{AB} = \frac{\tilde{D}_{AB}}{\sqrt{\tilde{p}_A(1-\tilde{p}_A)\tilde{p}_B(1-\tilde{p}_B)}}$$

is a correlation coefficient. As such, it can be transformed to a normal variable z by

$$z = \frac{1}{2}\ln\left(\frac{1+r}{1-r}\right)$$

Under the hypothesis of no correlation (i.e. no disequilibrium), it is approximately true that

$$z \sim N\left(0, \frac{1}{2n-3}\right)$$

for a sample of size $2n$ gametes. In other words, $(2n-3)z^2$ has a chi-square distribution with 1 d.f.

When linkage disequilibrium estimates are available from several samples, i, and $\bar{z}$ is the average of the z_i values, the identity

$$\sum_{i=1}^{m}(2n_i-3)z_i^2 = \sum_{i=1}^{m}(2n_i-3)(z_i-\bar{z})^2 + \bar{z}^2\sum_{i=1}^{m}(2n_i-3)$$

can be written as

$$X_m^2 = X_{m-1}^2 + X_1^2$$

The total chi-square, with m d.f., is partitioned into a component with $(m-1)$ d.f. for testing for homogeneity, and a component with 1 d.f. for testing for zero disequilibrium on the pooled data if the samples are homogeneous.

The same approach applies to the composite disequilibrium coefficient Δ_{AB} estimated from genotypic data. Provided there are no higher order disequilibria, the correlation becomes

$$r_{AB} = \frac{\tilde{\Delta}_{AB}}{\sqrt{[\tilde{p}_A(1-\tilde{p}_A)+\tilde{D}_A][\tilde{p}_B(1-\tilde{p}_B)+\tilde{D}_B]}}$$

and the appropriate sample sizes are now the numbers of individuals. These statistics were used to test for homogeneity of samples by Laurie-Ahlberg and Weir (1979). They worked with isozymes in *Drosophila melanogaster*, and calculations for loci *EST-C* and *EST-6* are shown in Table 3.10. The seven samples represented in that table are clearly not homogeneous.

Table 3.10 Testing for homogeneity of linkage disequilibrium estimates for data of Laurie-Ahlberg and Weir (1979).

Sample(i)	r_i	z_i	$n_i - 3$	$(n_i - 3)z_i$	$(n_i - 3)z_i^{2*}$	$(n_i - 3)(z_i - \bar{z})^2$
1	0.066	0.066	96	6.336	0.418	0.431
2	−0.053	−0.053	107	−5.671	0.301	0.289
3	0.429	0.459	95	43.605	20.015	20.102
4	−0.167	−0.169	92	−15.548	2.628	2.597
5	0.079	0.079	97	7.663	0.605	0.621
6	−0.116	−0.117	112	−13.104	1.533	1.507
7	−0.192	−0.194	125	−24.250	4.705	4.656
Total			724†	−0.969†	30.205	30.203

* These are single d.f. chi-squares, and indicate when disequilibria are significantly different from zero.
† Weighted average $\bar{z} = -0.969/724 = -0.001$.

As C. Zapata and G. Alvarez (personal communication) have pointed out, these tests are really for homogeneity of r values instead of homogeneity of D values. The two will not be equivalent when the populations have different allele frequencies. If attention is primarily on D, rather than r, then alternative methods should be used when allele frequencies differ. Bootstrapping within samples could be used to provide confidence intervals for the D's, for example.

SUMMARY

In studying associations between sets of alleles, whether at one or several loci, a convenient procedure is to define a measure of association as the difference between the joint frequency of the set and the frequency that would result if there was no association. For large samples, the maximum likelihood estimates of these measures of association are normally distributed, and this allows the construction of test statistics based on the expected variance of the estimates under the hypothesis of no association. For smaller samples, exact tests can be constructed that reject the hypothesis when the observed or less likely outcomes have a very low probability.

EXERCISES

Exercise 3.1

Suppose that 10 individuals are scored for a locus with two alleles A, a and 6 AA, 3 Aa, 1 aa individuals are found. Test the sample for Hardy-Weinberg equilibrium with both a chi-square and an exact test.

Exercise 3.2

Set up a likelihood ratio test of the hypothesis of gametic linkage equilibrium at two loci for a sample of gametes.

Exercise 3.3

Suppose 20 individuals are scored for two loci, each with two alleles, and the following genotypic counts are found:

	BB	Bb	bb
AA	9	2	1
Aa	3	2	1
aa	1	0	1

Test the sample for linkage disequilibrium, using the gametic measure D_{AB} (assuming HWE) and using the composite measure Δ_{AB} (not assuming HWE).

Exercise 3.4

A locus has three alleles, A_1, A_2, A_3 in frequencies p_1, p_2, p_3. Three disequilibrium coefficients can be defined in terms of the homozygote frequencies:

$$D_{11} = P_{11} - p_1^2,\ D_{22} = P_{22} - p_2^2,\ D_{33} = P_{33} - p_3^2$$

Find expressions for the maximum possible values of these homozygote disequilibria. How large must the sample be for the overall test of HWE to have 90% power and 5% significance when all three alleles are equally frequent and the disequilibria are at half their maximum values?

Exercise 3.5

For a set of $2n$ gametes, indexed by i, define the indicator variables x_i, y_i to take the values 1, 0 according to whether or not the ith gamete carries alleles A, B, respectively. What is the sample correlation coefficient of these two indicator variables?

Exercise 3.6

Verify the G values shown in Table 3.9.

Chapter 4

Diversity

INTRODUCTION

The study of evolution is that of characterizing the extent and causes of genetic variation. For the present discussion, different ways of measuring variation will be considered. The simplest descriptors are just the frequencies of alleles or genotypes, but emphasis in this chapter will be given to heterozygosity and gene diversity. The frequency of heterozygotes is important, since each heterozygote carries different alleles and represents the existence of variation. There are situations, however, such as in selfing species, where variation results from the continued presence of different homozygotes, and gene diversity is then a more appropriate measure. Sampling properties for each of these two measures are covered in this chapter.

HETEROZYGOSITY

A simple measure of genetic variation in a population is the amount of heterozygosity observed, and this is often reported for a single locus or as an average over several loci. Properties of the statistic will be investigated in both situations.

If n_{luv} is the observed count of A_uA_v heterozygotes, $u \neq v$, at locus l in a sample of size n, then the sample heterozygote frequency at this locus is

$$\tilde{H}_l = \sum_u \sum_{v \neq u} \frac{n_{luv}}{n}$$

If m loci are scored, the average heterozygosity is

$$\tilde{H} = \frac{1}{m} \sum_{l=1}^{m} \tilde{H}_l$$

Within-Population Variance of Heterozygosity

As $\tilde{H}_l$ is defined from counts that are multinomially distributed over samples from the same population, it has mean and variance

$$\begin{aligned} \mathcal{E}(\tilde{H}_l) &= H_l \\ \mathrm{Var}(\tilde{H}_l) &= \frac{1}{n}H_l(1-H_l) \end{aligned} \tag{4.1}$$

where H_l is the population frequency of heterozygotes at locus l. Note that H_l refers to the frequency of heterozygotes in the one population being sampled.

It is helpful to introduce indicator variables, as was done for allele frequencies in Chapter 2. Define x_{jl} for locus l in individual j as

$$x_{jl} = \begin{cases} 1 & \text{if individual is heterozygous at locus } l \\ 0 & \text{otherwise} \end{cases}$$

without regard to what the (heterozygous) genotype is at that locus. The sample heterozygosity is the average of the indicator variables

$$\tilde{H}_l = \frac{1}{n}\sum_{j=1}^{n} x_{jl}$$

with properties determined by the expected values of functions of the x's. Taking expectations over all samples for one population

$$\begin{aligned} \mathcal{E}(x_{jl}) &= H_l \\ \mathcal{E}(x_{jl}^2) &= H_l \\ \mathcal{E}(x_{jl}x_{j'l}) &= H_l^2 \end{aligned}$$

with the last expression reflecting the independence of individuals within a sample. Using the indicator variables leads to the same mean and variance of $\tilde{H}_l$ as given in Equations 4.1.

For average heterozygosity, the observed value remains unbiased

$$\mathcal{E}(\tilde{H}) = H = \frac{1}{m}\sum_{l} H_l$$

but the variance requires account to be taken of the covariance between heterozygosities at different loci, l and l', caused by them both depending on the frequency of double heterozygotes $\tilde{H}_{ll'}$ at those loci. This covariance is

$$\text{Cov}(\tilde{H}_l, \tilde{H}_{l'}) = \frac{1}{n}(H_{ll'} - H_l H_{l'})$$

as may be found using the expectation

$$\mathcal{E}(x_{jl}x_{jl'}) = H_{ll'}$$

i.e.,

$$\begin{aligned} \text{Cov}(\tilde{H}_l, \tilde{H}_{l'}) &= \mathcal{E}\left(\frac{1}{n}\sum_j x_{jl}\frac{1}{n}\sum_{j'} x_{j'l'}\right) - H_l H_{l'} \\ &= \mathcal{E}\left[\frac{1}{n^2}\left(\sum_j x_{jl}x_{jl'} + \sum_j \sum_{j' \neq j} x_{jl}x_{j'l'}\right)\right] - H_l H_{l'} \\ &= \frac{1}{n^2}\left[nH_{ll'} + n(n-1)H_l H_{l'}\right] - H_l H_{l'} \\ &= \frac{1}{n}(H_{ll'} - H_l H_{l'}) \end{aligned} \tag{4.2}$$

The variance of average heterozygosity within populations is, therefore,

$$\begin{aligned} \text{Var}(\tilde{H}) &= \frac{1}{m^2}\left[\sum_l \text{Var}(\tilde{H}_l) + \sum_l \sum_{l' \neq l} \text{Cov}(\tilde{H}_l, \tilde{H}_{l'})\right] \\ &= \frac{1}{nm^2}\sum_l H_l(1 - H_l) \\ &\quad + \frac{1}{nm^2}\sum_l \sum_{l' \neq l}(H_{ll'} - H_l H_{l'}) \end{aligned} \tag{4.3}$$

This expression illustrates the general result that the variance of an average of single-locus statistics must involve two-locus parameters. In itself, this provides a justification for determining the properties of two-locus measures.

It may be thought possible to use the variance among the sample heterozygosities at individual loci to estimate the variance of average heterozygosity, but this ignores the covariance between heterozygosities and it also ignores the differences among mean values of the individual heterozygosities. The sample variance of single-locus heterozygosities is

$$\begin{aligned} s_H^2 &= \frac{1}{m-1}\sum_l (\tilde{H}_l - \tilde{H})^2 \\ &= \frac{1}{m}\sum_l \tilde{H}_l^2 - \frac{1}{m(m-1)}\sum_l \sum_{l' \neq l} \tilde{H}_l \tilde{H}_{l'} \end{aligned}$$

which has expectation over all samples from that population of

$$\mathcal{E}(s_H^2) = \frac{1}{m}\sum_l \left[H_l^2 + \frac{1}{n}H_l(1-H_l)\right] - \frac{1}{m(m-1)}\sum_l \sum_{l' \neq l} \left[H_l H_{l'} + \frac{1}{n}(H_{ll'} - H_l H_{l'})\right] \quad (4.4)$$

This equation used the results

$$\mathcal{E}(\tilde{H}_l^2) = H_l^2 + \text{Var}(\tilde{H}_l)$$
$$\mathcal{E}(\tilde{H}_l \tilde{H}_{l'}) = H_l H_{l'} + \text{Cov}(\tilde{H}_l, \tilde{H}_{l'})$$

If the sample heterozygosities at each locus were independent and had the same distribution, the variance of the average heterozygosity would be $1/m$ times the variance of one of them. In other words, s_H^2/m would provide an estimate of $\text{Var}(\tilde{H})$. Indeed, for Equations 4.3 and 4.4, divided by m, to be equal, it is necessary that all loci have the same heterozygosity, $H_l = H$ for all l, and that the heterozygosities at different loci are independent, $H_{ll'} = H_l H_{l'}$ for all $l \neq l'$. In that case, the variance of the average heterozygosity reduces to a binomial form

$$\text{Var}(\tilde{H}) = \frac{H(1-H)}{nm}$$

Otherwise, writing $m_1 = m(m-1)$,

$$\mathcal{E}(\frac{s_H^2}{m}) - \text{Var}(\tilde{H}) = \frac{1}{m_1}\left[\sum_l (H_l - H)^2 - \frac{1}{n}\sum_l \sum_{l' \neq l}(H_{ll'} - H_l H_{l'})\right]$$
$$= \frac{1}{m_1}\left[\sum_l (H_l - H)^2 - \frac{1}{n}\sum_l \sum_{l' \neq l} \text{Cov}(\tilde{H}_l, \tilde{H}_{l'})\right] \quad (4.5)$$

Equation 4.5 shows that the sample variance of single-locus heterozygosities gives a biased estimate of the variance of average heterozygosity.

Total Variance of Heterozygosity

When variation between replicate populations is taken into account, expectations of the products of indicator variables in different individuals are no longer the products of the expectations. To accommodate the dependence between different sample members, caused by genetic sampling, write

$$\mathcal{E}(x_{jl} x_{j'l}) = M_l, \; j \neq j'$$

where M_l is the probability that two individuals in a population are heterozygous at locus l. This quantity is playing a role analogous to that of $P_{A/A}$ for the frequency with which two individuals both carry allele A, when the total variance of the frequency of A was being discussed in Chapter 2. The frequencies H and M refer to values expected over all populations. Using these expectations gives the mean and variance of $\tilde{H}_l$:

$$\begin{aligned}
\mathcal{E}(\tilde{H}_l) &= H_l \\
\text{Var}(\tilde{H}_l) &= \frac{1}{n^2}\mathcal{E}\left(\sum_j x_{jl}^2 + \sum_j \sum_{j \neq j'} x_{jl}x_{j'l}\right) - [\mathcal{E}(\tilde{H}_l)]^2 \\
&= (M_l - H_l^2) + \frac{1}{n}(H_l - M_l)
\end{aligned}$$

There is a between-population component to this variance that remains even when the sample size is very large, unless all individuals in the population are independent and $M_l = H_l^2$. This will not be the case in finite populations or when there is any other association between individuals.

Averaging over all m loci scored gives

$$\tilde{H} = \frac{1}{m}\sum_l \tilde{H}_l = \frac{1}{nm}\sum_j \sum_l x_{jl}$$

In addition to the two-locus heterozygosity $H_{ll'}$

$$\mathcal{E}(x_{jl}x_{jl'}) = H_{ll'}$$

which is the probability that a random individual is heterozygous at loci l and l', there is a need for a two-locus version of between-individual heterozygosity, M_l

$$\mathcal{E}(x_{jl}x_{j'l'}) = M_{ll'}$$

The quantity $M_{ll'}$ is the probability that two random individuals are heterozygous, one at locus l and one at locus l'. These new frequencies lead to

$$\begin{aligned}
\mathcal{E}(\tilde{H}) = H &= \frac{1}{m}\sum_l H_l \\
\text{Var}(\tilde{H}) &= \frac{1}{m^2n^2}\mathcal{E}\left(\sum_j \sum_l x_{jl}^2 + \sum_j \sum_{j' \neq j} \sum_l x_{jl}x_{j'l}\right. \\
&\quad \left. + \sum_j \sum_l \sum_{l' \neq l} x_{jl}x_{jl'} + \sum_j \sum_{j' \neq j} \sum_l \sum_{l' \neq l} x_{jl}x_{j'l'}\right) - H^2
\end{aligned}$$

$$\begin{aligned}\mathrm{Var}(\tilde{H}) &= \frac{1}{m^2}\left[\sum_l (M_l - H_l^2) + \sum_l \sum_{l' \neq l} (M_{ll'} - H_l H_{l'})\right] \\ &\quad + \frac{1}{m^2 n}\left[\sum_l (H_l - M_l) + \sum_l \sum_{l' \neq l} (H_{ll'} - M_{ll'})\right] \qquad (4.6)\end{aligned}$$

This last expression contains four terms that can be rearranged to show just how populations, loci, and individuals contribute to the variance of average heterozygosity. The rearrangement can be found by setting out the calculations slightly differently, in a framework similar to that used for analysis of variance.

Although the distributional properties of the various sums of squares of the indicator variables are not needed, the analysis of variance format simplifies calculations, and the method of moments is used to estimate components of variance. An index i is added to the heterozygosity indicator variable to denote the population being sampled, so that x_{ijl} is 1 when individual j from population i is heterozygous at locus l and is 0 otherwise. The observations fall in a nested scheme of individuals within populations, but as the same set of loci is scored in all individuals, loci are crossed with populations. The whole scheme has an analysis similar to that of a split-plot design (Steel and Torrie 1980, Chapter 16), with populations playing the role of whole-plot treatments and loci that of split-plot treatments. The analysis of variance format is shown in Table 4.1 for the case of m loci scored in n individuals taken from each of r populations.

In Table 4.1 sums of squares of indicator variables are constructed to reflect the sampling structure, and corresponding expected mean squares are written down as though the variables could be represented by the linear model

$$x_{ijl} = \alpha_i + \beta_{ij} + \gamma_l + (\alpha\gamma)_{il} + (\beta\gamma)_{ijl}$$

This equation has terms α_i for populations, β_{ij} for individuals within populations, γ_l for loci, $(\alpha\gamma)_{il}$ for populations by loci, and $(\beta\gamma)_{ijl}$ for loci by individuals within populations. The term for loci is regarded as a fixed effect since the same loci are repeatedly scored. All other effects must be regarded as being random if inferences are to be extended to all individuals and to all populations. Loci should also be regarded as random effects if inferences are to be extended to all loci in the genome, and then a mean would be added to the model.

The expected values of the random terms are 0, and γ_l equals H_l. The variances of α_i, β_{ij}, $(\alpha\gamma)_{ijl}$, and $(\beta\gamma)_{ijl}$ are σ_p^2, $\sigma_{i/p}^2$, σ_{pl}^2, and $\sigma_{li/p}^2$, respectively. Expectations of the terms in the sums of squares, defined in the footnote to Table 4.1, are

$$\begin{aligned}
\mathcal{E}(x_{ijl}^2) &= \sigma_p^2 + \sigma_{i/p}^2 + \gamma_l^2 + \sigma_{pl}^2 + \sigma_{l\,i/p}^2 \\
\mathcal{E}(x_{ij.}^2) &= m^2\sigma_p^2 + m^2\sigma_{i/p}^2 + (\sum_l H_l)^2 + m\sigma_{pl}^2 + m\sigma_{l\,i/p}^2 \\
\mathcal{E}(x_{i.l}^2) &= n^2\sigma_p^2 + n\sigma_{i/p}^2 + n^2H_l^2 + n^2\sigma_{pl}^2 + n\sigma_{l\,i/p}^2 \\
\mathcal{E}(x_{i..}^2) &= n^2m^2\sigma_p^2 + nm^2\sigma_{i/p}^2 + n^2(\sum_l H_l)^2 + n^2m\sigma_{pl}^2 + nm\sigma_{l\,i/p}^2 \\
\mathcal{E}(x_{..l}^2) &= rn^2\sigma_p^2 + rn\sigma_{i/p}^2 + r^2n^2H_l^2 + rn^2\sigma_{pl}^2 + rn\sigma_{l\,i/p}^2 \\
\mathcal{E}(x_{...}^2) &= rn^2m^2\sigma_p^2 + rnm^2\sigma_{i/p}^2 + r^2n^2(\sum_l H_l)^2 + rn^2m\sigma_{pl}^2 + rnm\sigma_{l\,i/p}^2
\end{aligned}$$

Table 4.1 Analysis of variance format for variable that indicates heterozygosity.*

Source	d.f.	Sum of Squares	Expected Mean Square
Populations	$r-1$	$SS_1 - C$	$\sigma_{l\,i/p}^2 + m\sigma_{i/p}^2 + n\sigma_{pl}^2 + mn\sigma_p^2$
Individuals within populations	$r(n-1)$	$SS_2 - SS_1$	$\sigma_{l\,i/p}^2 + m\sigma_{i/p}^2$
Loci	$m-1$	$SS_3 - C$	$\sigma_{l\,i/p}^2 + n\sigma_{pl}^2 + L$
Loci by populations	$(r-1)(m-1)$	$SS_4 - SS_1 - SS_3 + C$	$\sigma_{l\,i/p}^2 + n\sigma_{pl}^2$
Loci by individuals within populations	$r(n-1)(m-1)$	$SS_5 - SS_2 - SS_4 + SS_1$	$\sigma_{l\,i/p}^2$

*$x_{ij.} = \sum_l x_{ijl}$ $x_{i.l} = \sum_j x_{ijl}$ $x_{i..} = \sum_j \sum_l x_{ijl}$

$x_{..l} = \sum_i \sum_j x_{ijl}$ $x_{...} = \sum_i \sum_j \sum_l x_{ijl}$

$SS_1 = \frac{1}{mn}\sum_i x_{i..}^2$ $SS_2 = \frac{1}{m}\sum_i \sum_j x_{ij.}^2$ $SS_3 = \frac{1}{rn}\sum_l x_{..l}^2$

$SS_4 = \frac{1}{n}\sum_i \sum_l x_{i.l}^2$ $SS_5 = \sum_i \sum_j \sum_l x_{ijl}^2 = x_{...}$ $C = \frac{1}{mnr}x_{...}^2$

so that

$$\begin{aligned}
\mathcal{E}(SS_1) &= rnm\sigma_p^2 + rm\sigma_{i/p}^2 + rn(\sum_l H_l)^2/m + rn\sigma_{pl}^2 + r\sigma_{li/p}^2 \\
\mathcal{E}(SS_2) &= rnm\sigma_p^2 + rnm\sigma_{i/p}^2 + \frac{rn}{m}(\sum_l H_l)^2 + rn\sigma_{pl}^2 + rn\sigma_{li/p}^2 \\
\mathcal{E}(SS_3) &= nm\sigma_p^2 + m\sigma_{i/p}^2 + rn\sum_l H_l + nm\sigma_{pl}^2 + m\sigma_{i/p}^2 \\
\mathcal{E}(SS_4) &= rnm\sigma_p^2 + rm\sigma_{i/p}^2 + rn\sum_l H_l + rnm\sigma_{pl}^2 + rm\sigma_{li/p}^2 \\
\mathcal{E}(SS_5) &= rnm\sigma_p^2 + rnm\sigma_{i/p}^2 + rn\sum_l H_l + rnm\sigma_{pl}^2 + rnm\sigma_{li/p}^2 \\
\mathcal{E}(C) &= nm\sigma_p^2 + m\sigma_{i/p}^2 + \frac{rn}{m}(\sum_l H_l)^2 + n\sigma_{pl}^2 + \sigma_{li/p}^2
\end{aligned}$$

and these expressions lead to the expected mean squares in Table 4.1. The quantity L

$$L = \frac{1}{m-1}\sum_l (H_l - H)^2, \ m > 1$$

is analogous to a variance among heterozygosities.

For a single locus, the split-plot component of the analysis disappears and the only variance components are those for populations and individuals within populations. In that case the components are

$$\text{Populations} : \sigma_p^2 = M_l - H_l^2$$

$$\text{Individuals within populations} : \sigma_{i/p}^2 = H_l - M_l$$

For more than one locus, however, the four components are

$$\text{Populations} : \sigma_p^2 = \frac{1}{m(m-1)}\sum_l \sum_{l' \neq l}(M_{ll'} - H_l H_{l'})$$

$$\text{Individuals within populations} : \sigma_{i/p}^2 = \frac{1}{m(m-1)}\sum_l \sum_{l'}(H_{ll'} - M_{ll'})$$

Populations by loci :

$$\sigma_{pl}^2 = \frac{1}{m}\sum_l (M_l - H_l^2) - \frac{1}{m(m-1)}\sum_l \sum_{l' \neq l}(M_{ll'} - H_l H_{l'})$$

Loci by individuals within populations :

$$\sigma^2_{li/p} = \frac{1}{m}\sum_l (H_l - M_l) - \frac{1}{m(m-1)}\sum_l \sum_{l'} (H_{ll'} - M_{ll'})$$

Variances of means can be expressed very simply in terms of the four variance components. For the heterozygosity at a single locus l in one population, the value depends on which population is sampled and which individuals within that population are observed. Therefore there are two components to the variance: the component for populations and the component for individuals within populations:

$$\begin{aligned} \text{Var}(\tilde{H}_l) &= \text{Var}\left(\frac{1}{n}\sum_j x_{ijl}\right) \\ &= \sigma^2_p + \frac{1}{n}\sigma^2_{i/p} \end{aligned} \qquad (4.7)$$

As the sample size n increases, the effect of the variation between individuals becomes less and the only contribution to the variance of heterozygosity is due to that between populations.

If heterozygosity is reported as an average over several loci, the value depends on the population, the loci scored, and the individuals sampled. All four components of variance are needed:

$$\begin{aligned} \text{Var}(\tilde{H}) &= \text{Var}\left(\frac{1}{nm}\sum_j \sum_l x_{ijl}\right) \\ &= \sigma^2_p + \frac{1}{n}\sigma^2_{i/p} + \frac{1}{m}\sigma^2_{pl} + \frac{1}{mn}\sigma^2_{li/p} \end{aligned} \qquad (4.8)$$

Equation 4.8 allows the planning of surveys of populations in such a way as to achieve a desirable level of variance of sample heterozygosity. If the variance components are known, then adjustments can be made to the number, m, of loci scored and the number, n, of individuals sampled for the variance to be the size required. No juggling of sample sizes, however, can remove the effects of variation between replicate populations. Note that the variances in Equations 4.6 and 4.8 are the same.

A complete discussion of the effects of mating system, allele frequency distributions, and unbalanced data is given by Weir et al. (1990), but note that for finite random mating populations the variance components σ^2_p and $\sigma^2_{i/p}$ are quite small, and the variance of average heterozygosity is reduced most quickly by increasing the number of loci scored per individual. For infinite random mating populations in which there is a constant amount of selfing, the variance components σ_p^2 and σ_{pl}^2 are zero, and then the best strategy for reducing variance is to increase the number of individuals sampled.

As in the within-population situation, it may be desired to use the variation of heterozygosity among loci to estimate the total variance of average heterozygosity. Once again this procedure ignores the differences between heterozygosities at different loci, and it assumes that single-locus heterozygosities are independent. To see this, suppose that the sample variance of single-locus heterozygosities, as before, is written as

$$s_H^2 = \frac{1}{m-1}\sum_l(\tilde{H}_l - \tilde{H})^2$$

and if each locus had the same expected heterozygosity, the variance of the mean heterozygosity $\tilde{H}$ could be estimated by s_H^2/m. The same arguments as in the within-population case lead to

$$\mathcal{E}(\frac{s_H^2}{m}) - \text{Var}(\tilde{H}) = \frac{1}{m_1}\left[\sum_l(H_l - H)^2 - \frac{1}{n}\sum_l\sum_{l'\neq l}\text{Cov}(\tilde{H}_l, \tilde{H}_{l'})\right]$$

This expression, in which $m_1 = m(m-1)$, has the same form as Equation 4.5, although the frequencies and covariances now apply to all replicate populations, and not just to samples from a single population. They therefore involve the M frequencies.

Although the variance among loci does not provide an unbiased estimate of the total variance of average heterozygosity, it is all that is available when data are taken from only a single population. Such data do not allow the proper estimation of the between-population component of variance.

For large datasets, involving more than one population, analysis of heterozygosity is most conveniently handled by a statistical computer package. The data are recoded to consist of ones for every heterozygous locus, and zeros for every homozygous locus. The resulting set of ones and zeros is analyzed as being from a split-plot experiment. Statistical packages generally allow estimation of variance components from analyses of such designs.

GENE DIVERSITY

An alternative measure of variation, often referred to loosely as average heterozygosity, but more properly known as ***gene diversity***, is formed from the sum of squares of allele frequencies. It is a more appropriate measure of variability for inbred populations where there are very few heterozygotes, but there may be several different homozygous types. For random mating populations, it will be close in value to heterozygosity.

Since gene diversity is calculated from allele frequencies, it can be found from tables that give only these frequencies and do not give genotypic frequencies. Although this may be an advantage in some situations, it is well to remember that the data are generally collected at the genotypic level. Sampling properties of gene diversity will depend on the genotype as well as the allele frequencies.

If p_{lu} is the frequency of the uth allele at the lth locus, the gene diversity at this locus is

$$D_l = 1 - \sum_u p_{lu}^2$$

and, as an average over m loci,

$$D = 1 - \frac{1}{m}\sum_l \sum_u p_{lu}^2$$

Sample allele frequencies in these equations provide maximum likelihood estimates (MLE's) of gene diversity:

$$\hat{D} = 1 - \frac{1}{m}\sum_l \sum_u \tilde{p}_{lu}^2$$

Within-Population Variance of Gene Diversity

Among samples from a single population, the expected values of squared allele frequencies have been calculated in Chapter 2, and these allow the expected value of gene diversity at a locus to be found as

$$\begin{aligned}\mathcal{E}(\tilde{D}_l) &= 1 - \sum_u p_{lu}^2 - \frac{1}{2n}\sum_u p_{lu}(1-p_{lu})(1+f) \\ &= \left(1 - \frac{1+f}{2n}\right) D_l\end{aligned}$$

There is a small bias factor of $(2n-1)/2n$ for noninbred populations, but a more serious bias otherwise. The presence of the f term indicates that the expected value of D depends on genotype as well as allele frequencies.

The variance requires the sum of variances and covariances of squared frequencies of alleles at the same locus. From Fisher's variance formula (Chapter 2 and Weir 1989a), approximately

$$\begin{aligned}\text{Var}(\tilde{p}_{lu}^2) &= \frac{1}{n}2p_{lu}^3(1-p_{lu})(1+f) \\ \text{Cov}(\tilde{p}_{lu}^2, \tilde{p}_{lv}^2) &= \frac{1}{n}2p_{lu}^2 p_{lv}^2(1+f), \quad v \neq u\end{aligned}$$

so that

$$
\begin{aligned}
\operatorname{Var}(\hat{D}_l) &= \sum_u \operatorname{Var}(\tilde{p}_{lu}^2) + \sum_u \sum_{v \neq u} \operatorname{Cov}(\tilde{p}_{lu}^2, \tilde{p}_{lv}^2) \\
&= \frac{2(1+f)}{n} \left[\sum_u p_{lu}^3 - \left(\sum_u p_{lu}^2 \right)^2 \right]
\end{aligned}
$$

For multiple loci, there is no problem in calculating allele frequencies, and the genotypic frequencies at each separate locus are multinomially distributed. There is a covariance between sample allele frequencies at different loci, caused by the frequencies being estimated from the same individuals, and hence there is also a covariance between squared sample frequencies. Multinomial moments show that the covariance between frequencies $\tilde{p}_{lu}$ and $\tilde{p}_{l'v}$ of alleles u and v at loci l and l', respectively, is given by the composite coefficient of linkage disequilibrium for these two alleles

$$
\operatorname{Cov}(\tilde{p}_{lu}, \tilde{p}_{l'v}) = \frac{1}{n} \Delta_{lu,l'v}
$$

and Fisher's formula gives the covariance between squared allele frequencies at different loci to be approximately

$$
\operatorname{Cov}(\tilde{p}_{lu}^2, \tilde{p}_{l'v}^2) = \frac{2}{n} p_{lu} p_{l'v} \Delta_{lu,l'v}
$$

This last expression provides the covariance of diversities at different loci:

$$
\operatorname{Cov}(\tilde{D}_l, \tilde{D}_{l'}) = \frac{2}{n} \sum_u \sum_v p_{lu} p_{l'v} \Delta_{lu,l'v}
$$

The variance of average gene diversity over samples from the same population then follows as

$$
\begin{aligned}
\operatorname{Var}(\tilde{D}) &= \sum_l \sum_u \operatorname{Var}(\tilde{p}_{lu}^2) + \sum_l \sum_u \sum_{v \neq u} \operatorname{Cov}(\tilde{p}_{lu}^2, \tilde{p}_{lv}^2) \\
&\quad + \sum_l \sum_{l' \neq l} \sum_u \sum_v \operatorname{Cov}(\tilde{p}_{lu}^2, \tilde{p}_{l'v}^2) \\
&= \frac{2}{m^2 n} \sum_l (1+f_l) \left[\sum_u p_{lu}^3 - \left(\sum_u p_{lu}^2 \right)^2 \right] \\
&\quad + \frac{2}{m^2 n} \sum_l \sum_{l' \neq l} \sum_u \sum_v p_{lu} p_{l'v} \Delta_{lu,l'v} \qquad (4.9)
\end{aligned}
$$

Because data are collected on genotypes, this variance depends on the association f_l between alleles at locus l as well the associations $\Delta_{lu,l'v}$ between alleles at different loci. Simply taking the variance of single-locus diversities would ignore the between-locus associations, as well as ignoring the differences in expected diversity between loci.

When gene diversity is calculated for loci that all have two alleles, Equation 4.9 simplifies. If p_l is the frequency of one of the two alleles at locus l, and $\Delta_{ll'}$ is the composite linkage disequilibrium between these identified alleles at loci l and l'

$$\begin{aligned}\mathrm{Var}(\tilde{D}) &= \frac{2}{m^2n}\sum_l\left[(1+f_l)p_l(1-p_l)(1-2p_l)^2\right] \\ &\quad + \frac{2}{m^2n}\sum_l\sum_{l'\neq l}(1-2p_l)(1-2p_{l'})\Delta_{ll'} \\ &= \frac{2}{m^2n}\sum_l(D_l-\frac{1}{2}H_l)(1-2p_l)^2 \\ &\quad + \frac{2}{m^2n}\sum_l\sum_{l'\neq l}(1-2p_l)(1-2p_{l'})\Delta_{ll'}\end{aligned} \tag{4.10}$$

This expression is zero when $p_l = 0.5$, indicating the need to retain the terms of order n^{-2} that are ignored by Fisher's formula.

Total Variance of Gene Diversity

When expectations are taken over replicate populations as well as replicate samples from the same population, the argument that led to Equation 2.13 shows that

$$\mathcal{E}(\tilde{p}_{lu}^2) = p_{lu}^2 + p_{lu}(1-p_{lu})\theta + \frac{1}{2n}p_{lu}(1-p_{lu})(1+F-2\theta)$$

and the total expectation of gene diversity is

$$\mathcal{E}(\tilde{D}_l) = D_l\left[(1-\theta) - \frac{1}{2n}(1+F-2\theta)\right]$$

Cockerham (1967) defined a group coancestry coefficient

$$\theta_L = \theta + (F-\theta)/n + (1-F)/2n$$

that refers to the identity of a random pair of alleles among the $2n$ alleles of n individuals. This coefficient allows the expected gene diversity to be written simply as

$$\mathcal{E}(\tilde{D}_l) = (1-\theta_L)D_l$$

The total variance requires the expectations of squares and products of squared allele frequencies. In Chapter 2, it was necessary to introduce the joint frequency $P_{A|A}$ for the total expectation of $\tilde{p}_A^2$. Here, joint frequencies for three or four alleles are needed. Dropping the locus subscript l, the expectations for alleles u and v are (Weir 1989a)

$$
\begin{aligned}
\mathcal{E}(\tilde{p}_u^4) &= P_{u|u|u|u} + \frac{3}{n}(P_{u|u|u} + P_{uu|u|u} - 2P_{u|u|u|u}) \\
\mathcal{E}(\tilde{p}_u^2\tilde{p}_v^2) &= P_{u|u|v|v} + \frac{1}{2n}\Big(P_{u|v|v} + P_{u|u|v} + 4P_{uv|u|v} + \\
&\quad + P_{uu|v|v} + P_{vv|u|u} - 12P_{u|u|v|v}\Big), \quad v \neq u
\end{aligned}
$$

A quantity such as $P_{uu|v|v}$ is the frequency with which one individual carries two copies of allele u and two other individuals each has a copy of allele v. The total variance of sample gene diversity at one locus then becomes

$$
\begin{aligned}
\mathrm{Var}(\tilde{D}_l) &= \left[\sum_u\sum_v P_{u|u|v|v} - \left(\sum_u P_{u|u}\right)^2\right] \\
&\quad + \frac{1}{n}\left[2\sum_u P_{u|u|u} - \sum_u P_{uu}\sum_u P_{u|u} + 2\left(\sum_u P_{u|u}\right)^2\right. \\
&\quad \left. + \sum_u\sum_v (2P_{uv|u|v} + P_{uu|v|v} - 6P_{u|u|v|v})\right]
\end{aligned}
$$

For diversity averaged over loci, the total expectations necessary are quite complicated. The notation for joint frequencies of alleles at different loci separates alleles in different individuals with a vertical rule and alleles on different gametes within individuals by a slash. With this convention, the expectations needed are

$$
\begin{aligned}
\mathcal{E}(\tilde{p}_{lu}\tilde{p}_{l'v}) &= P_{lu|l'v} + \frac{1}{2n}(P_{lu,lv} + P_{lu/l'v} - 2P_{lu|l'v}) \\
\mathcal{E}(\tilde{p}_{lu}^2\tilde{p}_{l'v}^2) &= P_{lu|lu|l'v|l'v} + \frac{1}{2n}\Big[P_{lu|lu|l'v} + P_{lu|l'v|l'v} \\
&\quad + 4(P_{lu,l'v|lu|l'v} + P_{lu/l'v|lu|l'v}) \\
&\quad - 12P_{lu|lu|l'v|l'v}\Big]
\end{aligned}
$$

The total variance for gene diversity is a very cumbersome expression.

Estimating the Variance of Diversity

For data from a single population, the variance of gene diversity is estimated by substituting observed frequencies into Equation 4.9. A more convenient computing formula for the within-population variance is

$$\begin{aligned}\mathrm{Var}(\tilde{D}) &= \frac{2}{m^2 n}\left\{\sum_l \left[\sum_u (p_{lu}^3 + p_{lu}^2 P_{lu,lu} - 2p_{lu}^4)\right.\right.\\ &\quad \left. + \frac{1}{2}\sum_u \sum_{v \neq u} p_{lu}p_{lv}(P_{lu,lv} - 4p_{lu}p_{lv})\right]\\ &\quad \left. + \sum_l \sum_{l'} \sum_u \sum_v p_{lu}p_{l'v}\Delta_{lu,l'v}\right\}\end{aligned}$$

For two alleles this becomes

$$\begin{aligned}\mathrm{Var}(\tilde{D}) &= \frac{2}{m^2 n}\left\{\sum_l \left[p_{l1}^3 + p_{l2}^3 - 2(p_{l1}^2 + p_{l2}^2)^2\right.\right.\\ &\quad \left. + p_{l1}^2 P_{l1,l1} + p_{l2}^2 P_{l2,l2} + p_{l1}p_{l2}P_{l1,l2}\right]\\ &\quad \left. + \sum_l \sum_{l'} (1 - 2p_{l1})(1 - 2p_{l'1})\Delta_{l1,l'1}\right\} \qquad (4.11)\end{aligned}$$

For populations that have Hardy-Weinberg genotypic proportions at each locus, the genotypic frequencies can be replaced by products of allele frequencies and then Equations 4.10 and 4.11 both become

$$\begin{aligned}\mathrm{Var}(\tilde{D}) &= \frac{2}{m^2 n}\left\{\sum_l \left[p_{l1}p_{l2}(p_{l1} - p_{l2})^2\right]\right.\\ &\quad \left. + \sum_l \sum_{l'} (1 - 2p_{l1})(1 - 2p_{l'1})\Delta_{l1,l'1}\right\}\end{aligned}$$

Estimation of the total variance requires data from more than one population. It is often suggested that different loci can play the role of replicate populations, however, so that the variance among diversities at different loci could serve as an estimate of the total variance. The same strategy was discussed for heterozygosity in the previous section, and it has merit in some situations.

With this approach, the total variance of $\tilde{D}$ is taken to be $1/m$ times the variance among the $\tilde{D}_l$'s. If this variance is written as s_D^2, then

$$s_D^2 = \frac{1}{(m-1)} \sum_l (\tilde{D}_l - \tilde{D})^2$$

but taking total expectations shows that, writing $m_1 = m(m-1)$,

$$\mathcal{E}(\frac{s_D^2}{m}) = \text{Var}(\tilde{D}) + \frac{1}{m_1}\left[\sum_l (D_l - D)^2 + \sum_l \sum_{l' \neq l} \text{Cov}(\tilde{D}_l, \tilde{D}_{l'})\right]$$

Just as for heterozygosity, s_D^2/m can serve as an estimator of the total variance of average diversity only when each locus has the same expected diversity and when the diversities at different loci have zero covariance. These conditions are met for random-mating populations and independent loci at equilibrium for drift and mutation (Weir 1989a). A good indication of when the approach is not valid would be provided by evidence of linkage disequilibrium between loci. In the case of a plant population with mixed selfing and random outcrossing, the covariance between diversities at two loci is directly proportional to the (composite) linkage disequilibrium between the loci, and s_D^2/m should not be used to estimate the total variance of average diversity if the disequilibria are found to be significantly different from zero.

Example

Gene diversity is a better indication of variability than heterozygosity for selfing species, as will now be demonstrated for some data on esterase loci *A*, *B*, and *C* (*Est*1, *Est*2, and *Est*4) in generation 4 of Barley Composite Cross V (Weir et al. 1972). Table 4.2 shows two-locus counts for each of the three pairs of loci, obtained by collapsing the alleles at each locus into the most common one versus the rest.

From Table 4.2, the heterozygosities at one and two loci are:

l	$\tilde{H}_l$	l, l'	$\tilde{H}_{l,l'}$
A	0.0640	A, B	0.0049
B	0.0381	A, C	0.0219
C	0.0916	B, C	0.0146

and the average over loci is $\tilde{H} = 0.0646$. The estimate of the variance of this average, over replicate samples from the same population, is found from Equation 4.3.

$$\text{Var}(\tilde{H}) = \frac{1}{m^2}\left[\sum_l \text{Var}(\tilde{H}_l) + \sum_l \sum_{l' \neq l} \text{Cov}(\tilde{H}_l, \tilde{H}_{l'})\right]$$

Table 4.2 Two-locus genotypic counts for three esterase loci in barley data of Weir et al. (1972).

	BB	Bb	bb	Total
AA	520	3	20	543
Aa	72	6	1	79
aa	502	38	72	612
Total	1094	47	93	1234

	CC	Cc	cc	Total
AA	375	25	143	543
Aa	44	27	8	79
aa	391	61	160	612
Total	810	113	311	1234

	CC	Cc	cc	Total
BB	770	83	241	1094
Bb	4	18	25	47
bb	36	12	45	93
Total	810	113	311	1234

$$\begin{aligned}\mathrm{Var}(\tilde{H}) \;\hat{=}\; & \frac{1}{9}\left[\frac{0.0640 \times 0.9360}{1234} + \frac{0.0381 \times 0.9619}{1234} + \frac{0.0916 \times 0.9084}{1234}\right.\\ & + \frac{0.0049 - 0.0640 \times 0.0381}{1234} + \frac{0.0219 - 0.0640 \times 0.0916}{1234}\\ & \left. + \frac{0.0146 - 0.0381 \times 0.0916}{1234}\right] = 0.000018846\end{aligned}$$

The estimated standard deviation of the average heterozygosity is, therefore, 0.0043. In contrast, the standard deviation found from the square root of the variance among the three single-locus heterozygosities is 0.0154. For the within-population variance, the difference among single-locus heterozygosities and the covariances between them make s_H^2/m an inappropriate estimator.

The total variance, applicable to inferences for more than one population, cannot be estimated from data on one population. If counts from several populations were available, the analysis of variance framework of Table 4.1 would allow estimation of the components of variance, and hence of the average heterozygosity. The variance among single-locus heterozygosities is also inappropriate for the total variance.

For gene diversity, the single-locus estimates and the pairwise composite linkage disequilibria are

l	$\tilde{D}_l$	l, l'	$\hat{\Delta}_{l,l'}$
A	0.4984	A, B	0.0510
B	0.1710	A, C	0.0117
C	0.4182	B, C	0.0540

The diversities are substantially larger than the heterozygosities, reflecting the fact that this population has a great deal of polymorphism, but that it is almost entirely homozygous. The significant linkage disequilibria between loci A, B and B, C precludes the between-locus variance of diversity being of any help in estimating the variance of average diversity, within or between populations.

The average gene diversity is 0.3625, with a within-population variance estimated by using the following frequencies:

l	u	$\tilde{p}_{lu}$	$\tilde{P}_{lu,lu}$	u, v	$\tilde{P}_{lu,lv}$
A	1	0.4720	0.4400	1,2	0.0640
	2	0.5280	0.4959		
B	1	0.9056	0.8784	1,3	0.0381
	2	0.0944	0.0754		
C	1	0.7022	0.6564	2,3	0.0916
	2	0.2978	0.2520		

The variance is estimated from Equation 4.11 as

$$\mathrm{Var}(\tilde{D}) \;\hat{=}\; \frac{2}{9 \times 1234}\Big[(0.4720)^3 + (0.5280)^3 - 2(0.5016)^2$$

$$\begin{aligned}
& + (0.4720)^2(0.4400) + (0.5280)^2(0.4959) \\
& + (0.4720)(0.5280)0.0640 \\
& + (0.9056)^3 + (0.0944)^3 - 2(0.8290)^2 \\
& + (0.9056)^2(0.8784) + (0.0944)^2(0.0754) \\
& + (0.9056)(0.0944)0.0381 \\
& + (0.7022)^3 + (0.2978)^3 - 2(0.5818)^2 \\
& + (0.7022)^2(0.6564) + (0.2978)^2(0.2520) \\
& + (0.7022)(0.2978)0.0916 \\
& + (0.0560)(-0.8122)(0.0510) \\
& + (0.0560)(-0.4044)(0.0117) \\
& +(-0.8112)(-0.4044)(0.0540)] \\
= \; & 0.000030767
\end{aligned}$$

and the standard deviation estimated as 0.0055. There is no estimate of the total variance.

SUMMARY

Variation in populations can be characterized with heterozygosity or gene diversity, the latter being more appropriate for inbred populations. The variance of these measures when averaged over loci needs to take account of different levels of variation at different loci, and the associations of variation levels at different loci.

EXERCISES

Exercise 4.1

Consider the following data set:

	Individual	Locus 1	Locus 2	Locus 3
Population 1	1	*BB*	*AA*	*AA*
	2	*BD*	*AC*	*AA*
	3	*BB*	*AB*	*AA*
	4	*DD*	*AA*	*AA*
	5	*BB*	*AB*	*AA*
Population 2	1	*BD*	*BB*	*AA*
	2	*BB*	*AD*	*AA*
	3	*BB*	*AB*	*AA*
	4	*DE*	*AA*	*AA*
	5	*BE*	*AC*	*AA*

a. Within each population, estimate the average heterozygosity across the three loci, and estimate the variance of this average using Equation 4.3. Compare this estimated variance with one-third the variance among the three single-locus heterozygosities.

b. Perform an analysis of variance of heterozygosity using the format of Table 4.1. Replace any negative estimated variance components by zero and then estimate the total variance to attach to the average heterozygosity for any population, using Equation 4.8.

Exercise 4.2

Using the same data as in Exercise 4.1, estimate the gene diversities for each locus and each population. Set up an analysis like that in Table 4.1 for these six gene diversities, and hence estimate the total variance for the average gene diversity in a population.

Exercise 4.3

Modify Equations 4.2, 4.3 to allow for unequal sample sizes at each locus.

Chapter 5

Population Structure

INTRODUCTION

This chapter considers ways of characterizing genetic variation within and between populations. Analyses are established for the case in which data are available in samples from different populations, or different subdivisions of the same population. The discussion of variation in such data, quantified with F-statistics, leads naturally into a treatment of genetic distance.

As discussed in Chapter 1, unless a population is invariant for the loci being studied, different samples from the population will show different levels of genetic variation. This is simply a consequence of the ***statistical sampling*** that results in each sample having a different set of individuals. Analyses need to be set up that allow statements to be made about the population, based on the sample at hand, with this sampling variation accommodated.

In population genetics there is another level of sampling to be considered. Each generation of a population is formed by the union of gametes chosen from among those produced by the previous generation. There is a ***genetic sampling*** process here that would cause the population to look different if the formation of a new generation was replicated. Population genetic theory depends on the concept of replicate populations that are maintained under the same conditions, but that will differ because of genetic sampling. It is possible to derive variances for statistics of interest that include both types of variation.

One use for such total variances is in predicting future values. From a specified population, it is possible to predict the expected value of a statistic, such as allele frequency or heterozygosity or linkage disequilibrium, in a future population, but not to specify the exact value of the statistic. For a neutral allele in a finite population, for example, it is known that the allele frequencies are *expected* to remain constant, although in any particular population the frequency may have drifted to any value from zero to one. Statements about the statistic in some future sample must therefore take account of the variation between replicate populations, as well as between replicate samples from any one population.

Difficulty arises in that the magnitude of between-population variation cannot be estimated with a sample from a single population. One way around this problem is sometimes afforded by the availability of several loci in the data set. To the extent that different loci may be regarded as being independent, they may also be regarded as playing the role of separate populations. The loci have each been exposed to the genetic sampling forces between generations, but the specific descent history (or coalescent) for each locus can be different.

The distinction between statistical and genetic sampling can also be phrased in terms of ***fixed*** and ***random*** effects, to show how the intended scope of inference affects the sampling properties of genetic statistics. If there is interest only in the one particular population sampled, then prior genetic sampling is not of consequence. It is necessary to take account only of the statistical sampling for repeated samples from this one ("fixed") population. Future samples would be taken from the population as it is presently constituted. Comparisons between different ("fixed") populations can be phrased in terms of means, and it will be shown that procedures of permutation and numerical resampling are of use in drawing inferences.

A different situation arises when the sample is to be used to make inferences about a collection of populations, or maybe even about the species as a whole. There is then less interest in the particular population sampled, which can now be regarded as being random. Future samples may very well be drawn from a different population, so that both statistical and genetic sampling variation needs to be considered. The distinction between fixed and random effects arises in statistics. In the analysis of variance context it is easier to detect differences between means in a fixed effects situation because interaction terms are not included in the mean square used in the denominator for the F-test statistic, leading generally to smaller mean squares and larger test statistics. It is only one specific set of means (fixed effects) that is being compared, and not some population of means (random effects) for which the means at hand are just a sample. The random genetic model considers each population to be a replicate sample of the evolutionary process.

FIXED POPULATIONS

With data collected for genotypes, the first descriptors of a population are the genotypic frequencies. When the population is sampled in such a way that every member of the population has an equal chance of being sampled, and individuals are sampled independently, the genotypic counts are multinomially distributed. Under the fixed-population approach, different populations for the same species are compared simply by comparing frequencies. When Hardy-Weinberg equilibrium (HWE) cannot be assumed, this requires the comparison of genotypic frequencies. Under HWE, however, the data may be regarded as being multinomial allele frequencies and comparisons made at the allele level.

Contingency Tables

The most straightforward procedure to test for equal genotypic frequencies over populations is to use contingency table chi-square tests. With v alleles at a locus, the genotypic counts in each of r samples are arranged in a $v(v+1)/2 \times r$ contingency table and a chi-square statistic with $\{[v(v+1)/2]-1\} \times (r-1)$ degrees of freedom is calculated. In practice, the method has problems when some cells have small expected counts, under the hypothesis of equal frequencies in each population, since this can give test statistics that are spuriously large, and it may be necessary to combine the least frequent classes. This problem increases with the number of alleles, but even for two alleles a sample of size 100 is expected to have only four individuals in the aa class when allele frequencies are $p_A = 0.8$, $p_a = 0.2$. Conventional wisdom says that goodness-of-fit chi-square tests should not be performed on classes with expected counts less than five, although this is probably too conservative. The test statistic may have an approximate chi-square distribution even with some small expected counts in the table. Doubts about the effect of small expected counts are best resolved by looking at the contribution of each class to the test statistic.

When HWE is assumed, it is sufficient to compare the allele frequency arrays in each of the populations. The contingency table is then only $v \times r$, and problems of small expected counts are less likely, although they may still be considerable for many-allele systems.

Numerical Resampling

An alternative to the contingency table approach is provided by numerical resampling (Chapter 2). This is a means of making inferences about population allele frequencies from the sample frequencies. Variances can be estimated and confidence intervals can be constructed, both referring to repeated sampling from the same populations. Briefly, numerical resampling mimics the drawing of new samples from each population by drawing them from each sample. Only bootstrapping will be considered here since it provides very much more information about the parameter being estimated than does jackknifing. It leads to an estimate of the sampling distribution of the estimate.

For bootstrapping, from the original set of n observations a new sample of the same size n is constructed by random sampling with replacement. In place of the single estimate from the original sample, bootstrapping provides as many new estimates as desired. Whereas this collection of new estimates could be used to provide an estimated variance for the original estimate, it is more informative to work with the whole distribution of estimates. For example, a 95% confidence interval for the parameter being estimated can be constructed as the limits between which the middle 95% of the bootstrap estimates lie.

Bootstrapping within each of the r samples will therefore provide confidence intervals for the population allele frequencies, without having to invoke the binomial theorem and, therefore, without having to assume Hardy-Weinberg equilibrium. Two populations can be judged to have different allele frequencies if the estimated frequencies have nonoverlapping confidence intervals. How wide should these confidence intervals be in order to have a specified confidence in the statement that the populations have different frequencies?

A very conservative answer that does not invoke normality (or any other distribution) can be based on Chebyshev's inequality, which states that the probability of a random variable being more than k standard deviations from its mean is less than $1/k^2$. For population I with frequency p_{i_I} for allele A_i, if the sample frequency $\tilde{p}_{i_I}$ has variance σ^2, this means that

$$\Pr(|\tilde{p}_{i_I} - p_{i_I}| \geq k\sigma) \ \leq \ 1/k^2$$

There is a similar equation for the sample frequency $\tilde{p}_{i_{II}}$ in population II. Under the hypothesis that the two populations have the same allele frequencies, $p_{i_I} = p_{i_{II}}$, and the same variances of sample frequencies, Chebyshev's inequality applied to the variable $\tilde{p}_{i_I} - \tilde{p}_{i_{II}}$ gives

$$\Pr(|\tilde{p}_{i_I} - \tilde{p}_{i_{II}}| \geq K\sigma\sqrt{2}) \ \leq \ 1/K^2$$

because the variance of the difference of two independent frequencies is the sum of their variances. For this last probability to be 0.05, corresponding to a 95% confidence interval on the difference of frequencies, it is necessary that $K^2 = 20$. However, if this corresponds to non-overlapping confidence intervals for the two separate allele frequencies, then $2k\sigma = K\sigma\sqrt{2}$ and $k^2 = 10$. The two separate intervals are 90% confidence intervals.

Permutation Tests

A distribution-free approach can also be based on permutation. For two populations, extending the deck-of-cards analogy mentioned in Chapter 3, the deck now has a card for each allele in each of the two populations. The hypothesis of equality of frequencies in the two populations is equivalent to alleles being allocated independently into two samples. After the permutation is completed, the first $2n_I$ cards are taken to represent the first sample, and the remaining $2n_{II}$ to represent the second sample, where n_I, n_{II} were the original sample sizes. The proportion of permutations in which the difference in allele A_i counts between the two samples is as great or greater than the original observed value provides the significance level for rejecting the hypothesis of equality. This test is local to allele A_i.

If HWE is assumed, a global test over all alleles could be based on the joint probability of the two arrays

$$\Pr(\{n_{i_I}\}, \{n_{i_{II}}\}) = \left(\frac{(2n_I)!}{\prod_i n_{i_I}!} \prod_i p_{i_I}^{n_{i_I}}\right) \left(\frac{(2n_{II})!}{\prod_i n_{i_{II}}!} \prod_i p_{i_{II}}^{n_{i_{II}}}\right)$$

where $n_{i_I}, n_{i_{II}}$ are the counts for allele A_i in the two samples. Making this joint probability conditional on the total counts of alleles $(\{n_{i_I} + n_{i_{II}}\})$

$$\Pr(\{n_{i_I}\}, \{n_{i_{II}}\} | \{n_{i_I} + n_{i_{II}}\}) = \frac{(2n_I)!(2n_{II})!}{(2n_I + 2n_{II})!} \frac{\prod_i (n_{i_I} + n_{i_{II}})!}{\prod_i n_{i_I}! \prod_i n_{i_{II}}!} \quad (5.1)$$

under the hypothesis of equality of allele frequencies $p_{i_I} = p_{i_{II}}$. The total set of alleles is permuted and then divided into two samples of sizes $2n_I, 2n_{II}$, and the proportion of the resulting conditional probabilities less than or equal to the observed value provides the significance value. Permutation tests for comparing populations were discussed by Fisher (1936b) and by Roff and Bentzen (1989), although Manly (1991) gave more efficient randomization techniques.

As an example, suppose two samples of five individuals have the following allele counts for alleles A, a:

	Sample I	Sample II
Allele A	$n_{A_I} = 8$	$n_{A_{II}} = 2$
Allele a	$n_{a_I} = 2$	$n_{a_{II}} = 8$
	$2n_I = 10$	$2n_{II} = 10$

Table 5.1 Possible allocations of 10 A and 10 a alleles into two samples of size 10 alleles, along with conditional probabilities.

Number of A alleles		
Sample I	Sample II	Probability
0	10	0.0000
1	9	0.0005
2	8	0.0110
3	7	0.0779
4	6	0.2387
5	5	0.3438
6	4	0.2387
7	3	0.0779
8	2	0.0110
9	1	0.0005
10	0	0.0000

In this simple case, it is not necessary to perform random permutations because there are only 11 possible pairs of samples when total allele counts are held constant. The possible pairs, along with the conditional probabilities are displayed in Table 5.1. The observed outcome is one of the more unusual under the null hypothesis of equal frequencies in each population, and the significance level is 0.0230 (as is also found from the chi-square test that gives $X^2 = 7.2$). For larger samples and many alleles, random permutations will be necessary when the number of possible pairs of samples is too large to enumerate, just as for the Hardy-Weinberg tests in Chapter 3.

F Statistics

It is often the case that a single statistic, written as F_{ST}, is calculated to compare populations. For a set of r populations with sample allele frequencies $\tilde{p}_i (i = 1, 2, \ldots, r)$ for some allele A, the statistic could be defined as

$$\begin{aligned} F_{ST} &= \frac{\sum_i (\tilde{p}_i - \bar{p})^2/(r-1)}{\bar{p}(1-\bar{p})} \\ &= \frac{s^2}{\bar{p}(1-\bar{p})} \end{aligned}$$

where $\bar{p} = \sum_i \tilde{p}_i / r$ is the average sample frequency of the allele over the samples and s^2 is the sample variance. If the samples were of unequal sizes

n_i, weighted means and variances should be used and then

$$F_{ST} = \frac{\sum_i n_i(\tilde{p}_i - \bar{p})^2/(r-1)\bar{n}}{\bar{p}(1-\bar{p})}$$

with $\bar{p} = \sum_i n_i\tilde{p}_i / \sum_i n_i, \bar{n} = \sum n_i/r$.

It is certainly true that this quantity increases as the sample allele frequencies diverge, but it is difficult to assess the significance of divergence. Using the F_{ST} statistic to test for allele frequency differences would require knowledge of its sampling properties. In the fixed-population framework, it is possible to relate F_{ST} to the contingency-table chi-square test statistic. Suppose attention is focused on one allele A and its alternative allele(s) $\bar{A}$. If the sample frequency of allele A in the ith sample is $\tilde{p}_i$, and the sample size is n_i, the chi-square statistic for comparing allele counts over populations is

$$\begin{aligned} X^2 &= \sum_i \frac{(n_i\tilde{p}_i - n_i\bar{p})^2}{n_i\bar{p}} + \sum_i \frac{[n_i(1-\tilde{p}_i) - n_i(1-\bar{p})]^2}{n_i(1-\bar{p})} \\ &= \sum_i \frac{n_i(\tilde{p}_i - \bar{p})^2}{\bar{p}(1-\bar{p})} \\ &= (r-1)\bar{n}F_{ST} \end{aligned}$$

(If F_{ST} had been defined with r instead of $r-1$ in the numerator, $X^2 = r\bar{n}F_{ST}$ would have resulted.)

There seems little point, however, in calculating F_{ST} in order to test for population differentiation as that can be done directly with the chi-square test or alternatives such as the permutation test. Under the hypothesis that the allele frequencies are the same over populations, the X^2 statistic has an expected value of $(r-1)$ so that the expected value of F_{ST} is $1/\bar{n}$ rather than zero. The estimate is slightly biased, and the dependence of expected value on sample size is not desirable. Of course, for very large sample sizes F_{ST} is expected to be close to zero when populations have the same frequencies.

Analysis of Variance

Comparison of several means in a statistical setting is often performed by an analysis of variance. Different variances are computed, which, when the means are the same, are expected to be estimating the same quantity. For normally distributed variables, the ratio of between-population to within-population variances has an F distribution under the hypothesis of equal population means. The same approach is not strictly applicable here, but this is a convenient place to introduce the concepts and notation that are

Table 5.2 Analysis of variance layout for variable indicating allele A in fixed populations.

Source	d.f.	Sum of Squares	Expected Mean Square*
Between populations	$r-1$	$\sum_i \frac{x_{i.}^2}{n_i} - \frac{x_{..}^2}{n.}$	$\frac{1}{r-1}\sum_i k_1 p_{Ai}(1-p_{Ai})$
		$=\sum_i n_i(\tilde{p}_{Ai}-\tilde{p}_{A\cdot})^2$	$+\frac{1}{r-1}\sum_i n_i(p_{Ai}-\bar{p}_{A\cdot})^2$
Within populations	$\sum_i(n_i-1)$	$\sum_{i=1}^r \sum_{j=1}^{n_i} x_{ij}^2 - \sum_i \frac{x_{i.}^2}{n_i}$	$\frac{1}{\sum_i(n_i-1)}\sum_i k_2 p_{Ai}(1-p_{Ai})$
	$=n.-r$	$=\sum_i n_i\tilde{p}_{Ai}(1-\tilde{p}_{Ai})$	

$^*k_1 = \left(1 - \frac{n_i}{\sum_i n_i}\right), \quad k_2 = (n_i - 1)$

needed later for random populations. For allele j in the sample from population i, a variable x_{ij} can be defined by

$$x_{ij} = \begin{cases} 1 \text{ if allele is } A \\ 0 \text{ if allele is not } A \end{cases}$$

The mean of the x_{ij} values for population i is just the allele frequency p_{Ai} in that population, so that the hypothesis of equal allele frequencies among populations would appear to be addressed by an analysis of variance of the x's. The analysis of variance table has the structure shown in Table 5.2. In this table, the average of the sample allele frequencies over the r samples is

$$\tilde{p}_{A\cdot} = \frac{\sum_i n_i \tilde{p}_{Ai}}{\sum_i n_i}$$

and the average of the population frequencies is

$$\bar{p}_{A\cdot} = \frac{\sum_i n_i p_{Ai}}{\sum_i n_i}$$

The expected values in Table 5.2 are calculated on the basis of the n_i alleles from population i being independent. When the r populations all have the same allele frequencies, $p_{Ai} = \bar{p}_{A\cdot} = p_A$, the expected mean squares for between- and within-populations are both $p_A(1-p_A)$. Although an analysis of variance layout is used here, it is not claimed that the ratio of the two mean squares has the F distribution since the underlying variables x_{ij} are not normally distributed. (It may well be that the F distribution is adequate, however.) The purpose for presenting Table 5.2 in the fixed model framework is that it serves to clarify the difference between the fixed and random models.

In the fixed model, the only variation being considered is that between samples from the same population. This is the variation giving rise to the distribution of the test statistic, and is to be contrasted to the random model case considered next.

RANDOM POPULATIONS

Under the random model, the populations sampled may be considered to represent the species and therefore to have a common evolutionary history. Even though the populations may have been distinct for some time, the analysis is built on the assumption that there is a single reference population. The expectations implied by taking means and variances now refer to repeated samples from the populations *and* to replicate populations. In the absence of disturbing forces, such as differing amounts of selection in different populations, all populations are expected to have the same allele frequencies.

Underlying the analysis of differentiation in the random model is the notion that genetic sampling causes different alleles in a population to be dependent, or related. Even though individuals, or alleles, may be sampled randomly, the process of taking expectations must recognize that they are dependent through their shared ancestry. Another essential concept for the analysis is that the relationships between various alleles are relative to the least-related alleles in the data. It is generally assumed that the least-related alleles are independent, since the data do not allow measures of relationship to be estimated otherwise.

Interest can be centered on the extent to which different populations within the species have differentiated over time. The action of evolutionary forces, or genetic sampling, will result in intraspecific differentiation, and this differentiation is conveniently quantified with the F statistics of Wright (1951), or the analogous measures of Cockerham (1969, 1973). These quantities measure the degree of relatedness of various pairs of alleles. Cockerham (1969) described the three basic quantities in the situation when diploid individuals are sampled from a series of populations as follows: the overall inbreeding coefficient F, Wright's F_{IT}, is the correlation of alleles within in-

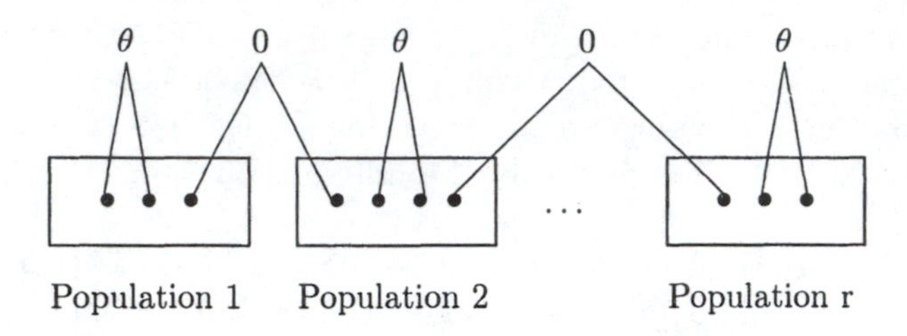

Figure 5.1 Relationships between pairs of alleles; θ within and 0 between populations.

dividuals over all populations, the coancestry θ, Wright's F_{ST}, is the correlation of alleles of different individuals in the same population, and f, Wright's F_{IS}, is the correlation of alleles within individuals within one population. If the populations are assumed to have been isolated since an ancestral population, alleles in one population are independent of those in another. This model is illustrated in Figure 5.1, where different alleles within a population are related to an extent θ and alleles in different populations are not related. If conditions within populations are different, meaning that θ values differ among populations, then the estimation procedures discussed below provide some kind of average θ value. The pairing of alleles within individuals, necessitating the addition of measure F, is not shown in Figure 5.1. There may, of course, be nonzero relationships between alleles in different populations, especially when there is migration between populations. In that case the estimates of θ are a combination of within- and between-population measures (Cockerham and Weir 1987).

Haploid Data

If data are available on alleles directly, then the analysis is in terms of allele frequencies. For haploid data, there is only one F statistic as shown in Figure 5.1. It measures the relationship between different alleles in the same population, relative to the zero relationship between alleles of different populations. The quantity is written here as θ.

Shared ancestry means that the expected value of a squared sample frequency from a sample of size n_i is

$$\mathcal{E}(\tilde{p}_{Ai}^2) = p_A^2 + p_A(1-p_A)\theta + \frac{1}{n_i}p_A(1-p_A)(1-\theta)$$

where p_A is the frequency in the ancestral population and θ, or F_{ST}, is the coancestry coefficient. The derivation of this expression was given in Chapter 2, following Equation 2.14 with $F = \theta$. Another way of contrasting the fixed and random populations approaches is to note that the fixed approach variances are within populations, whereas the random approach variances are for total variation.

Although the allele frequency p_A is assumed to remain at its ancestral value, for finite populations the coancestry will increase over time as inbreeding accrues in each population. In other words, θ measures the extent of differentiation between populations. It is worth stressing that between-population differentiation goes hand in hand with relatedness of alleles within populations. As individuals become more related within populations, the independent populations are expected to become more differentiated.

Estimation of θ can proceed by the method of moments, with the various statistics being conveniently organized in an analysis of variance format (Table 5.3), which removes the need to know the expected population frequencies p_A. (Other methods of estimation were reviewed by Spielman et al. 1977 and Jiang 1987). The different expectation framework from the fixed population analysis causes Table 5.3 to have different expected mean squares from Table 5.2. The two tables differ in that Table 5.3 has a common expectation for each $\tilde{p}_{Ai}$. The more substantial difference between the two tables, however, is that Table 5.3 explicitly allows different alleles in a sample to be related to an extent measured by θ. In Table 5.2, the sampled alleles are all independent. Note that in Table 5.3, the quantity n_c is like an average sample size but it also involves the variance of sample sizes over populations.

From the method of moments, the parameters are estimated by equating observed and expected mean squares. Writing the observed mean square for between populations as *MSP* and for alleles within populations as *MSG*:

$$\begin{aligned}
p_A(1-p_A)(1-\theta) &\hat{=} \frac{1}{\sum_i(n_i-1)}\sum_i n_i\tilde{p}_{Ai}(1-\tilde{p}_{Ai}) \\
&= MSG \\
p_A(1-p_A)(1-\theta) + n_c p_A(1-p_A)\theta &\hat{=} \frac{1}{r-1}\sum_i n_i(\tilde{p}_{Ai}-\tilde{p}_{A\cdot})^2 \\
&= MSP
\end{aligned}$$

Table 5.3 Analysis of variance layout for variable indicating allele A in random populations.

Source	d.f.	Sum of Squares	Expected Mean Square*
Between populations	$r-1$	$\sum_{i=1}^{r} \frac{x_{i.}^2}{n_i} - \frac{x_{..}^2}{n.}$	$p_A(1-p_A)[(1-\theta)+n_c\theta]$
		$=\sum_{i=1}^{r} n_i(\tilde{p}_{Ai}-\tilde{p}_{A\cdot})^2$	$=\sigma_G^2+n_c\sigma_P^2$
Within populations	$\sum_{i=1}^{r}(n_i-1)$	$\sum_{i=1}^{r}\sum_{j=1}^{n_i} x_{ij}^2 - \sum_i \frac{x_{i.}^2}{n_i}$	$p_A(1-p_A)(1-\theta)$
	$=n.-r$	$=\sum_i n_i\tilde{p}_{Ai}(1-\tilde{p}_{Ai})$	$=\sigma_G^2$

$$^*n_c = \frac{1}{r-1}\left(\sum_{i=1}^{r} n_i - \frac{\sum_i n_i^2}{\sum_i n_i}\right)$$

(Recall that $\hat{=}$ means "is estimated by.")

The expected mean squares in Table 5.3 have also been written in terms of two ***components of variance***. For alleles within populations:

$$\sigma_{\mathrm{G}}^2 = p_A(1-p_A)(1-\theta)$$

This component can be estimated, by the method of moments, as

$$\hat{\sigma}_{\mathrm{G}}^2 = MSG$$

A component for populations is defined as

$$\sigma_{\mathrm{P}}^2 = p_A(1-p_A)\theta$$

and is estimated as

$$\hat{\sigma}_{\mathrm{P}}^2 = \frac{1}{n_c}(MSP - MSG)$$

Note that the two components of variance have been defined in terms of the expected mean squares in Table 5.3. They were not defined as variances, in spite of the conventional σ^2 notation. Instead, the component σ_{P}^2 is actually a covariance (for alleles in the same population) and there is no requirement that this quantity be positive.

The two variance components, within and between populations, reflect all the factors that lead to variation in allele frequencies. The sum of these two components involves only the allele frequency p_A, and this allows the unknown quantity $p_A(1-p_A)$ to be eliminated. An estimate of θ can then be found as

$$\begin{aligned} \hat{\theta} &= \frac{\hat{\sigma}_P^2}{\hat{\sigma}_P^2 + \hat{\sigma}_G^2} \\ &= \frac{MSP - MSG}{MSP + (n_c - 1)MSG} \end{aligned} \tag{5.2}$$

This can also be expressed in terms of the variance of allele frequencies over populations

$$s_A^2 = \frac{1}{(r-1)\bar{n}} \sum_i n_i(\tilde{p}_{Ai} - \tilde{p}_{A\cdot})^2$$

where $\bar{n} = \sum_{i=1}^r n_i/r$. The estimator is a ratio of two functions of the data, whose expectations differ by the factor θ. The ratio may be denoted

$$\hat{\theta} = \frac{T_1}{T_2}$$

with

$$\begin{aligned} T_1 &= s_A^2 - \frac{1}{\bar{n}-1}\left[\tilde{p}_{A\cdot}(1-\tilde{p}_{A\cdot}) - \frac{r-1}{r}s_A^2\right] \\ T_2 &= \frac{n_c - 1}{\bar{n}-1}\tilde{p}_{A\cdot}(1-\tilde{p}_{A\cdot}) + \left[1 + \frac{(r-1)(\bar{n}-n_c)}{\bar{n}-1}\right]\frac{s_A^2}{r} \end{aligned}$$

For a large number of large samples, when both $1/\bar{n}$ and $1/r$ can be ignored, this estimate reduces to

$$\hat{\theta} = \frac{s_A^2}{\tilde{p}_{A\cdot}(1-\tilde{p}_{A\cdot})}$$

which is just the estimator F_{ST} discussed above in the fixed-model section. Note, however, that the estimators have been derived under very different models. Nothing was implied about evolutionary forces causing population differentiation in the fixed model. There is a difference between arriving at this equation as an approximate expression for the estimator of a parameter, and taking the equation as the definition of a quantity of interest.

Estimation of θ has been presented in terms of one of the alleles, A, at a locus. If there are only two alleles at the locus, then the same estimate would be obtained if the other allele is used. For more than two alleles, however, a different *estimate* may result with every allele. Since the *parameter* θ is the same for every allele when there are no disturbing forces, all these estimates refer to the same quantity and an appropriate average is needed to give the best single estimate. For the uth allele, the estimate could be written as

$$\hat{\theta}_u = \frac{T_{1u}}{T_{2u}}$$

and then an overall estimate with the desirable properties of low bias and small variance is given by combining the information from all v alleles

$$\hat{\theta} = \frac{\sum_{u=1}^{v} T_{1u}}{\sum_{u=1}^{v} T_{2u}}$$

There is an additional extension to cover the case where several loci are scored. Once again, under a model with no disturbing forces, every allele at every locus provides an estimate of the same quantity. Indexing the loci by l and alleles within loci by u, then the individual estimates are

$$\hat{\theta}_{lu} = \frac{T_{1lu}}{T_{2lu}}$$

and the overall estimate from all m loci is

$$\hat{\theta} = \frac{\sum_{l=1}^{m} \sum_{u=1}^{v} T_{1lu}}{\sum_{l=1}^{m} \sum_{u=1}^{v} T_{2lu}} \tag{5.3}$$

With θ estimated, there is a quantification of the degree of divergence among a set of r populations. Equal allele frequencies in all populations will cause θ to be zero, and for a pair of populations θ may serve as a measure of genetic distance. Under the pure drift model, $\ln(1-\theta)$ will increase linearly with the time since divergence of the two populations from the ancestral population. When all populations are fixed at all loci scored, the estimator is undefined, since the equation becomes the ratio of zeros. Indeed, unless there is additional information on the molecular structure of the fixed alleles, there is then no information in the present data on how long the populations have been fixed or on the time of divergence of the populations.

Making inferences about θ beyond simply estimating it may be accomplished by numerical resampling. Confidence intervals follow from bootstrapping. Because θ is a parameter appropriate for the random model, numerical resampling cannot be performed by resampling alleles at a locus within populations, as was done in the fixed model. Resampling must mimic both the genetic sampling that causes replicate populations to differ, and the sampling of alleles for observation from each population. Two possibilities are suggested.

In the first place bootstrapping may be done over loci. For a study in which m loci are scored, this provides confidence intervals for θ. Instead of drawing all the data for sets of m loci, each bootstrap sample is found by drawing a set of m pairs of values $\sum_u T_{1lu}, \sum_u T_{2lu}$ with replacement from the m calculated values, and the combined estimate formed from this new collection of T's. As before, the middle 95% of these estimates provides a 95% confidence interval. The hypothesis that θ has some specified value can be rejected, at the 5% significance level, if the confidence interval does not contain that value. Data sets with overlapping (90%) confidence intervals provide no evidence that the corresponding θ values differ (at the 5% level).

It is also possible to bootstrap populations, and this may be done for each locus separately. In that way the estimates of θ for each locus could be compared. Loci that did not give overlapping confidence intervals for θ indicate the presence of disturbing forces. Bootstrapping over populations requires that there are several populations, just as resampling over loci supposes that several loci have been scored. In practice, many populations or loci appear to be necessary. The problem with bootstrapping over populations is that the structure of the original sample is not preserved, and the resulting distribution may not apply to the number and divergence of the sampled populations. A detailed discussion was given by Dodds (1986).

The hypothesis that $\theta = 0$ may also be addressed by permutation tests, in a way similar to the comparison of allele frequencies in the fixed-population analysis (Raymond and Rousset 1995). Figure 5.1 makes it clear that alleles have the same zero relationship within and between populations when the hypothesis is true. If the alleles in the total dataset are permuted into new population groupings, in such a way that preserves the sample sizes, then the new estimate of θ should be from the same distribution as the original estimate if $\theta = 0$. The hypothesis would be rejected at the 5% level if the original estimate was among the least likely 5% of the estimates found from permuted data. Note that permutation changes both the between- and within-population mean squares, MSG and MSP. This causes changes both to the estimate of θ and to the estimate of the sum of variance components $\sigma_G^2 + \sigma_P^2 = p_A(1 - p_A)$.

It is necessary to comment on the possibility that the estimate of θ may be negative. There are two situations that are likely to give this outcome. It may be that the true value of θ is positive but small. Since the estimate has fairly low bias, it is about as probable to be below as above the true value. Estimates less than the true value will often be less than zero when the true value is close to zero. The second situation is that the parameter may be negative. In statistical language, this corresponds to a negative ***intraclass correlation*** (not a negative variance). In genetic language, it means that alleles are more related between than within populations. A more complete discussion of biological causes for a negative component of variance between populations, $\sigma^2_{\mathrm{P}} < 0$, was given by Cockerham (1973) for the analysis of diploid data. Any mating system, such as the avoidance of self-mating (e.g., any system with separate sexes), that allows alleles to be more alike between individuals than within individuals can cause this phenomenon, as will become evident in the next section.

Diploid Data

When observations are made on diploid individuals, the analysis should be performed at the diploid level. The same general approach is followed, but it is now possible to estimate the degree of relationship F between alleles within individuals, as well as the degree θ between alleles of different individuals. The haploid analysis essentially dropped the distinction between F and θ. Differentiation between populations is still measured in terms of θ, reflecting the relatedness of individuals within populations, but the analysis also provides an estimate of the overall inbreeding coefficient F. In Wright's notation, $F = F_{IT}, \theta = F_{ST}$. The degree of inbreeding within populations, f, or F_{IS}, can be expressed as

$$f = \frac{F - \theta}{1 - \theta}$$

The expected value of the squared sample allele frequencies now must reflect the two levels of relatedness of different alleles within populations — both a consequence of prior genetic sampling:

$$\mathcal{E}(\tilde{p}^2_{Ai}) = p^2_A + p_A(1 - p_A)\theta + \frac{1}{2n_i}p_A(1 - p_A)(1 + F - 2\theta)$$

Once again, the estimation procedure follows naturally from an analysis of variance layout. There are now three sources of variation: populations, individuals within populations, and alleles within individuals. The sums of squares are constructed with allele *and* genotypic frequencies, as shown in Table 5.4. The first form of each sum of squares is the most convenient for computations with data.

The expected mean squares have also been written in terms of variance components

$$\sigma^2_{\mathrm{P}} = p_A(1 - p_A)\theta \text{ for populations}$$

$$\sigma^2_{\mathrm{I}} = p_A(1 - p_A)(F - \theta) \text{ for individuals within populations}$$

Table 5.4 Analysis of variance layout for genotypic data in random populations.

Source	d.f.	Sum of Squares	Expected Mean Square*
Between populations	$r-1$	$2\sum_{i=1}^{r} n_i(\tilde{p}_{Ai}-\tilde{p}_{A\cdot})^2$	$p_A(1-p_A)[(1-F)$
		$=2(r-1)\bar{n}s_A^2$	$+2(F-\theta)+2n_c\theta]$
			$=\sigma_G^2+2\sigma_I^2+2n_c\sigma_P^2$
Individuals in populations	$\sum_{i=1}^{r}(n_i-1)$	$\sum_{i=1}^{r} n_i(\tilde{p}_{Ai}+\tilde{P}_{AAi}-2\tilde{p}_{Ai}^2)$	$p_A(1-p_A)[(1-F)$
	$=n.-r$	$=2r\bar{n}\tilde{p}_{A\cdot}(1-\tilde{p}_{A\cdot})-\frac{1}{2}r\bar{n}\tilde{H}_{A\cdot}$	$+2(F-\theta)]$
		$-2(r-1)\bar{n}s_A^2$	$=\sigma_G^2+2\sigma_I^2$
Alleles in individuals	$\sum_{i=1}^{r} n_i$	$\sum_{i=1}^{r} n_i(\tilde{p}_{Ai}-\tilde{P}_{AAi})$	$p_A(1-p_A)(1-F)$
	$=n.$	$=\frac{1}{2}r\bar{n}\tilde{H}_{A\cdot}$	$=\sigma_G^2$

$$^*n_c=\frac{1}{r-1}\left(\sum_{i=1}^{r} n_i-\frac{\sum_i n_i^2}{\sum_i n_i}\right)$$

and

$$\sigma_{\mathrm{G}}^2 = p_A(1-p_A)(1-F) \text{ for alleles within individuals}$$

From the method of moments, unbiased estimates of the components are functions of the observed mean squares *MSP*, *MSI*, and *MSG* for populations, individuals, and alleles, respectively:

$$\begin{aligned}
\hat{\sigma}_{\mathrm{G}}^2 &= MSG \\
\hat{\sigma}_{\mathrm{I}}^2 &= \frac{1}{2}(MSI-MSG) \\
\hat{\sigma}_{\mathrm{P}}^2 &= \frac{1}{2n_c}(MSP-MSI)
\end{aligned}$$

In Table 5.4, the sample sizes n_i are for the numbers of individuals in each sample, whereas in Tables 5.2 and 5.3 they were the numbers of alleles. Also in Table 5.4, $\tilde{P}_{AAi}$ is the frequency of AA homozygotes in the ith sample, and $\tilde{H}_{A\cdot}$ is the frequency of heterozygous individuals that have allele A, averaged over all samples. In other words

$$\begin{aligned}\tilde{H}_{A\cdot} &= \frac{\sum_i n_i \tilde{H}_{Ai}}{\sum_i n_i} \\ &= \frac{1}{\sum_i n_i} \sum_i 2n_i(\tilde{p}_{Ai} - \tilde{P}_{AAi})\end{aligned}$$

Measures of differentiation can be estimated as

$$\begin{aligned}\hat{F} &= \frac{\hat{\sigma}_{\mathrm{P}}^2 + \hat{\sigma}_{\mathrm{I}}^2}{\hat{\sigma}_{\mathrm{P}}^2 + \hat{\sigma}_{\mathrm{I}}^2 + \hat{\sigma}_{\mathrm{G}}^2} \\ &= 1 - \frac{2n_c MSG}{MSP + (n_c - 1)MSI + n_c MSG} \\ &= 1 - \frac{S_3}{S_2}\end{aligned}$$

$$\begin{aligned}\hat{\theta} &= \frac{\hat{\sigma}_{\mathrm{P}}^2}{\hat{\sigma}_{\mathrm{P}}^2 + \hat{\sigma}_{\mathrm{I}}^2 + \hat{\sigma}_{\mathrm{G}}^2} \\ &= \frac{MSP - MSI}{MSP + (n_c - 1)MSI + n_c MSG} \\ &= \frac{S_1}{S_2}\end{aligned}$$

These are probably the most convenient computing formulas, but it is possible to give explicit expressions for the terms S_1, S_2, and S_3:

$$\begin{aligned}S_1 &= s_A^2 - \frac{1}{\bar{n} - 1}\left[\tilde{p}_{A\cdot}(1 - \tilde{p}_{A\cdot}) - \frac{r-1}{r} s_A^2 - \frac{1}{4}\tilde{H}_{A\cdot}\right] \\ S_2 &= \tilde{p}_{A\cdot}(1 - \tilde{p}_{A\cdot}) - \frac{\bar{n}}{r(\bar{n} - 1)}\left\{\frac{r(\bar{n} - n_c)}{\bar{n}}\tilde{p}_{A\cdot}(1 - \tilde{p}_{A\cdot})\right. \\ &\quad \left. - \frac{1}{\bar{n}}[(\bar{n} - 1) + (r - 1)(\bar{n} - n_c)]s_A^2 - \frac{r(\bar{n} - n_c)}{4\bar{n}n_c}\tilde{H}_{A\cdot}\right\} \\ S_3 &= \frac{n_c}{2\bar{n}}\tilde{H}_{A\cdot}\end{aligned}$$

It is worth stressing that, with data collected for genotypes, estimates of the F statistics F, θ are functions of genotypic frequencies.

These estimates have all been presented in terms of one particular allele A. In practice, several alleles at several loci will be available, and each will provide an estimate of the same parameters under a strictly neutral model. To combine estimates over all alleles, numerators and denominators are combined separately, as in the haploid case. For the uth allele at the lth locus, the estimates may be expressed as

$$\hat{F}_{lu} = 1 - \frac{S_{3lu}}{S_{2lu}}$$

$$\hat{\theta}_{lu} = \frac{S_{1lu}}{S_{2lu}}$$

and then the combined estimates follow from

$$\hat{F} = 1 - \frac{\sum_l \sum_u S_{3lu}}{\sum_l \sum_u S_{2lu}}$$

$$\hat{\theta} = \frac{\sum_l \sum_u S_{1lu}}{\sum_l \sum_u S_{2lu}}$$

Also, as in the haploid case, bootstrapping over many loci provides the means for making inferences about the parameters F and θ, while bootstrapping over many populations may allow comparisons between loci. Permutations could be for individuals among populations, or for alleles among individuals and populations, depending on the hypotheses to be tested.

Note that large numbers of large samples allow approximate expressions to be found for the estimates from each allele:

$$\hat{F} = 1 - \frac{\tilde{H}_{A\cdot}}{\tilde{p}_{A\cdot}(1 - \tilde{p}_{A\cdot})}$$

$$\hat{\theta} = \frac{s_A^2}{\tilde{p}_{A\cdot}(1 - \tilde{p}_{A\cdot})}$$

but the routine use of computers for data analysis makes this level of approximation unnecessary.

The estimates of F and θ can be combined to give a single estimate of the within-population coefficient f

$$\hat{f} = \frac{\hat{F} - \hat{\theta}}{1 - \hat{\theta}}$$

A separate estimate can be obtained for each of the populations using the methods described at the end of Chapter 2.

Effects of Evolutionary Forces

A random model allows statements to be made about evolutionary forces from estimated values of θ. Once an evolutionary model is specified, in principle it is possible to determine the distribution of θ values and then to compare this distribution to empirical distributions. In practice, it is difficult to derive algebraic results. The methods of Cockerham and Weir (1983) lead to expected variances of θ under models where the mating system and mutation or migration process are specified, but this is a long way from a complete distribution. The problems of determining the distributions of allele frequencies over populations were referred to in Chapter 2.

For (monoecious) populations mating at random, alleles are equally related whether they are within or between individuals. In this case $F = \theta$ or $f = 0$. Estimates of F and θ which differ significantly therefore may indicate departures from random mating within populations. Any avoidance of mating between relatives will cause F to be less than θ and f to be negative. In the language of variance components, the component for individuals within populations will then be negative. Recall that this component is actually the difference of two positive quantities. It is not being claimed that there is a variance which is negative. Different patterns of differences for the two estimates of F and θ at different loci indicate that there are forces other than nonrandom mating affecting these loci.

The effects of selection on the F statistics were also detailed by Cockerham (1973). Different selective forces in different populations, tending to increase their differences, will increase the value of θ. Within a population, Lewontin and Cockerham (1959) showed that selection at a locus gives negative f values unless the viability of a heterozygote is less than or equal to the geometric mean of the viabilities of the two homozygotes.

If forces such as mutation are involved, then the allele frequencies no longer remain constant over time, and it is not appropriate to regard p_A as a constant in the expectation of $\tilde{p}_{Ai}^2$. Instead, for haploid data, pairs of alleles can be compared within and between populations. In the notation of Cockerham and Weir (1987), Q_2 is the probability that two different alleles within a population are the same, and Q_3 is the probability that two alleles, one in each of two separate populations, are the same. These probabilities can be expressed in terms of indicator variables, letting x_{uij} take the value 1 if the jth allele from the ith population is A_u, and the value 0 otherwise. The similarity probabilities are therefore defined from the sum over all alleles

$$Q_2 = \sum_u \mathcal{E}(x_{uij}x_{uij'}), \ j' \neq j \qquad (5.4)$$

$$Q_3 = \sum_u \mathcal{E}(x_{uij}x_{ui'j'}), \ i' \neq i \qquad (5.5)$$

The quantities Q_2, Q_3 can be related to the variance components from the analysis of variance layout. Subscripting the variance components by u for allele A_u,

$$1 - Q_2 = \sum_u \sigma^2_{\mathrm{G}_u}$$
$$Q_2 - Q_3 = \sum_u \sigma^2_{\mathrm{P}_u}$$

The expected mean squares in Table 5.3, after summing over alleles, therefore can be written as

$$\mathcal{E}(\sum_u MSP_u) = (1 - Q_2) + n_c(Q_2 - Q_3)$$
$$\mathcal{E}(\sum_u MSG_u) = (1 - Q_2)$$

These similarity measures, in turn, can be related to descent measures, θ_2 for alleles within populations and θ_3 for alleles between populations. Until now populations have been assumed to be independent, so that there was no need for a θ_3, and θ_2 was written as θ (see Figure 5.1). The only estimable quantity is

$$\beta = \frac{\theta_2 - \theta_3}{1 - \theta_3} = \frac{Q_2 - Q_3}{1 - Q_3}$$

which is analogous to $f = F_{IS}$. The analogy should not be carried too far, since f was defined in terms of the relationship of pairs of alleles within and between individuals of the same population, whereas β refers to pairs of alleles within and between different populations. Cockerham and Weir (1987) discussed how β depends on the population size, number of populations, mutation rate and migration rate.

Without migration, $\theta_3 = 0$ and β reduces to θ, which had been introduced to accommodate drift. Regardless of the forces acting, β serves as a measure of differentiation between populations. However, Weir and Basten (1990) showed that, in a model with drift and mutation between a finite number of alleles, β could discriminate among populations for only a relatively short time.

With the measures Q_2, Q_3 defined in terms of similarity over all alleles at a locus, it may be more convenient to estimate them in the following way. Two statistics, X, Y, are defined in terms of sample allele frequencies $\tilde{p}_{ui}$ for allele A_u in the sample of size n_i alleles from population i

$$X = \sum_u \sum_{i=1}^{r} n_i \tilde{p}^2_{ui} \tag{5.6}$$

$$Y = \sum_u (\sum_{i=1}^{r} n_i \tilde{p}_{ui})^2 \tag{5.7}$$

Manipulating the mean squares in Table 5.3 provides the following unbiased estimates of the similarities

$$\hat{Q}_2 = \frac{X - r}{r(\bar{n} - 1)} \tag{5.8}$$

$$\hat{Q}_3 = \frac{1}{r(r-1)\bar{n}n_c}\left(Y - \frac{\bar{n}(n_c - 1)}{\bar{n} - 1}X\right) + \frac{\bar{n} - n_c}{n_c(\bar{n} - 1)}\left(1 - \frac{1}{r-1}X\right) \tag{5.9}$$

with great simplification when the samples are all of equal size, $n_i = n = \bar{n} = n_c$

$$\hat{Q}_2 = \frac{X - r}{r(n - 1)} \tag{5.10}$$

$$\hat{Q}_3 = \frac{1}{r(r-1)n^2}(Y - nX) \tag{5.11}$$

It may be more convenient, in the multiple alleles case, to use the statistics from Equations 5.6 and 5.7 in Equations 5.8 and 5.9 and then estimate θ from

$$\hat{\theta} = \frac{\hat{Q}_2 - \hat{Q}_3}{1 - \hat{Q}_3}$$

than to use the analysis of variance format in Table 5.3.

Suppose that there is no migration, but there is mutation of the ***infinite-alleles*** type, whereby every mutation is to a new allelic type. In a finite population of size N, an equilibrium will be established between the loss of variation by drift and the introduction of variation by mutation. At such an equilibrium, if the mutation rate is μ, the value of θ is given by

$$\theta = \frac{1}{1 + 4N\mu}$$

(Kimura and Crow 1964). Note that different populations are considered here to be independent.

It has often been suggested that migration rates could be estimated from θ values (e.g., Slatkin 1985). There is an infinite-island model, corresponding to the infinite-alleles mutation model. Each generation, any allele sampled from a population has probability m of having migrated from any one of an infinite number of other populations. When these various "islands" are of finite size N, an equilibrium is again established between loss of variation due to drift within islands, and gain of variation by migration from other islands. The equilibrium value of θ is simply

$$\theta = \frac{1}{1 + 4Nm}$$

While this suggests a means of estimating Nm, there are complications in practice because the infinite-island model is unrealistic. As soon as a finite number n of islands is postulated, there is a nonzero probability that two islands receive migrant alleles from the same island. The island populations are not independent and it is necessary to distinguish between θ_2 and θ_3. The quantity β can be estimated, and was shown by Cockerham and Weir (1987) to have expectation in an equilibrium population of

$$\beta = \frac{\rho d}{2N(1 - \rho d) + \rho d}$$

where for mutation rate μ, $\rho = (1 - \mu)^2, d = (1 - m\alpha)^2, \alpha = n/(n - 1)$. For a large number of islands

$$\beta \approx \frac{1}{1 + 4N\mu + 4Nm}$$

which reduces to the $1/(1 + 4Nm)$ value given above when the mutation rate is small. Cockerham and Weir (1993) showed that $[(1/\hat{\beta}) - 1]$ performs satisfactorily as an estimator of $4Nm$. A related discussion, of using F_{ST} estimated from DNA sequence data to estimate gene flow, was given by Hudson et al. (1992).

The drawback of using θ or related quantities to estimate migration rates is that it is assumed that the pattern of allele frequencies among populations is due to gene flow. In the absence of direct observations on gene flow, this may be the only way of estimating migration rates, but it does mean that any other evolutionary scenario that would give the same pattern of allele frequencies cannot be eliminated.

POPULATION SUBDIVISION

It is often the case that individuals are sampled in a nested classification such as transects and sites within transects (Barker et al. 1986), or even more complex structures such as drainages, rivers within drainages, and sites within rivers (P. Kukuk, personal communication). Each recognizable level of such hierarchies adds another level to the degrees of relatedness of alleles. Appropriate nested analysis of variance layouts suggest the procedures for estimating the various measures of differentiation. Details will be given now for three- and four-level hierarchies.

Table 5.5 Analysis of variance layout for a three-level sampling hierarchy in random populations.

Source	d.f.	M.S.	Expected M.S.*
Between populations	$r-1$	*MSP*	$(1-F)+2(F-\theta_S)$ $+2n_{c1}(\theta_S-\theta_P)+2n_{c2}\theta_P$
Subpopulations in populations	$\sum_{i=1}^r (s_i-1)$ $= s.-r$	*MSS*	$(1-F)+2(F-\theta_S)$ $+2n_{c3}(\theta_S-\theta_P)$
Individuals in subpopulations	$\sum_{i=1}^r \sum_{j=1}^{s_i} (n_{ij}-1)$ $= n..-s.$	*MSI*	$(1-F)+2(F-\theta_S)$
Alleles in individuals	$\sum_{i=1}^r \sum_{j=1}^{s_i} n_{ij}$ $= n..$	*MSG*	$(1-F)$

*To be multiplied by $p_A(1-p_A)$. n_{c1}, n_{c2}, n_{c3} are defined in the text.

Three-Level Hierarchy

Suppose that n_{ij} individuals are sampled from the jth subpopulation of the ith population, and that there are s_i subpopulations sampled from the ith population. There are still r populations. Pairs of alleles may be related by virtue of being within individuals, between individuals within subpopulations, or between subpopulations within populations. The three degrees of relatedness are quantified by F, θ_S, and θ_P, respectively, and these parameters give the measures of differentiation among individuals, subpopulations, or populations. Alleles in different populations are assumed to be unrelated and all alleles sampled have the same expectation, p_A, of being of type A. The analysis of variance layout is given in Table 5.5.

Although no explicit use of them will be made, components of variance can be defined

$$\sigma_P^2 = p_A(1-p_A)\theta_P \text{ for populations}$$

$$\sigma_S^2 = p_A(1-p_A)(\theta_S-\theta_P) \text{ for subpopulations within populations}$$

$$\sigma_I^2 = p_A(1-p_A)(F-\theta_S) \text{ for individuals within subpopulations}$$

and

$$\sigma_G^2 = p_A(1-p_A)(1-F) \text{ for alleles within individuals}$$

Calculations of the mean squares in Table 5.5 make use of the following sums:

$$n_{i\cdot} = \sum_{j=1}^{s_i} n_{ij}, \; n_{\cdot\cdot} = \sum_{i=1}^{r} n_{i\cdot}, \; s_{\cdot} = \sum_{i=1}^{r} s_i$$

$$n_{c1} = \frac{1}{r-1}\sum_{i=1}^{r}\sum_{j=1}^{s_i}\left[\frac{(n_{\cdot\cdot}-n_{i\cdot})n_{ij}^2}{n_{i\cdot}n_{\cdot\cdot}}\right]$$

$$n_{c2} = \frac{1}{r-1}\left(n_{\cdot\cdot} - \frac{1}{n_{\cdot\cdot}}\sum_{i=1}^{r} n_{i\cdot}^2\right)$$

$$n_{c3} = \frac{1}{s_{\cdot}-r}\left[n_{\cdot\cdot} - \sum_{i=1}^{r}\sum_{j=1}^{s_i}\left(\frac{n_{ij}^2}{n_{i\cdot}}\right)\right]$$

and mean allele frequencies

$$\tilde{p}_{Ai\cdot} = \frac{1}{n_{i\cdot}}\sum_{j=1}^{s_i} n_{ij}\tilde{p}_{Aij}$$

$$\tilde{p}_{A\cdot\cdot} = \frac{1}{n_{\cdot\cdot}}\sum_{i=1}^{r}\sum_{j=1}^{s_i} n_{ij}\tilde{p}_{Aij} = \frac{1}{n_{\cdot\cdot}}\sum_{i=1}^{r} n_{i\cdot}\tilde{p}_{Ai\cdot}$$

where $\tilde{p}_{Aij}$ is the allele frequency of A in subpopulation j of population i.

The four mean squares use the allele frequencies and the heterozygote frequencies: $\tilde{H}_{Aij}$ is the frequency in the sample from subpopulation j of population i of heterozygous individuals that contain allele A.

$$MSP = \frac{2}{r-1}\sum_{i=1}^{r} n_{i\cdot}(\tilde{p}_{Ai\cdot} - \tilde{p}_{A\cdot\cdot})^2$$

$$MSS = \frac{2}{s_{\cdot}-r}\sum_{i=1}^{r}\sum_{j=1}^{s_i} n_{ij}(\tilde{p}_{Aij} - \tilde{p}_{Ai\cdot})^2$$

$$MSI = \frac{2}{n_{\cdot\cdot}-s_{\cdot}}\sum_{i=1}^{r}\sum_{j=1}^{s_i} n_{ij}[\tilde{p}_{Aij}(1-\tilde{p}_{Aij}) - \frac{1}{4}\tilde{H}_{Aij}]$$

$$MSG = \frac{1}{2n_{\cdot\cdot}}\sum_{i=1}^{r}\sum_{j=1}^{s_i} n_{ij}\tilde{H}_{Aij}$$

The method of moments provides estimators for each of the three measures of differentiation:

$$\hat{F} = 1 - \frac{R_4}{R_2}$$

$$\hat{\theta}_S = \frac{R_3}{R_2}$$

$$\hat{\theta}_P = \frac{R_1}{R_2}$$

where

$$R_1 = \frac{MSP - MSI}{2n_{c2}} - \frac{n_{c1}(MSS - MSI)}{2n_{c2}n_{c3}}$$

$$R_2 = \frac{MSP - MSI}{2n_{c2}} + \frac{(n_{c2} - n_{c1})(MSS - MSI)}{2n_{c2}n_{c3}} + \frac{MSI + MSG}{2}$$

$$R_3 = \frac{MSP - MSI}{2n_{c2}} + \frac{(n_{c2} - n_{c1})(MSS - MSI)}{2n_{c2}n_{c3}}$$

$$R_4 = MSG$$

The combination over alleles and loci proceeds as before, as does numerical resampling over loci or populations.

There is an alternative to using these algebraic expressions for the various mean squares and subsequent estimates of measures of differentiation. Every individual in the data set could be coded as 1,1 for AA homozygotes, 1,0 for heterozygotes for allele A, and 0,0 for all other genotypes. This is equivalent to introducing the indicator variable x_{ijkl} for the lth allele ($l = 1, 2$), in the kth individual ($k = 1, 2, \ldots, n_{ij}$), from the jth subpopulation ($j = 1, 2, \ldots, s_i$), of the ith population ($i = 1, 2, \ldots, r$). The variable takes the value 1 if that allele is of type A, and the value 0 otherwise. Table 5.5 can be produced by a statistical computer package as a nested analysis of variance for the x's, and such packages generally allow the estimation of components of variance, even for unbalanced data. With the components of variance estimated, it is not difficult to recover the measures of differentiation.

Four-Level Hierarchy

As a final example, consider the case of individuals (indexed by l) sampled from subsubpopulations (indexed by k) within subpopulations (indexed by

Table 5.6 Analysis of variance layout for a four-level sampling hierarchy in random populations.

Source	d.f.*	Mean Square	Expected Mean Square†
Between populations	$r-1$	*MSP*	$(1-F)+2(F-\theta_{SS})$ $+2n_{c1}(\theta_{SS}-\theta_{\rm S})+2n_{c2}(\theta_{\rm S}-\theta_{\rm P})$ $+2n_{c3}\theta_{\rm P}$
Subpopulations in populations	$s.-r$	*MSS*	$(1-F)+2(F-\theta_{SS})$ $+2n_{c4}(\theta_{SS}-\theta_{\rm S})+2n_{c5}(\theta_{\rm S}-\theta_{\rm P})$
Subsubpopula– tions in subpopulations	$t..-s.$	*MSSS*	$(1-F)+2(F-\theta_{SS})$ $+2n_{c6}(\theta_{SS}-\theta_{\rm S})$
Individuals in subsubpop– ulations	$u...-t..$	*MSI*	$(1-F)+2(F-\theta_{SS})$
Alleles in individuals	$u...$	*MSG*	$(1-F)$

†To be multiplied by $p_A(1-p_A)$. $n_c's$ are defined in the text.
*$s. = \sum_{i=1}^{r} s_i$, $t.. = \sum_{i=1}^{r}\sum_{j=1}^{s_i} t_{ij}$, $u... = \sum_{i=1}^{r}\sum_{j=1}^{s_i}\sum_{k=1}^{t_{ij}} u_{ijk}$

j), within populations (indexed by i). There are r populations, s_i subpopulations from the ith population, t_{ij} subsubpopulations from the jth subpopulation of the ith population, and u_{ijk} individuals sampled from that subsubpopulation. The analysis of variance layout is shown in Table 5.6.

The variance components are

$$\sigma^2_{\rm P} = p_A(1-p_A)\theta_{\rm P} \text{ for populations}$$

$$\sigma^2_{\rm S} = p_A(1-p_A)(\theta_{\rm S}-\theta_{\rm P}) \text{ for subpopulations in populations}$$

$$\sigma^2_{\rm SS} = p_A(1-p_A)(\theta_{\rm SS}-\theta_{\rm S}) \text{ for subsubpopulations in subpopulations}$$

$$\sigma^2_{\rm I} = p_A(1-p_A)(F-\theta_{\rm SS}) \text{ for individuals in subsubpopulations}$$

$$\sigma^2_{\mathrm{G}} = p_A(1-p_A)(1-F) \text{ for alleles within individuals}$$

The coefficients for these components in the mean squares are

$$\begin{aligned} n_{c1} &= (C_2 - C_3)/(r-1) \quad , \quad n_{c2} = (C_4 - C_5)/(r-1) \\ n_{c3} &= (u_{...} - C_6)/(r-1) \quad , \quad n_{c4} = (C_1 - C_2)/(s_{.} - r) \\ n_{c5} &= (u_{...} - C_5)/(s_{.} - r) \quad , \quad n_{c6} = (u_{...} - C_1)/(t_{..} - s_{.}) \end{aligned}$$

where

$$C_1 = \sum_{i=1}^{r}\sum_{j=1}^{s_i}\left[(\sum_{k=1}^{t_{ij}} u_{ijk}^2)/u_{ij.}\right] \quad , \quad C_2 = \sum_{i=1}^{r}\left[(\sum_{j=1}^{s_i}\sum_{k=1}^{t_{ij}} u_{ijk}^2)/u_{i..}\right]$$

$$C_3 = (\sum_{i=1}^{r}\sum_{j=1}^{s_i}\sum_{k=1}^{t_{ij}} u_{ijk}^2)/u_{...} \quad , \quad C_4 = \sum_{i=1}^{r}\left[(\sum_{j=1}^{s_i} u_{ij.}^2)/u_{i..}\right]$$

$$C_5 = (\sum_{i=1}^{r}\sum_{j=1}^{s_i} u_{ij.}^2)/u_{...} \quad , \quad C_6 = (\sum_{i=1}^{r} u_{i..}^2)/u_{...}$$

and dots indicate summation over subscripts for s, t, u.

It is informative to express the mean squares in a way that reveals the nature of the corresponding sums of squares:

$$\begin{aligned} MSP &= (S_1 - C)/(r-1) \\ MSS &= (S_2 - S_1)/(s_{.} - r) \\ MSSS &= (S_3 - S_2)/(t_{..} - s_{.}) \\ MSI &= (S_4 - S_3)/(u_{...} - t_{..}) \\ MSG &= (S_5 - S_4)/(u_{...}) \end{aligned}$$

The analysis may be formed in terms of indicator variables x_{ijklm} for the mth allele in the lth individual of the kth subsubpopulation from the jth subpopulation of the ith population. As previously, these variables take the value 1 if that allele is allele A, and have the value 0 otherwise. A statistical computer package will produce the structure of Table 5.6 if the genotypes are recoded as 1,1 for AA; 1,0 for A heterozygotes; and 0,0 for all other genotypes at that locus. Alternatively, a program can be written that uses the indicator variables, and for which the above sums of squares are

$$C = x_{.....}^2/2u_{...}$$

$$
\begin{aligned}
S_1 &= \sum_{i=1}^{r} \left(x_{i....}^2 / 2u_{i..} \right) \\
S_2 &= \sum_{i=1}^{r} \sum_{j=1}^{s_i} \left(x_{ij...}^2 / 2u_{ij.} \right) \\
S_3 &= \sum_{i=1}^{r} \sum_{j=1}^{s_i} \sum_{k=1}^{t_{ij}} \left(x_{ijk..}^2 / 2u_{ijk} \right) \\
S_4 &= \sum_{i=1}^{r} \sum_{j=1}^{s_i} \sum_{k=1}^{t_{ij}} \sum_{l=1}^{u_{ijk}} \left(x_{ijkl.}^2 / 2 \right) \\
S_5 &= \sum_{i=1}^{r} \sum_{j=1}^{s_i} \sum_{k=1}^{s_{ij}} \sum_{l=1}^{u_{ijk}} \sum_{m=1}^{2} x_{ijklm}^2 = x_{.....}
\end{aligned}
$$

It is also possible to express the sums of squares in terms of average allele frequencies in various levels of the hierarchy. Within one subsubpopulation, the sample frequencies of A and heterozygotes for A are $\tilde{p}_{Aijk}$ and $\tilde{H}_{Aijk}$. Within subpopulations, within populations, or over the whole data set, the average sample allele frequencies are

$$
\begin{aligned}
\tilde{p}_{Aij\cdot} &= \sum_{k=1}^{t_{ij}} u_{ijk} \tilde{p}_{Aijk} / \sum_{k=1}^{t_{ij}} u_{ijk} \\
\tilde{p}_{Ai\cdot\cdot} &= \sum_{j=1}^{s_i} t_{ij} \tilde{p}_{Aij\cdot} / \sum_{j=1}^{s_i} t_{ij} \\
\tilde{p}_{A\cdots} &= \sum_{i=1}^{r} s_i \tilde{p}_{Ai\cdot\cdot} / \sum_{i=1}^{r} s_i
\end{aligned}
$$

The sums of squares become

$$
\begin{aligned}
S_1 - C &= 2 \sum_{i=1}^{r} s_i (\tilde{p}_{Ai\cdot\cdot} - \tilde{p}_{A\cdots})^2 \\
S_2 - S_1 &= 2 \sum_{i=1}^{r} \sum_{j=1}^{s_i} t_{ij} (\tilde{p}_{Aij\cdot} - \tilde{p}_{Ai\cdot\cdot})^2 \\
S_3 - S_2 &= 2 \sum_{i=1}^{r} \sum_{j=1}^{s_i} \sum_{k=1}^{t_{ij}} u_{ijk} (\tilde{p}_{Aijk} - \tilde{p}_{Aij\cdot})^2
\end{aligned}
$$

$$S_4 - S_3 = 2\sum_{i=1}^{r}\sum_{j=1}^{s_i}\sum_{k=1}^{t_{ij}} u_{ijk}[\tilde{p}_{Aijk}(1-\tilde{p}_{Aijk}) - \frac{1}{4}\tilde{H}_{Aijk}]$$

$$S_5 - S_4 = \frac{1}{2}\sum_{i=1}^{r}\sum_{j=1}^{s_i}\sum_{k=1}^{t_{ij}} u_{ijk}\tilde{H}_{Aijk}$$

GENETIC DISTANCE

When genetic data are available from several populations, it is natural to ask how genetically similar are the populations. Generally, genetic distance is thought of as being related to the time since the populations being compared diverged from a single ancestral population. This, in turn, requires a genetic model specifying the processes such as mutation and drift causing the populations to diverge. Genetic distances must therefore satisfy criteria different than those for geometric distances, but the latter kind will be reviewed first.

The ***geometric distance*** d_{PQ} between points P and Q in Euclidean space is a quantity that satisfies three axioms:

- $d_{PQ} \geq 0, d_{PP} = 0$

  distances are not negative

- $d_{PQ} = d_{QP}$

  distances are symmetric

- $d_{PQ} + d_{QR} \geq d_{PR}$

  distances satisfy the triangle inequality

The most common distance satisfying these axioms is the ***Euclidean distance***, defined in terms of the coordinates $\{p_i\}$, $\{q_i\}$ of points P and Q in n-dimensional space as

$$d_{PQ} = \sqrt{\sum_{i=1}^{n}(p_i - q_i)^2}$$

Genetic distances are designed to express the genetic differences between two populations as a single number (Smith 1977). If there are no differences, the distance could be set to zero, whereas if the populations have no alleles in common at any locus the distance may be set equal to its maximum value, say 1. Distances may be regarded simply as data reduction devices,

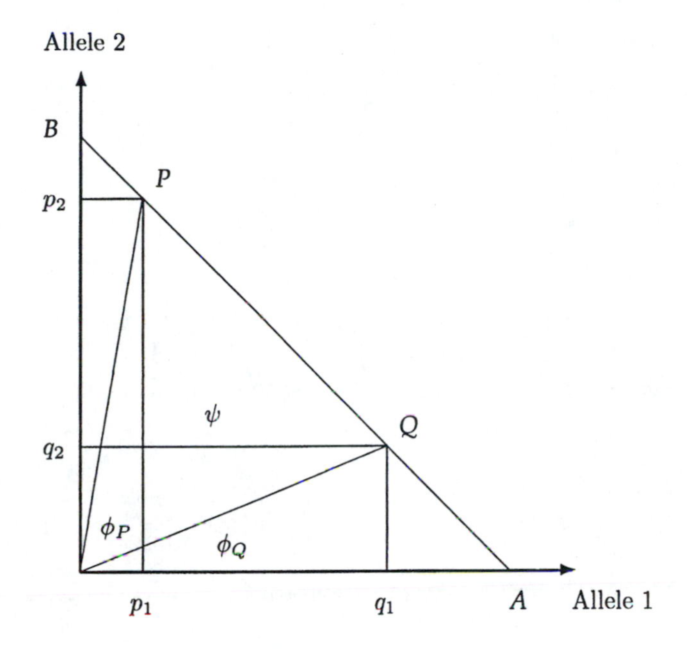

Figure 5.2 Representation of two populations in allele frequency space. Each population is represented by a point on the line from A:(1,0) to B:(0,1).

or as a means of comparing pairs of extant populations, or as the basis for constructing evolutionary histories for the populations. These different goals require different distances and, especially for inferring phylogenies, require genetic models. A review was given by Reynolds (1981), and some of the genetic distances that have been proposed are now considered.

Geometric Distances

Simply as a means of data reduction, a population may be represented as a point in v-dimensional space on the basis of frequencies of the v alleles at a locus. All populations lie on the hyperplane defined by $\sum_{u=1}^{v} p_u = 1$. This is illustrated in two dimensions in Figure 5.2.

The Euclidean distance between the populations in Figure 5.2 is expressed in terms of the allele frequencies. For a locus with two alleles, if population P has allele frequencies p_1 and $p_2 = 1 - p_1$ and population Q has frequencies q_1, q_2, then the Euclidean distance between the populations, based on that locus, is

$$d_{PQ} = \sqrt{[(p_1 - q_1)^2 + (p_2 - q_2)^2]}$$

or, for v alleles,

$$d_{PQ} = \sqrt{\left[\sum_{u=1}^{v} (p_u - q_u)^2\right]}$$

An alternative measure is based on the angle between the lines joining the origin to the points representing the populations. If ϕ_P and ϕ_Q are the angles between the allele-1 axis and these two lines, the angle ψ between the two lines is $(\phi_P - \phi_Q)$ and has cosine

$$\begin{aligned}
\cos(\psi) &= \cos(\phi_P - \phi_Q) \\
&= \cos(\phi_P)\cos(\phi_Q) + \sin(\phi_P)\sin(\phi_Q) \\
&= \frac{\sum_{u=1}^{2} p_u q_u}{\sqrt{\sum_{u=1}^{2} p_u^2}\sqrt{\sum_{u=1}^{2} q_u^2}}
\end{aligned}$$

A more appropriately scaled measure would be $1 - \cos(\psi)$, which takes the value 0 when P and Q have the same frequencies, $P \equiv Q$, and the value 1 when there are no alleles in common so that $\sum_u p_u q_u = 0$.

These geometric distances can also be established when the square roots of allele frequencies are used as coordinates. Populations now lie on the surface of a hypersphere with radius 1 as illustrated in Figure 5.3 for two alleles. The cosine of ψ with this scaling is

$$\cos(\psi) = \sum_u \sqrt{(p_u q_u)}$$

and $1 - \cos(\psi)$ may again be used as a distance measure, as may ψ itself, or as may the chord length $\sqrt{[2 - 2\cos(\psi)]}$ between points P and Q. Such distance measures were used by Cavalli-Sforza and Bodmer (1971) .

If a sample of size n, drawn from a population with frequencies p_u, has frequencies $\tilde{p}_u$, the geometric distance between sample and population can be based on

$$\begin{aligned}
\cos(\psi) &= \sum_u \sqrt{p_u \tilde{p}_u} \\
&= \sum_u p_u \sqrt{1 + \frac{\tilde{p}_u - p_u}{p_u}}
\end{aligned}$$

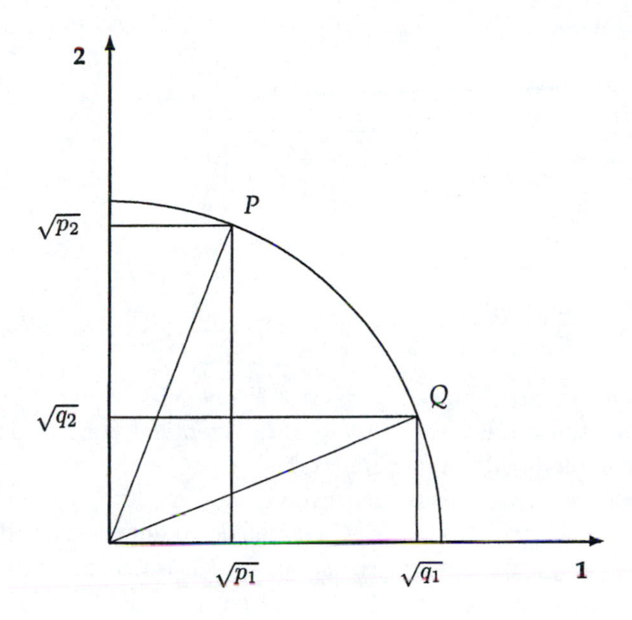

Figure 5.3 Representation of two populations in square-root allele frequency space. Populations represented by points on a quarter-circle centered on the origin.

$$\approx \sum_u p_u \left[1 + \frac{1}{2}\left(\frac{\tilde{p}_u - p_u}{p_u}\right) - \frac{1}{8}\left(\frac{\tilde{p}_u - p_u}{p_u}\right)^2 \right]$$
$$= 1 - \frac{1}{8}\frac{X^2}{n}$$

where X^2 is the chi-square goodness-of-fit statistic

$$X^2 = \sum_u \frac{(n\tilde{p}_u - np_u)^2}{np_u}$$

for testing agreement of sample to population frequencies. If the sample allele counts $n\tilde{p}_u$ are multinomially distributed, then X^2 is distributed as chi-square with $(v-1)$ d.f. It is also possible to express $\cos(\psi)$ as

$$\cos(\psi) = 1 - 2\sin^2\left(\frac{\psi}{2}\right)$$
$$\approx 1 - 2\left(\frac{\psi}{2}\right)^2$$

to establish

$$\psi^2 \approx \frac{1}{4n}X^2$$

A similar argument holds for the distance between two populations:

$$\begin{aligned}
\cos(\psi) &= \sum_u \sqrt{p_u q_u} \\
&= \frac{1}{2}\sum_u (p_u + q_u)\sqrt{\left[1 - \left(\frac{p_u - q_u}{p_u + q_u}\right)^2\right]}
\end{aligned}$$

so that

$$\psi^2 \approx \frac{1}{2}\sum_u \left[\frac{(p_u - q_u)^2}{p_u + q_u}\right]$$

Population frequencies are replaced by sample frequencies to provide sample distances. This distance takes the value of 0 for identical populations, and the value 1 for completely distinct populations.

None of these geometric distances involves any evolutionary concepts other than the general one that more similar allele frequencies imply more genetic similarity. To give a meaning to "similarity" it is necessary to invoke an evolutionary model and, for example, to say that more similar populations are fewer generations removed from a common ancestral population.

Coancestry as Distance

For a finite population of size N mating at random, including a random amount of selfing, the inbreeding and coancestry coefficients are equal, $F = \theta$, and these values change over time according to

$$\theta_{t+1} = \frac{1}{2N} + \left(1 - \frac{1}{2N}\right)\theta_t$$

If the initial value θ_0 is zero

$$\theta_t = 1 - \left(1 - \frac{1}{2N}\right)^t$$

This shows that an estimate of θ from two or more populations of the same size and same history that have been diverging because of drift will give information about t, the time since the populations diverged. Specifically

$$\begin{aligned}
d &= -\ln(1 - \theta) \\
&= -t\ln\left(1 - \frac{1}{2N}\right) \\
&\approx \frac{t}{2N}
\end{aligned}$$

is an appropriate distance for divergence due to drift. Estimation can proceed by the variance-component method discussed earlier in this chapter. The analysis shown in Table 5.3 is used since it is being assumed $F = \theta$. For equal-sized samples from two populations in which the frequencies of allele A_u at locus l are $\tilde{p}_{lu_1}$ and $\tilde{p}_{lu_2}$, the estimator becomes

$$\hat{\theta} = \frac{\sum_l \left\{\frac{1}{2}\sum_u(\tilde{p}_{lu_1} - \tilde{p}_{lu_2})^2 - \frac{1}{2(2n-1)}\left[2 - \sum_u(\tilde{p}^2_{lu_1} + \tilde{p}^2_{lu_2})\right]\right\}}{\sum_l (1 - \sum_u \tilde{p}_{lu_1}\tilde{p}_{lu_2})} \quad (5.12)$$

This corrects a typographical error in the expression of Reynolds et al. (1983). Note that the estimator $\hat{\theta}$ explicitly involves the sample size n, although the parameter θ being estimated does not.

If only one locus is scored, and the two populations are fixed for different alleles, each sample has only one nonzero frequency, but for different u values. The estimated distance is then equal to 1. The populations cannot diverge any further. If the two samples have no alleles in common but each has more than one allele, the estimated distance is less than 1, reflecting the fact that drift has not yet caused fixation for one allele per locus. When each of the two samples has only the same single allele at a single scored locus, the estimator is undefined, since Equation 5.12 becomes 0/0. This result occurs because it cannot be determined from observing two samples fixed for the same allele whether the two populations have just become homozygous or have been homozygous for some time that may have extended back to the ancestral population. It is unlikely that this will cause a problem in practice, since more than one locus will be used in constructing the estimator and it is unlikely that all loci will be invariant for the same allele in both samples. If they were, however, there is no information in the data about genetic distance between those populations.

Nei's Genetic Distance

The most widely used measure of genetic distance was proposed by Nei (1972). With the same notation as in the previous section, a "genetic identity" I is defined between the two populations as

$$I = \frac{\sum_l \sum_u \tilde{p}_{lu_1}\tilde{p}_{lu_2}}{\sqrt{\sum_l \sum_u \tilde{p}^2_{lu_1} \sum_l \sum_u \tilde{p}^2_{lu_2}}}$$

with some similarity to the geometric measure $\cos(\psi)$. When a bias correction (Nei 1978) is included

$$I = \frac{(2n-1)\sum_l \sum_u \tilde{p}_{lu_1}\tilde{p}_{lu_2}}{\sqrt{\sum_l(2n\sum_u \tilde{p}^2_{lu_1} - 1)\sum_l(2n\sum_u \tilde{p}^2_{lu_2} - 1)}} \quad (5.13)$$

and Nei's standard genetic distance D is

$$D = -\ln(I)$$

The quantity I is a ratio of the proportions of alleles that are alike between (q) and within (Q) populations, and Equation 5.13 can be written in terms of the sample proportions $\tilde{Q}$ and $\tilde{q}$:

$$I = \frac{\tilde{q}_{12}}{\sqrt{\tilde{Q}_1\tilde{Q}_2}}$$

Cockerham (1984) derived transition equations for the corresponding parameters Q and q. For a random mating population of finite size N, and for loci at which there is mutation away from any allele at rate μ per generation and mutation to any other specific allele at rate ν ($\nu = 0$ for the ***infinite allele*** model), in generation t

$$Q_t = Q^* - (Q^* - Q_0)\lambda_Q^t$$
$$q_t = q^* - (q^* - q_0)\lambda_q^t$$

where $Q_0 = q_0$ are the values at time zero, before the two populations diverged, and Q^*, q^* are the final, or equilibrium, values given by

$$Q^* = \frac{1 + 4N\nu}{1 + 4N\mu + 4N\nu} \ , \ q^* = \frac{\nu}{\mu + \nu}$$

The two rates of change are

$$\lambda_Q = 1 - 2\mu - 2\nu - \frac{1}{2N} \ , \ \lambda_q = 1 - 2\mu - 2\nu$$

all supposing that squares and products of μ, ν, and $1/2N$ and smaller terms can be ignored. The statistics $\tilde{Q}$ and $\tilde{q}$ are unbiased for the parameters Q and q. To see what genetic parameters are being estimated by Nei's distance, approximate the expectation of the ratio in Equation 5.13 by

$$\mathcal{E}(I) = \frac{q}{Q} = \frac{q^* - (q^* - Q_0)\lambda_q^t}{Q^* - (Q^* - Q_0)\lambda_Q^t}$$

It is reasonable to assume the ancestral population was in equilibrium for the joint forces of drift and mutation, and that each population separately remains in equilibrium over time, $Q_t = Q^* = Q_0$. Then

$$\mathcal{E}(I) = \frac{q^*}{Q^*} + \left(1 - \frac{q^*}{Q^*}\right)\lambda_q^t$$

Although this has a finite final value, $-\ln I$ does increase approximately linearly with time until t becomes moderately large (Weir and Basten 1990). There is greater simplification with the infinite alleles model, for then $q^* = 0$ and

$$\mathcal{E}(I) = \lambda_q^t$$

so that Nei's distance has an expected value of

$$\mathcal{E}(D) \approx 2\mu t$$

which, as desired, is linearly related to the time since divergence.

This discussion of Nei's distance and the coancestry distance has included explicit statements of the underlying genetic models. Nei's distance is appropriate for long-term evolution when populations diverge because of drift and mutation. The distance is proportional to the time since divergence in the special case of the infinite alleles mutation model and equilibrium in the ancestral population. The coancestry distance is appropriate for divergence due to drift only, and no assumptions need be made about the ancestral population. Neither distance should be used in the situation for which it was not intended.

When there is no mutation, the proportions of like alleles change over time according to

$$\begin{aligned} Q_t &= 1-(1-Q_0)\left(\frac{2N-1}{2N}\right)^t \\ q_t &= Q_0 \end{aligned}$$

with Q_0 equal to the sum of squares of allele frequencies in the ancestral population. In this drift situation Nei's distance is estimating

$$\mathcal{E}(D) \approx \ln\left[\frac{1}{Q_0} - \left(\frac{1}{Q_0}-1\right)\left(1-\frac{1}{2N}\right)^t\right]$$

which is confounded by the unknown Q_0. This dependence on initial allele frequencies is also a feature of Rogers' (1972) distance

$$D_R = \frac{1}{m}\sum_l \left[\frac{1}{2}\sum_u (\tilde{p}_{1lu} - \tilde{p}_{2lu})^2\right]^{1/2}$$

Goodman (1972) pointed out that equal geometric distance may not indicate equal biological divergence, and a similar lack of correspondence between distance and degree of divergence can occur if mutation-based distances are applied to drift situations.

An alternative estimate of θ for the pure drift situation is provided by one of the geometric measures, as proposed by Balakrishnan and Sanghvi (1968). If locus l has v_l alleles, their measure G^2 is

$$G^2 = \frac{1}{\sum_l (v_l - 1)} \sum_l \sum_u \left[\frac{(\tilde{p}_{lu_1} - \tilde{p}_{lu_2})^2}{\tilde{p}_{lu_1} + \tilde{p}_{lu_2}} \right]$$

Taking expectations for each allele and summing over all alleles in the case of equal sample sizes leads to

$$\mathcal{E}(G^2) = \frac{1}{2n} + \frac{2n-1}{2n}\theta$$

so that $-\ln(1 - G^2)$ will estimate θ for large n.

Variance of Distance Estimates

All of the distance estimators discussed here are ratios of quadratic functions of allele frequencies and the methods of numerical resampling appear to be the best for providing estimates of sampling variances. Resampling over loci is expected to mimic the genetic sampling process, and so will lead to appropriate variances when distances are being used in a random population framework.

If only the variances among repeated samples from the same populations are wanted, then the delta method discussed in Chapter 2 could be used. Because the distances are not homogeneous functions of degree zero in the genotypic counts, Fisher's variance formula cannot be used. Of course, numerical resampling over individuals (or alleles) within the two populations is also appropriate in this fixed population framework.

SUMMARY

When observations are collected from several populations, it is possible to estimate measures of population structure. Specifically, measures of association between pairs of alleles in various levels of a hierarchy can be estimated. When data are available from only a single population it is not possible to estimate the effects of genetic sampling between populations unless different loci can be regarded as being independent. If inferences are to be made on only the population(s) sampled, it is not necessary to invoke the variance-based measures represented by F statistics. Analyses of mean allele frequencies are appropriate.

A useful quantification of the structure in a population is provided by genetic distances between components of the population. Again, care is needed to match the distance to the intended scope of inference.

EXERCISES

Exercise 5.1

The following set of data are gametes, or haplotypes, at five loci A, B, C, D, E in samples from six populations. Each line represents one gamete.

Popn.	A	B	C	D	E	Popn.	A	B	C	D	E
1	4	3	3	3	4	4	4	4	3	4	4
1	4	4	3	3	4	4	4	4	3	3	4
1	4	4	3	3	4	4	4	4	3	3	4
1	4	4		3	4	4	4	4	3	4	4
1	4	4	2	4	4	4	4	4	4	3	4
1	4	4	4	4	4	4	4	4	4	4	4
1	4	4	3	4	4						
1	4	4	3	4	4						
1	4	4		3	4	5	4	4	4	1	4
1	4	4	3	3	4	5	4	4	4	3	4
						5	4	4	2	3	4
2	4	4	2	2	4	5	4	4	3	3	4
2	4	3	4	3	4	5	4	4	3	3	4
2	4	4	3	3	4	5	4	4	4	4	4
2	4	4	4	4	4	5	4	4	3	3	4
2	4	3	4	4	4	5	4	4	4		4
2	4	4	4	2	4						
2	4	4	3	3	4						
2	4	4	4	4	4	6	4	4	4	3	4
						6	4	4	3	3	4
3	4	4	4	3	4	6	4	4	2	2	4
3	4	4	4	3	4	6	4	4	3	1	4
3	4	3	4	4	4	6	4	4	4	4	4
3	4	4	4	4	4	6	4	4	4	4	4
3	4	4	3	1	4	6	4	4	4	1	4
3	4	4	3	3	4						

For each locus, test for differences in allele frequencies among the six populations using a fixed-population chi-square test statistic. For a random-population analysis, estimate θ for each locus, and over all loci.

Exercise 5.2

For the data in Exercise 5.1, use an exact test to test the hypothesis that populations from which samples 1 and 2 were drawn have the same allele frequencies at locus B.

Exercise 5.3

A locus with two alleles, A, a, has been scored in four populations. The four sets of allele frequencies are listed below. Ignoring sample size corrections, calculate the Euclidean distance, Nei's standard distance and the coancestry-based distance between populations 1 and 2 and between populations 3 and 4. Discuss the differences among these three pairs of distances.

Population	A	a	Population	A	a
1	0.55	0.45	3	0.95	0.05
2	0.45	0.55	4	0.85	0.15

Exercise 5.4

The two sets of data on the next page are the 11-locus genotypes for two populations (Ghana and N'goye) of the mosquito *Aedes aegypti* (J. Powell, personal communication). Using all the data, estimate the three components of variance (between populations, between individuals within populations, and between alleles within individuals) for each allele, and hence estimate F and θ.

Ghana

1	2	3	4	5	6	7	8	9	10	11
11	11	11	12	11	34	12	11	11	11	11
11	11	11	12	11	22	12	11	11	11	11
11	11	11	12	11	23	11	11	11	11	11
11	11	11	11	11	11	11	11	11	11	11
11	11	11	12	11	34	12	11	11	11	
11	11	11	11	11	22	11	11	11	11	
11	11	11	11	11	11	11	11	11	11	
11	11	11	12	11	24	12	11	11	11	
11	11	11	11	12	11	11	11	11	11	
11	11	11	11	11	12	11	11	11	11	11
11	11	11	12	11	24	11	11	11	11	11
11	11	11	12	11	24	12	11	11	11	11
11	11	11	12	11	34	11	11	11	11	11
11	11	11	11	11	11	11	11	11	11	
11	11	11	12	11	22	12	11	11	11	
11	11	11	12	11	34	11	11	11	11	11
11	11	11	12	11	23	12	11	11	11	11
11	11	11	12	11	24	11	11	11	11	11
11	11	11	12	11	22	11	11	11	11	11
11	11	11	12	11	34	11	11	11	11	11
11	11	11	11	12	11	11	11	11	11	11
11	11	11	12	11	23	12	11	11	11	11
11	11	11	12	11	34	11	11	11	11	11
11	11	11	11	11	11	11	11	11	11	11
11	11	11	11	11	34	11	11	11	11	11
11	11	11	11	11	24	12	11	11	11	11
11	11	11	11	11	11	11	11	11	11	11
11	11	11	11	11	34	12	11	11	11	11
11	11	11	11	11	24	12	11	11	11	11
11	11	11	11	11	24	11	11	11	11	11
11	11	11	11	11	23	11	11	11	11	11
11	11	11	11	11	22	11	11	11	11	11
11	11	11	11	11	23	11	11	11	11	11
11	11	11	11	11	11	11	11	11	11	11
11	11	11	11	11	34	12	11	11	11	11
11	11	11	11	11	24	11	11	11	11	11
11	11	11	11	11	34	11	11	11	11	11
11	11	11	11	11	23	12	11	11	11	11
11	11	11	11	11	11	11	11	11	11	11
11	11	11	11	11	23	11	11	11	11	11

N'goye

1	2	3	4	5	6	7	8	9	10	11
11	11	11	11	11	11	13	11	11	11	11
11	11	11	11	11	11	11	11	11	11	11
11	11	11	11	11	33	11	11	11	11	11
11	11	11	11	11	24	11	11	11	11	11
11	11	11	11	12	11	13	11	11	11	11
11	11	11	11	11	11	11	11	11	11	11
11	11	11	11	11	14	12	11	11	11	11
11	11	11	11	11	12	13	11	11	11	11
11	11	11	12	11	13	23	11	11	11	11
11	11	11	11	11	14	12	11	11	11	11
11	11	11	11	11	11	11	11	11	11	11
11	11	11	11	11	11	33	11	11	11	11
11	11	11	11	11	11	12	11	11	11	11
11	11	11	11	11	24	13	11	11	11	11
11	11	11	11	11	11	11	11	11	11	11
11	11	11	11	11	12	11	11	11	11	11
11	11	11	11	11	11	11	11	11	11	11
11	11	11	11	11	12	11	11	11	11	11
11	11	11	11	11	11	33	11	11	11	11
11	11	11	11	11	13	12	11	11	11	11
11	11	11	11	11	12	11	11	11	11	11
11	11	11	11	11	11	22	11	11	11	11
11	11	11	11	11	11	13	11	11	11	11
11	11	11	11	11	14	13	11	11	11	11
11	11	11	11	11	12	13	11	11	11	11
11	11	11	11	11	11	13	11	11	11	11
11	11	11	11	11	14	11	11	11	11	11
11	11	11	11	11	11	12	11	11	11	11
11	11	11	11	11	14	12	11	11	11	11
11	11	11	12	11	12	12	11	11	11	11
11	11	11	11	11	11	13	11	11	11	11
11	11	11	12	11	12	11	11	11	11	11
11	11	11	11	12	12	11	11	11	11	11
11	11	11	12	11	11	11	11	11	11	11
11	11	11	11	11	12	11	11	11	11	11
11	11	11	11	11	11	11	11	11	11	11
11	11	11	11	11	12	11	11	11	11	11
11	11	11	11	11	12	11	11	11	11	11
11	11	11	11	11	11	11	11	11	11	11
11	11	11	11	11	11	33	11	11	11	11
11	11	11	11	11	24	11	11	11	11	11
11	11	11	11	11	12	11	11	11	11	11
11	11	11	11	11	11	23	11	11	11	11
11	11	11	11	11	12	13	11	11	11	11
11	11	11	11	11	11	11	11	11	11	11
11	11	11	11	11	14	11	11	11	11	11
11	11	11	12	11	34	12	11	11	11	11
11	11	11	22	11	23	11	11	11	11	11
11	11	11	11	11	11	33	11	11	11	11
11	11	11	11	11	12	11	11	11	11	11

Chapter 6

Individual Identification

INTRODUCTION

Discrete genetic markers are being used to an increasing extent to identify individuals, especially humans (Weir 1995). The contexts are quite varied, with the most established being that of paternity testing. Examination of genetic profiles of mother, child, and alleged father can lead to statements about the probability of the genetic evidence if the alleged father is the actual father. Similar reasoning can be used to identify mothers, or to reunite families separated by war or other acts. Examinations of genetic profiles need not be confined to living people, and in fact they are often used for inheritance disputes or identification of remains from war or other disasters. In these cases the profiles from remains are compared to those of family members.

The other major use of genetic profiles is in comparing biological samples, such as blood or semen stains serving as evidentiary samples from a crime, with the profiles from people suspected of having contributed those samples. Although matching of profiles between a person and a sample does not prove a common source, let alone guilt, such matches can constitute powerful evidence.

There are several issues involved in the analysis of genetic data used for individual identification, including the estimation of profiles at many loci and the accommodation of population substructure. Many of the complexities are handled best through the use of Bayes' theorem, as has been shown in a series of papers by Evett (e.g. Evett 1987, 1990, 1992, 1995). A thorough exploration of the issues was recently published by Aitken (1995).

BAYES' THEOREM

Bayes' theorem was discussed at the end of Chapter 2, and it made use of conditional probabilities of one event given another. A conditional probability may also refer to the probability of an event E when some conditions C hold. If $\bar{C}$ means not-C:

$$\begin{aligned}\Pr(E) &= \Pr(E \cap \Omega) \\ &= \Pr[E \cap (C \cup \bar{C}] \\ &= \Pr[(E \cap C) \cup (E \cap \bar{C})] \\ &= \Pr(E \cap C) + \Pr(E \cap \bar{C}) \\ &= \Pr(E|C)\Pr(C) + \Pr(E|\bar{C})\Pr(\bar{C})\end{aligned}$$

After event E has been observed, the probability of the conditions C is

$$\begin{aligned}\Pr(C|E) &= \frac{\Pr(C \cap E)}{\Pr(E)} \\ &= \frac{\Pr(E|C)\Pr(C)}{\Pr(E)} \\ &= \frac{\Pr(E|C)\Pr(C)}{\Pr(E|C)\Pr(C) + \Pr(E|\bar{C})\Pr(\bar{C})}\end{aligned}$$

This last expression is a restatement of Bayes' theorem. Before the event E there are ***prior probabilities*** $\Pr(C)$ and $\Pr(\bar{C})$ of conditions C and $\bar{C}$. After the event E there are ***posterior probabilities*** $\Pr(C|E)$ and $\Pr(\bar{C}|E)$.

A useful quantity is the ratio of posterior probabilities:

$$\frac{\Pr(C|E)}{\Pr(\bar{C}|E)} = \frac{\Pr(E|C)}{\Pr(E|\bar{C})} \times \frac{\Pr(C)}{\Pr(\bar{C})}$$

The ratio of posterior probabilities (the posterior odds) is L times the ratio of prior probabilities (the prior odds) of conditions C and $\bar{C}$, where the likelihood ratio

$$L = \frac{\Pr(E|C)}{\Pr(E|\bar{C})}$$

expresses how probable event E is when conditions C hold, compared to when conditions $\bar{C}$ hold.

The important concept here is that L compares the probabilities of an event under two explanations without having to invoke any prior probability for these two explanations. If $L > 1$, it is more probable that the event will occur when C holds than it will occur when $\bar{C}$ holds.

This concept is used in both paternity and forensic applications. For paternity testing, L is called the ***paternity index***. Event E represents the evidence, generally profiles of mother, child, and alleged father. Condition C represents the alleged father being the biological father of the child. In forensic applications, E may be the evidence of a match between the profiles of a suspect and an evidentiary stain. Condition C is that the stain was left by the suspect.

INBREEDING AND RELATEDNESS

In both paternity and forensic settings, there is often a need to accommodate related people in the calculations. Relatedness is a consequence of sharing identical alleles. Alleles that have descended from a single ancestral allele are said to be ***identical by descent, ibd***. Genetic drift, resulting from random mating in finite populations, leads to a build-up of the probability of ibd and this has been quantified in this book by the inbreeding coefficient F, for alleles within the same individual, and the coancestry coefficient θ, for alleles in different individuals.

Drift is not the only way in which inbreeding can accrue. Whenever two parents are related, meaning that they have a chance of sharing ibd alleles, their offspring are inbred. More formally, if θ_{XY} is the coancestry coefficient of individuals X and Y, then the inbreeding coefficient of their offspring I is

$$F_I = \theta_{XY} \tag{6.1}$$

which follows directly from the definitions of the two parameters.

There is another relationship of importance between F and θ. If two alleles, a and b, are taken at random from individual X (to go to two offspring), then the probability that these two alleles are ibd is written as θ_{XX}. Now suppose that X received alleles x_1 and x_2 from its parents. Each of a, b is equally likely to be copies of each of x_1, x_2:

$$\begin{aligned}
\theta_{XX} &= \Pr(a \equiv b) \\
&= \frac{1}{4}[\Pr(x_1 \equiv x_1) + \Pr(x_1 \equiv x_2) + \Pr(x_2 \equiv x_1) + \Pr(x_2 \equiv x_2)] \\
\theta_{XX} &= \frac{1}{2}(1 + F_X) \qquad (6.2)
\end{aligned}$$

Note that the coancestry θ_{XY} for distinct individuals X, Y can be regarded as giving the probability of ibd for a random pair of alleles received *by* X, Y or received *from* X, Y, whereas θ_{XX} is for two alleles received *from* X. For the two alleles received *by* X, the coancestry measure is written as $\theta_{\ddot{X}}$ and is just the inbreeding coefficient F_X of X.

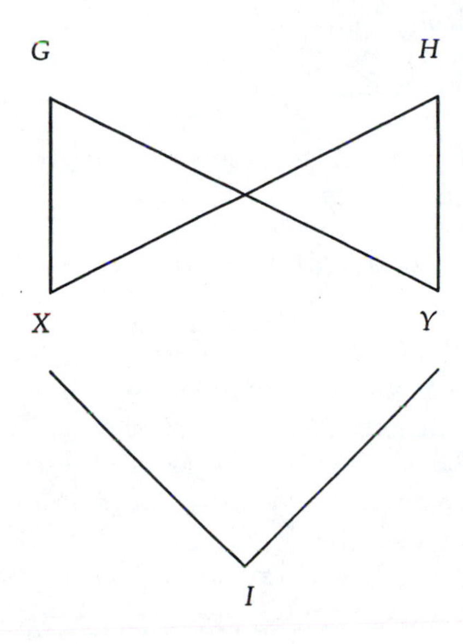

Figure 6.1 *I* is offspring of full sibs *X, Y*

The relations in Equations 6.1 and 6.2 allow the inbreeding coefficient to be determined in any pedigree. If I is the offspring of full sibs X and Y, who had parents G and H (Figure 6.1):

$$\begin{aligned} F_I &= \theta_{XY} \\ &= \frac{1}{4}[\theta_{GG} + \theta_{GH} + \theta_{HG} + \theta_{HH}] \\ &= \frac{1}{8}(1 + F_G) + \frac{1}{2}\theta_{GH} + \frac{1}{8}(1 + F_H) \\ &= \frac{1}{4} \end{aligned}$$

with the last line following if G, H are not inbred and are unrelated. In other words, there is a 25% chance that the offspring of noninbred full sibs is inbred, and this serves as a measure of relatedness of the sibs.

An alternative procedure is to use a general result that follows from the kind of argument just presented for full sibs. If the parents X, Y of an individual I have a common ancestor A, then the contribution of A to the inbreeding of I depends on the numbers of generations separating these individuals. Specifically, if there are n individuals in the path from X to A and back to Y, there is a contribution of $(1/2)^n(1 + F_A)$. If parents X, Y have several unrelated common ancestors A_i, with n_i members in the path linking them to A_i, then

$$F_I = \sum_i \left(\frac{1}{2}\right)^{n_i} (1 + F_{A_i})$$

For the case of full sibs in Figure 6.1, there are two paths: XGY and XHY, each with three individuals. Therefore

$$\begin{aligned} F_I &= \left(\frac{1}{2}\right)^3 (1 + F_G) + \left(\frac{1}{2}\right)^3 (1 + F_H) \\ &= \frac{1}{4} \end{aligned}$$

Additional descent measures are needed to express the joint genotypic frequencies of pairs of relatives. Whereas the genotypic frequency of possibly inbred individuals requires the use of a descent measure defined for pairs of alleles, the genotypic frequencies for pairs of relatives require descent measures for up to four alleles being ibd. Suppose individuals X and Y have alleles a, b and c, d at locus A. The various possible ibd relations among these four alleles are shown in Table 6.1. The fifteen probabilities in that table add to one.

The descent status of the four alleles a, b, c, d puts constraints on the possible genotypes of X and Y. If all four alleles were ibd, for example, both individuals would need to be of the same homozygous genotype. The frequencies with which X, Y have various genotypes at locus A are shown in Table 6.2. Although two alleles, A, a are indicated, a may be regarded as representing all alleles other than A.

As an example, consider the possible genotype pairs for two sibs, X, Y that have noninbred and unrelated parents. The descent measure values are shown in Table 6.1, and the probabilities for any of the six possible pairs of genotypes (regardless of order) are:

$$\begin{array}{llcll} AA, AA & p_A^2(1+p_A)^2/4 & , & aa, aa & p_a^2(1+p_a)^2/4 \\ AA, Aa & p_A^2 p_a(1+p_A) & , & Aa, aa & p_A p_a^2(1+p_a) \\ AA, aa & p_A^2 p_a^2/2 & , & Aa, Aa & p_A p_a(1+p_A+p_a+2p_A p_a)/2 \end{array}$$

These are not the same as simply multiplying together the frequencies of each of the two genotypes. Relatives are more likely than random pairs of people from the population to have the same genotype.

When there is no possibility of alleles within individuals being ibd, the number of descent measures reduces from 15 to 7, and symmetry may further reduce the number to three: the probabilities that two individuals share zero, one or two alleles ibd. Because there will be situations in which inbreeding is possible, however, it seems preferable to stay with the full set of descent measures in general and reduce to a lower number in specific cases.

Table 6.1 Descent relations among alleles for two individuals X and Y with alleles a, b and c, d.

Alleles ibd*	Probability	
	General	Full sibs†
none	δ_0	1/4
$a \equiv b$	δ_{ab}	0
$c \equiv d$	δ_{cd}	0
$a \equiv c$	δ_{ac}	1/4
$a \equiv d$	δ_{ad}	0
$b \equiv c$	δ_{bc}	0
$b \equiv d$	δ_{bd}	1/4
$a \equiv b \equiv c$	δ_{abc}	0
$a \equiv b \equiv d$	δ_{abd}	0
$a \equiv c \equiv d$	δ_{acd}	0
$b \equiv c \equiv d$	δ_{bcd}	0
$a \equiv b, c \equiv d$	$\delta_{ab.cd}$	0
$a \equiv c, b \equiv d$	$\delta_{ac.bd}$	1/4
$a \equiv d, b \equiv c$	$\delta_{ad.bc}$	0
$a \equiv b \equiv c \equiv d$	δ_{abcd}	0

*Alleles not specified are not ibd.
†a, c from mother; b, d from father

Conditional Genotypic Frequencies

It is often the case that the probability of one individual X having a specified genotype, A_iA_j for example, is needed when another individual Y is known to have that type. From the definition of conditional probabilities

$$\Pr(X = A_iA_j|Y = A_iA_j) = \frac{\Pr(X = A_iA_j, Y = A_iA_j)}{\Pr(Y = A_iA_j)}$$

The joint genotypic frequencies derived in the previous section are evidently relevant. Generalizing the expressions in Table 6.2, the probabilities with which $X = (ab), Y = (cd)$ have the same genotype come from the following expressions. In these equations $\Pr(ij; kl)$ means $\Pr(a = A_i, b = A_j; c = A_k, d = A_l)$.

$$\begin{aligned}\Pr(ii; ii) = {} & \delta_o p_i^4 + (\delta_{abc} + \delta_{abd} + \delta_{acd} + \delta_{bcd})p_i^2 \\ & + (\delta_{ab} + \delta_{ac} + \delta_{ad} + \delta_{bc} + \delta_{bd} + \delta_{cd})p_i^3 \\ & + (\delta_{ab.cd} + \delta_{ac.bd} + \delta_{ad.bc})p_i^2 + \delta_{abcd}p_i\end{aligned}$$

Table 6.2 Joint genotypic probabilities for two individuals $X(a,b)$ and $Y(c,d)$

X	Y	Probability
AA	AA	$\delta_{abcd}p_A + (\delta_{abc} + \delta_{abd} + \delta_{acd} + \delta_{bcd})p_A^2$ $+(\delta_{ab.cd} + \delta_{ac.bd} + \delta_{ad.bc})p_A^2 + (\delta_{ab} + \delta_{ac} + \delta_{ad} + \delta_{bc} + \delta_{bd} + \delta_{cd})p_A^3 + \delta_0 p_A^4$
AA	Aa	$(\delta_{abc} + \delta_{abd})p_A p_a + (2\delta_{ab} + \delta_{ac} + \delta_{bc} + \delta_{ad} + \delta_{bd})p_A^2 p_a + 2\delta_0 p_A^3 p_a$
Aa	AA	$(\delta_{acd} + \delta_{bcd})p_A p_a + (2\delta_{cd} + \delta_{ac} + \delta_{bc} + \delta_{ad} + \delta_{bd})p_A^2 p_a + 2\delta_0 p_A^3 p_a$
AA	aa	$\delta_{ab.cd}p_A p_a + \delta_{ab}p_A p_a^2 + \delta_{cd}p_A^2 p_a + \delta_0 p_A^2 p_a^2$
aa	AA	$\delta_{ab.cd}p_A p_a + \delta_{cd}p_A p_a^2 + \delta_{ab}p_A^2 p_a + \delta_0 p_A^2 p_a^2$
Aa	Aa	$2(\delta_{ac.bd} + \delta_{ad.bc})p_A p_a + (\delta_{ac} + \delta_{ad} + \delta_{bc} + \delta_{bd})(p_A p_a^2 + p_A^2 p_a) + 4\delta_0 p_A^2 p_a^2$
aa	Aa	$(\delta_{abc} + \delta_{abd})p_A p_a + (2\delta_{ab} + \delta_{ac} + \delta_{bc} + \delta_{ad} + \delta_{bd})p_a^2 p_A + 2\delta_0 p_a^3 p_A$
Aa	aa	$(\delta_{acd} + \delta_{bcd})p_A p_a + (2\delta_{cd} + \delta_{ac} + \delta_{bc} + \delta_{ad} + \delta_{bd})p_a^2 p_A + 2\delta_0 p_a^3 p_A$
aa	aa	$\delta_{abcd}p_a + (\delta_{abc} + \delta_{abd} + \delta_{acd} + \delta_{bcd})p_a^2$ $+(\delta_{ab.cd} + \delta_{ac.bd} + \delta_{ad.bc})p_a^2 + (\delta_{ab} + \delta_{ac} + \delta_{ad} + \delta_{bc} + \delta_{bd} + \delta_{cd})p_a^3 + \delta_0 p_a^4$

$$
\begin{aligned}
\Pr(ij;ij) &= \delta_o p_i^2 p_j^2 + \delta_{ac} p_i p_j^2 + \delta_{bd} p_i^2 p_j + \delta_{ac.bd} p_i p_j \\
\Pr(ij;ji) &= \delta_o p_i^2 p_j^2 + \delta_{ad} p_i p_j^2 + \delta_{bc} p_i^2 p_j + \delta_{ad.bc} p_i p_j \\
\Pr(ji;ij) &= \delta_o p_i^2 p_j^2 + \delta_{bc} p_i p_j^2 + \delta_{ad} p_i^2 p_j + \delta_{ad.bc} p_i p_j \\
\Pr(ji;ji) &= \delta_o p_i^2 p_j^2 + \delta_{bd} p_i p_j^2 + \delta_{ac} p_i^2 p_j + \delta_{ac.bd} p_i p_j
\end{aligned}
$$

For X and Y taken at random from a population, there is symmetr among descent measures of the same order, and it is more convenient t employ summary measures. The probabilities that *any* two, three, or fou alleles are identical by descent are written as θ, γ, δ, and the probabilit that *any* two pairs are identical by descent is written as Δ. The relationshi between the two sets of measures for a random set of four alleles a, b, c, d is:

$$\begin{aligned}
\delta_o &= 1 - 6\theta + 8\gamma + 3\Delta - 6\delta \\
\delta_{ab} = \delta_{ac} = \delta_{ad} = \delta_{bc} = \delta_{bd} = \delta_{cd} &= \theta - 2\gamma - \Delta + 2\delta \\
\delta_{abc} = \delta_{abd} = \delta_{acd} = \delta_{bcd} &= \gamma - \delta \\
\delta_{ab.cd} = \delta_{ac.bd} = \delta_{ad.bc} &= \Delta - \delta \\
\delta_{abcd} &= \delta
\end{aligned}$$

The probabilities that X, Y have the same genotypes can then be written as

$$\begin{aligned}
\Pr(X = A_iA_i, Y = A_iA_i) &= p_i[\delta + (4\gamma + 3\Delta - 7\delta)p_i \\
&\quad + 6(\theta - 2\gamma - \Delta + 2\delta)p_i^2 \\
&\quad + (1 - 6\theta + 8\gamma + 3\Delta - 6\delta)p_i^3] \\
\Pr(X = A_iA_j, Y = A_iA_j) &= 4p_ip_j[(\Delta - \delta) + (\theta - 2\gamma - \Delta + 2\delta)(p_i + p_j) \\
&\quad + (1 - 6\theta + 8\gamma + 3\Delta - 6\delta)p_ip_j]
\end{aligned}$$

This general formulation makes no assumptions about the evolutionary forces that have shaped the population, other than supposing symmetry among ibd probabilities involving the same numbers of alleles.

PATERNITY TESTING

In paternity disputes it is necessary to determine whether a particular man is the father of a particular child. Classical considerations of such questions were limited to excluding a man from paternity of a child if the man did not have the child's paternal allele at some locus, or, if the paternal allele cannot be determined, if the man has neither of the child's alleles. The increasing availability of diagnostic loci has given rise to calculations of the probability of the mother-child-man genotypic array if that man is the father.

Paternity Exclusion Probability with Codominant Loci

It may be useful to characterize a genetic marker by its ability to exclude a "random" man from paternity. Such exclusion probabilities depend on allele frequencies for that locus, but do not depend on the genotypes in any particular case.

For a single autosomal locus with codominant alleles, those men that can be excluded as possible fathers for all possible mother-child combinations are shown in Table 6.3. The table also shows the probabilities of the various genotypes for the mothers and random men. The offspring probabilities are conditional on those of the mother. An assessment of how good this locus

Table 6.3 Paternity exclusion configurations at one locus with an arbitrary number of alleles.

Mother		Child		Excluded Man	
Type	Prob.[1]	Type	Probability[2]	Genotype	Probability[3]
A_uA_u	p_u^2	A_uA_u	p_u	$A_wA_x,\ w,x \neq u$	$(1-p_u)^2$
		A_uA_v	p_v	$A_wA_x,\ w,x \neq v$	$(1-p_v)^2$
A_uA_v $v \neq u$	$2p_up_v$	A_uA_u	$p_u/2$	$A_wA_x,\ w,x \neq u$	$(1-p_u)^2$
		A_vA_v	$p_v/2$	$A_wA_x,\ w,x \neq v$	$(1-p_v)^2$
		A_uA_v	$(p_u+p_v)/2$	$A_wA_x,\ w,x \neq u,v$	$(1-p_u-p_v)^2$
		$A_uA_y^*$	$p_y/2$	$A_wA_x,\ w,x \neq y$	$(1-p_y)^2$
		$A_vA_y^*$	$p_y/2$	$A_wA_x,\ w,x \neq y$	$(1-p_y)^2$

[1]Probability of mother's genotype
[2]Probability of child's genotype given mother's genotype
[3]Probability of excluded genotype
*$y \neq u, v$

is for being able to exclude a random man from paternity, assuming that he is not the father of the child, is given by summing the joint probabilities of all the mother-child-excluded man combinations shown in the table. The probability of an A_uA_u mother with an A_uA_u child is $p_u^2 \times p_u$, and this combination excludes all men that do not have an A_u allele. Such men have frequency $(1-p_u)^2$, so that the trio in the first line of Table 6.3 has a combined probability of $p_u^3(1-p_u)^2$. Adding such probabilities from all seven lines of the table gives the exclusion probability Q

$$Q = \sum_u p_u(1-p_u)^2 - \frac{1}{2}\sum_u \sum_{v \neq u} p_u^2 p_v^2(4 - 3p_u - 3p_v)$$

and will be maximized when all V alleles at the locus have frequency $1/V$.

$$Q_{\max} = 1 - \frac{2V^3 + V^2 - 5V + 3}{V^4}$$

The utility of a locus increases with the number of alleles, although even with 10 alleles the value of $Q_{\max}$ is only 0.79. Larger number of alleles are being found with VNTR loci (e.g., Chimera et al. 1989), and $V = 30$ is not unusual. With 30 equally frequent alleles, the value of Q increases to 0.9324.

Table 6.4 Paternity exclusion arrangements for the three-allele *ABO* system.

		Mother			
	Excluded Men	*A*	*B*	*AB*	*O*
Child	*A*	None	*B*, *O*	None	*B*
	B	*A*, *O*	None	None	*A*
	AB	*A*, *O*	*B*, *O*	O	Impossible
	O	*AB*	*AB*	Impossible	*AB*

Exclusion probabilities are increased with the use of several loci, since it is sufficient to exclude at even one out of several loci if mutation can be ignored. If Q_l is the exclusion probability at locus l, then the overall probability of exclusion follows from being able to exclude from at least one locus. In other words, Q is one minus the probability that none of the loci allows exclusion. If the Q_l are independent

$$Q = 1 - \prod_l (1 - Q_l)$$

For two loci, each with 10 equally frequent alleles, the value of Q increases to $[1-(1-0.79)^2] = 0.96$. In practice, exclusion is required for at least two loci to guard against the chance of mutation falsely indicating exclusion at one locus.

Paternity Exclusion Probabilities with Dominant Loci

Exclusion calculations can also be made for diagnostic loci that have dominant alleles, as illustrated in Table 6.4 for the *ABO* system. Adding over all mother-child-excluded man combinations, the probability of paternity exclusion from the *ABO* system is found to be

$$Q = p(1-p)^4 + q(1-q)^4 + pqr^2(3-r)$$

where p, q, and r are the population frequencies of the A, B, and O alleles.

Examples of exclusion probabilities Q were compiled by Chakraborty et al. (1974). They used a four-allele model for the *ABO* system (alleles

O, A_1, A_2, B) and reported Q values of 0.1774, 0.1342, and 0.1917 for populations they characterized as "Black," "White," and "Japanese" respectively. Greater discrepancy was found between populations for the isozyme marker *acid phosphatase* with three alleles. The Q values were then 0.1588, 0.2323, and 0.1340 for Black, White, and Japanese populations.

Paternity Index

Unlike the exclusion probability, the paternity index is based on the actual genotypes of the mother, child, and alleged father. Before the collection of any genetic evidence, there may be some prior beliefs about the probability that the alleged father is the true father: $\Pr(C)$ is the ***prior probability*** of paternity. The main concern is with the situation after the genotypes of mother, child, and alleged father are available. If E denotes this genetic evidence, Bayes' theorem provides

$$\frac{\Pr(\text{man is father}|E)}{\Pr(\text{man is not father}|E)} = \frac{\Pr(E|\text{man is father})}{\Pr(E|\text{man is not father})} \times \frac{\Pr(\text{man is father})}{\Pr(\text{man is not father})}$$

The left-hand side of this expression is the ratio of posterior probabilities, $\Pr(C|E)$ to $\Pr(\bar{C}|E)$, whereas the right-hand side is the product of the paternity index L and the ratio of prior probabilities $\Pr(C)$ to $\Pr(\bar{C})$:

$$\frac{\Pr(C|E)}{\Pr(\bar{C}|E)} = L \times \frac{\Pr(C)}{\Pr(\bar{C})}$$

Rearranging this equation gives the posterior probability $\Pr(C|E)$ as

$$\Pr(C|E) = \frac{L\Pr(C)}{L\Pr(C) + (1 - \Pr(C))}$$

When several independent loci are available, L may be formed as the product of the values for each locus. Note that $\Pr(C|E)$ here is the probability of an alleged man with a particular genotype being the father when genetic evidence is available from the mother, child, and himself.

Discussion of paternity indices is quite valid, but debate begins when particular values are given to the prior probabilities. They could be set to 0.5, for instance, so that before the genetic evidence is examined, the alleged father is said to have a fifty percent probability of being the father. Then $\Pr(C|E) = L/(L+1)$ and Gürtler (1956) suggested that the man be declared the father if $\Pr(C|E)$ exceeded 0.95 ($L > 19$) and be declared not the father if $\Pr(C|E)$ was less than 0.05 ($L < 0.053$). Some paternity laboratories report the paternity index along with the posterior probabilities that result when priors are set to 0.1, 0.5, or 0.9. It may be difficult to assign a prior acceptable to both sides in a paternity dispute, but if L is sufficiently large even a small prior will lead to a high posterior probability.

The paternity index L offers a valuable means of quantifying the weight of the genetic evidence. The simplest case is when all three individuals; mother, child and alleged father, have different heterozygous genotypes. The genotypes of mother and child allow the maternal allele of the child to be determined, and the paternal allele to be inferred. Only alleged fathers having that allele will not be excluded. For an offspring O of genotype A_1A_2 with a mother M having genotype A_1A_3, the obligate paternal allele is A_2, and an alleged father AF with this allele would not be excluded as being the father of the offspring. If the alleged father has genotype A_2A_4, the genetic evidence E is the three genotypes ($M = A_1A_3, O = A_1A_2, AF = A_2A_4$). Under the condition C that the alleged father is the father, the probability of the evidence is:

$$\begin{aligned}\Pr(E|C) = \Pr(O, M, AF|C) &= \Pr(O|M, AF, C)\Pr(M, AF|C) \\ &= \frac{1}{2} \times \frac{1}{2} \times \Pr(M, AF)\end{aligned}$$

The $1/4$ value is the probability of a child of individuals M, AF having the genotype of O.

Under the condition that some man, unrelated to the alleged father, is the father, then the probability of the evidence is:

$$\begin{aligned}\Pr(E|\bar{C}) = \Pr(O, M, AF|\bar{C}) &= \Pr(O|M, AF, \bar{C})\Pr(M, AF|\bar{C}) \\ &= \Pr(O|M)\Pr(M, AF) \\ &= \frac{1}{2} \times p_2 \times \Pr(M, AF)\end{aligned}$$

The $p_2/2$ value is the probability of a child of M and some random man having the genotype of O. This random man must have contributed allele A_2. The joint probability $\Pr(M, AF)$ of the genotypes of the mother and the alleged father does not depend on conditions C or $\bar{C}$. The paternity index is therefore

$$L = \frac{1}{2p_2}$$

showing that loci with an allele in frequency greater than 0.5 are not very useful in paternity testing. If the paternal allele has a frequency greater than 0.5, a random man is more likely to have transmitted that allele than is the alleged father for whom the probability is 0.5. Of course this argument ignores any nongenetic evidence.

Elston (1986a) and Majumder and Nei (1983) have pointed out that, at a single locus, L can have quite a restricted range of values, and Elston showed that the expected value of the posterior probability over all mother-child-alleged father combinations in the population is just the true value π. The arguments put forward by Majumder and Nei (1983) and Elston (1986a) are summarized in Table 6.5. For a two-allele locus, the various mother-child-alleged father combinations are given, and paternity indices are shown. Use of this table shows that the mean paternity index for fathers is always higher than that of nonfathers (Elston 1986b).

Paternity Index for Relatives

It may be the case that the alleged father is a relative of the true father, and the paternity index can be calculated to assess the strength of the genetic evidence against that alternative. Suppose the evidence E is that the obligate paternal allele is A_i. Explanation C is that the alleged father (AF) with genotype A_iA_j is the true father, and explanation $\bar{C}$ is that his brother (B) is the true father. If $X \rightarrow A_i$ means that individual X transmits allele A_i to an offspring, then

$$\begin{aligned}
\Pr(E|C) &\propto \Pr(AF \rightarrow A_i|AF = A_iA_j) \\
&= \frac{1}{2} \\
\Pr(E|\bar{C}) &\propto Pr(B \rightarrow A_i|AF = A_iA_j) \\
&= \Pr(B \rightarrow A_i, AF = A_iA_j)/\Pr(AF = A_iA_j) \\
&= \Big[\Pr(B = A_iA_i, AF = A_iA_j) + \frac{1}{2}\Pr(B = A_iA_j, AF = A_iA_j) \\
&\quad + \frac{1}{2}\sum_{k\neq i,j}\Pr(B = A_iA_k, AF = A_iA_j)\Big] / \Pr(AF = A_iA_j) \\
&= \frac{1}{4}p_i(1+p_i) + \frac{1}{8}(1+p_i+p_j+2p_ip_j) \\
&\quad + \frac{1}{8}(1-p_i-p_j)(1+2p_i) \\
&= \frac{1+2p_i}{4}
\end{aligned}$$

and the paternity index is

$$L = \frac{(1/2)}{(1+2p_i)/(4)} = \frac{2}{1+2p_i}$$

Table 6.5 Paternity index calculations for a two-allele locus as given by Elston (1986a).

E					
Mother	Child	Alleged Father	$\Pr(E\|C)$	$\Pr(E\|\bar{C})$	L
AA	AA	AA	p_A^4	p_A^5	$1/p_A$
		Aa	$p_A^3 p_a$	$2p_A^4 p_a$	$1/2p_A$
		aa	0	$p_A^3 p_a^2$	0
	Aa	AA	0	$p_A^4 p_a$	0
		Aa	$p_A^3 p_a$	$2p_A^3 p_a^2$	$1/2p_a$
		aa	$p_A^2 p_a^2$	$p_A^2 p_a^3$	$1/p_a$
Aa	AA	AA	$p_A^3 p_a$	$p_A^4 p_a$	$1/p_A$
		Aa	$p_A^2 p_a^2$	$2p_A^3 p_a^2$	$1/2p_A$
		aa	0	$p_A^2 p_A^3$	0
	Aa	AA	$p_A^3 p_a$	$p_A^3 p_a$	1
		Aa	$2p_A^2 p_a^2$	$2p_A^2 p_a^2$	1
		aa	$p_A p_a^3$	$p_A p_a^3$	1
	aa	AA	0	$p_A^3 p_a^2$	0
		Aa	$p_A^2 p_a^2$	$2p_A^2 p_a^3$	$1/2p_a$
		aa	$p_A p_a^3$	$p_A p_a^4$	$1/p_a$
aa	Aa	AA	$p_A^2 p_a^2$	$p_A^3 p_a^2$	$1/p_A$
		Aa	$p_A p_a^3$	$2p_A^2 p_a^3$	$1/2p_A$
		aa	0	$p_A p_a^4$	0
	aa	AA	0	$p_A^2 p_a^3$	0
		Aa	$p_A p_a^3$	$2p_A p_a^4$	$1/2p_a$
		aa	p_a^4	p_a^5	$1/p_a$

FORENSIC TESTING

Suppose a crime has been committed and profiles for crime scene material and suspect are found to match, and that it is known from other evidence that the suspect has not been excluded from guilt. How should the match be interpreted, and what light does it shed on the guilt or innocence of the suspect? There are only two possibilities: the suspect is either guilty or not guilty. For this discussion two related possibilities are considered:

either the suspect provided the crime scene material (event C) or someone else provided the material (event $\bar{C}$). The calculations are simplest when "someone else" refers to a person unrelated to the suspect, but specified degrees of relationship can be accommodated. Suppose further that the matching profile is of type A and that it can be assumed that the crime scene material was left by the perpetrator P. The suspect is labeled S, so event C is that S and P are the same person and $\bar{C}$ is that they are different people. Event E refers to the evidence that both S and P have profile A.

In speaking about probabilities, there is an implicit (frequentist) reference to repeated trials, and the long-term average proportion with which an event occurs is the probability of that event. In the present context, the repeated trials refer to sampling people from some population. This population is defined by the crime, and could be referred to as the population of potential perpetrators. Some crimes might narrow this population by precise descriptions, for example based on geographic location, while others may merely refer to the general population. The suspect belongs to the population of potential perpetrators, otherwise he or she could be excluded from being the perpetrator. The suspect's membership in this population, however, may have nothing to do with his or her particular attributes, such as ethnicity.

If the suspect acknowledges responsibility for the crime scene material, there is no need for any further analysis. If the suspect claims it is a coincidence that he or she has the same profile A as the perpetrator, however, it is necessary to establish some level of credence to this claim. It is helpful to compare the values of this probability under the two explanations, C and $\bar{C}$, and the comparison is made by the likelihood ratio

$$L = \frac{\Pr(S = A, P = A|C)}{\Pr(S = A, P = A|\bar{C})}$$

As Aitken (1995) points out, the analysis may be "scene-anchored" or "suspect-anchored." It is convenient here to anchor on the suspect, and this changes the appearance of the likelihood ratio

$$L = \frac{\Pr(P = A|S = A, C)\Pr(S = A|C)}{\Pr(P = A|S = A, \bar{C})\Pr(S = A|\bar{C})}$$

which makes use of the conditional probability of the perpetrator having the profile, given that it is known the suspect has that profile, $\Pr(S = A|P = A)$. In this notation, $S = A$ means that S has profile A. From the fact that the probability with which S has profile A is independent of C and $\bar{C}$, the likelihood ratio becomes

$$L = \frac{\Pr(P = A|S = A, C)}{\Pr(P = A|S = A, \bar{C})}$$

The strength of the evidence is presented by saying that the evidence is L times more likely to have occurred if S and P are the same person than if they are different people. Further statements can be made if prior probabilities $\Pr(C), \Pr(\bar{C})$ that S and P are the same person, are introduced:

$$\frac{\Pr(C|E)}{\Pr(\bar{C}|E)} = L \times \frac{\Pr(C)}{\Pr(\bar{C})}$$

If experimental error is ignored, the numerator in L has the value 1, and

$$L = \frac{1}{\Pr(P = A|S = A, \bar{C})}$$

The possibility of incorrect determinations of profiles for either S or P is also being ignored here.

The simplest thing is to assume the probabilities of S and P having profile A are independent when they are different people. Then

$$L = \frac{1}{\Pr(P = A)}$$

This is the basis for presenting the evidence of a match along with the value of the frequency with which a person taken at random from the population of potential perpetrators has the profile in question. Clearly independence will not hold if S and P are related, and the results presented earlier show that it does not strictly hold for two individuals in the same population.

Estimation of Profile Frequency

Although interpretation of the evidence of a DNA profile match requires the likelihood ratio, there may be interest in estimating the frequency with which the matching profile occurs in a population. This estimation is made difficult by the fact that profiles with a frequency of 10^{-6}, or less, are unlikely to be seen in any sample from the population. Each of the constituent alleles, however, is likely to be seen with a moderate frequency and the profile frequency can be estimated as the product of these constituent frequencies.

Suppose the profile A has alleles A_{l_1}, A_{l_2} at locus l. If allele A_{l_i} has sample frequency $\tilde{p}_{l_i}$, then the genotypic frequency P_l at locus l is estimated as

$$\tilde{P}_l = \begin{cases} \tilde{p}_{l_1}^2, & l_1 = l_2 \\ 2\tilde{p}_{l_1}\tilde{p}_{l_2}, & l_1 \neq l_2 \end{cases}$$

and the whole profile frequency P is estimated as

$$\tilde{P} = \prod_l \tilde{P}_l$$

Of course, this requires a demonstration that the hypothesis of allele frequencies being independent is not rejected, and the methods of Chapter 3 are needed. It has been shown (Maiste and Weir 1995, Zaykin et al. 1995), that exact tests are appropriate in this context.

An estimate of the profile frequency should be accompanied by an indication of its sampling properties. The variance of a product of independent quantities $\tilde{P}_l$ can be determined (Goodman 1957) approximately from

$$\text{Var}(\prod_l \tilde{P}_l) \doteq (\prod_l P_l)^2 \left(\prod_l \left[1 + \frac{\text{Var}(\tilde{P}_l)}{P_l^2} \right] - 1 \right)$$

and the individual variances $\text{Var}(\tilde{P}_l)$ were given in Chapter 2. If the Hardy-Weinberg law holds, for a sample of n individuals:

$$\begin{aligned} \text{Var}(\tilde{p}_{l_1}^2) &= \frac{2}{n} p_{l_1}^3 (1 - p_{l_1}) \\ \text{Var}(2\tilde{p}_{l_1}\tilde{p}_{l_2}) &= \frac{2}{n} p_{l_1} p_{l_2} (p_{l_1} + p_{l_2} - 4p_{l_1}p_{l_2}) \end{aligned}$$

If confidence intervals are desired, rather than variance estimates, problems of normal-theory confidence limits not applying to very small $\tilde{P}$ values can be avoided by using bootstrapping (Chapter 2) to provide confidence limits for the profile frequency.

At the beginning of this section, estimated frequencies of 10^{-6} were mentioned and even smaller estimates can result when many loci are included in the profile. Is it sensible to provide an estimate smaller than the entire world population? The answer goes back to the discussion of genetic and statistical sampling in Chapter 1. An estimate formed as the product of allele frequencies refers to the collection of all possible genotypes that may be formed from those alleles, whereas the current world population is itself only a sample from this whole collection. Even six 10-allele loci allow more genotypes than there are people in the world, and an estimate can be given of the probability of any genotype regardless of whether that genotype actually exists. A case could be made for declaring that a six-locus match for 10-allele loci is very likely to indicate a common source for the two matching profiles rather than giving an estimate of the frequency of that profile. Exceptions to this non-numeric approach may be provided by the cases of the contributors of matching profiles being related or belonging to the same population, and by the case of mixed stains.

Likelihood Ratio for Members of the Same Population

For members of different populations, profile frequencies are independent. When S and P belong to the same population, however, there can be dependence. The issue is to determine conditional probabilities $\Pr(S = A|P = A)$ using the measures of relatedness appropriate to the population to which S and P belong.

The probabilities that random members S, P of the same population have the same genotypes were found above to be

$$\begin{aligned}\Pr(S = A_iA_i, P = A_iA_i) &= p_i[\delta + (4\gamma + 3\Delta - 7\delta)p_i \\ &\quad + 6(\theta - 2\gamma - \Delta + 2\delta)p_i^2 \\ &\quad + (1 - 6\theta + 8\gamma + 3\Delta - 6\delta)p_i^3]\end{aligned}$$

$$\begin{aligned}\Pr(S = A_iA_j, P = A_iA_j) &= 4p_ip_j[(\Delta - \delta) + (\theta - 2\gamma - \Delta + 2\delta)(p_i + p_j) \\ &\quad + (1 - 6\theta + 8\gamma + 3\Delta - 6\delta)p_ip_j]\end{aligned}$$

Evaluation of these expressions requires values for the four descent measures, $\theta, \gamma, \delta, \Delta$. Theoretical values can be obtained once the evolutionary forces acting on a population are specified. Procedures of estimation of the four quantities, however, have not yet been formulated (Weir 1994) , but one approach is to approximate the measures γ, δ, Δ by functions of θ and then use estimates of θ. In essence, this is what Balding and Nichols (1994) did in arriving at the approximate expressions

$$\begin{aligned}\Pr(S = A_iA_i|P = A_iA_i) &= \frac{[p_i + \theta(2 - p_i)][p_i + \theta(3 - p_i)]}{(1 + \theta)(1 + 2\theta)} \\ \Pr(S = A_iA_j|P = A_iA_j) &= \frac{2[p_i + \theta(1 - p_i)][p_j + \theta(1 - p_j)]}{(1 + \theta)(1 + 2\theta)}\end{aligned} \qquad (6.3)$$

These expressions are exact for populations at equilibrium under evolutionary forces, and as approximations they often perform well before equilibrium. Numerical values from these equations are shown in Table 6.6 for a locus at which all alleles have the frequencies p_i.

Likelihood Ratio for Relatives

The forensic likelihood ratio is easily modified for the case where the perpetrator of a crime is related to the suspect. Although unrelated individuals have a very low probability of sharing the same genetic profile, the probability increases for relatives. At any locus, father and son have one identical

Table 6.6 Conditional genotypic frequencies.

p_i	θ	$\Pr(S_{ii}\|P_{ii})$	$\Pr(S_{ij}\|P_{ij})$
0.005	0.000	0.0000	0.0001
	0.001	0.0001	0.0001
	0.005	0.0002	0.0002
	0.010	0.0006	0.0004
	0.050	0.0084	0.0055
	0.100	0.0293	0.0189
0.01	0.000	0.0001	0.0002
	0.001	0.0002	0.0002
	0.005	0.0004	0.0004
	0.010	0.0009	0.0008
	0.050	0.0097	0.0065
	0.100	0.0316	0.0204
0.05	0.000	0.0025	0.0050
	0.001	0.0028	0.0052
	0.005	0.0038	0.0059
	0.010	0.0052	0.0069
	0.050	0.0210	0.0171
	0.100	0.0502	0.0350
0.1	0.000	0.0100	0.0200
	0.001	0.0105	0.0203
	0.005	0.0124	0.0216
	0.010	0.0148	0.0231
	0.050	0.0381	0.0372
	0.100	0.0749	0.0584

allele, for example, while brothers may have zero, one, or two identical alleles. In each case identical means that the two alleles are both (recent) copies of the same allele. They are identical by descent, ibd, and so must be the same. The chance of two relatives having the same genetic profile therefore depends on the chance of their nonidentical alleles being the same.

For forensic calculations, the situation may be that a suspect matches the evidence sample, and the probability of having obtained a match if the evidence was left by some (untyped) relative is required. It will not be known which, if either, of the two alleles at a locus will be identical to an allele of the relative. Suppose the one-locus profile in question has alleles A_i and

Table 6.7 Probability that the perpetrator P has a genotype given that a suspect S has that genotype, when they are related. Likelihood ratio (L) assumes allele frequencies of 0.1.

Genotype A	Relationship	$\Pr(P = A \mid S = A)$	L
A_iA_j	Full brothers	$(1 + p_i + p_j + 2p_ip_j)/4$	3.3
	Father and son	$(p_i + p_j)/2$	10.0
	Half brothers	$(p_i + p_j + 4p_ip_j)/4$	16.7
	Uncle and nephew	$(p_i + p_j + 4p_ip_j)/4$	16.7
	First cousins	$(p_i + p_j + 12p_ip_j)/8$	25.0
	Unrelated	$2p_ip_j$	50.0
A_iA_i	Full brothers	$(1 + p_i)^2/4$	3.3
	Father and son	p_i	10.0
	Half brothers	$p_i(1 + p_i)/2$	18.2
	Uncle and nephew	$p_i(1 + p_i)/2$	18.2
	First cousins	$p_i(1 + 3p_i)/4$	30.8
	Unrelated	p_i^2	100.0

A_j for which the population frequencies are p_i and p_j. The chance that the father or the son of the suspect is also of type A_iA_j is $(p_i + p_j)/2$. Other probabilities are listed in Table 6.7, along with the likelihood ratios for loci in which alleles have frequencies of 0.1. These calculations make use only of the specified degrees of relationship, and they ignore any relationship following from membership in the same population. Those additional calculations were given by Weir (1994) but they have little effect for most populations.

MIXED SAMPLES

General Situation

Evidentiary samples sometimes contain DNA from more than one person. This may be obvious from the fact that more than two alleles per locus are found, or it may be dictated from circumstances of the crime. The circumstances may also point to some people being known contributors to the sample, such as the victim in a rape case. The general treatment of samples from more than one contributor (Weir et al. 1996) makes the need for a likelihood ratio approach very clear.

Suppose the sample has a set of alleles $\{e\}$ at some locus. There may be known contributors to the profile, having alleles $\{k\}$ between them, where the set $\{k\}$ is wholly contained within the set $\{e\}$. Under at least one of the alternative explanations, it is necessary to account for alleles $\{u\}$ from unknown people. If there are x unknown contributors to the profile, then what is needed is the probability $P_x(\{u\}|\{e\})$ that these people have alleles $\{u\}$ between them, but they do not have any alleles not in $\{e\}$. The x unknown people may also have alleles in the set $\{k\}$. When there are no unknown contributors, the symbol P_ϕ will be used, and this probability is one.

Calculating Probabilities

Profiles with one allele If the sample profile has only one allele, of type a, then all contributors must be aa homozygotes. Assuming independence of alleles within loci, the probability that one unknown contributor has only allele a is

$$P_1(a|a) = p_a^2$$

The probability of a, and only a, from x contributors is

$$P_x(a|a) = p_a^{2x}$$

If a known contributor to the profile is aa, then there are no other alleles that need to be accounted for. The probability needed, for x unknowns, is $P_x(\phi|a)$, where ϕ indicates the empty set. These unknown people can have only allele a, so

$$P_x(\phi|a) = p_a^{2x}$$

Profiles with two alleles If the sample profile has alleles a, b, then all contributors must be of genotypes aa, ab or bb. The probability that one unknown contributor has allele a, and no allele other than a or b, is

$$\begin{aligned} P_1(a|ab) &= p_a^2 + 2p_ap_b \\ &= (p_a + p_b)^2 - p_b^2 \end{aligned}$$

The generalization to x contributors is

$$P_x(a|ab) = (p_a + p_b)^{2x} - p_b^{2x}$$

since the case of all x contributors being homozygous bb does not provide allele a. If known contributors already have alleles a, b, then unknowns need not have any specific alleles but cannot have alleles other than a, b, and

$$P_x(\phi|ab) = (p_a + p_b)^{2x}$$

Finally, if there are no known contributors, then both alleles a, b must be carried by the unknown contributors

$$P_x(ab|ab) = (p_a + p_b)^{2x} - p_a^{2x} - p_b^{2x}$$

since not all x contributors can be of the same homozygous type.

Profiles with three alleles If the sample profile has alleles a, b, c, then all contributors must be of genotype aa, ab, bb, ac, bc or cc. If known contributors already have alleles a, b, c, then unknowns need not have any specific alleles but cannot have alleles other than a, b, c, and

$$P_x(\phi|abc) = (p_a + p_b + p_c)^{2x}$$

The probability that x unknown contributors have allele a, and no allele other than a, b or c, is

$$P_x(a|abc) = (p_a + p_b + p_c)^{2x} - (p_b + p_c)^{2x}$$

The probability that x unknown contributors have alleles a, b, and no allele other than a, b or c, is

$$P_x(ab|abc) = (p_a + p_b + p_c)^{2x} - (p_b + p_c)^{2x} - (p_a + p_c)^{2x} + p_c^{2x}$$

Finally, if there are no known contributors, then all alleles a, b, c must be carried by the unknown contributors, which means that $x > 1$, and

$$\begin{aligned} P_x(abc|abc) = {} & (p_a + p_b + p_c)^{2x} - (p_a + p_b)^{2x} - (p_b + p_c)^{2x} \\ & - (p_a + p_c)^{2x} + p_a^{2x} + p_b^{2x} + p_c^{2x} \end{aligned}$$

since not all contributors can be of the same homozygous type.

Profiles with four alleles Similar reasoning provides the four-allele results:

$$\begin{aligned} P_x(\phi|abcd) &= (p_a + p_b + p_c + p_d)^{2x} \\ P_x(a|abcd) &= (p_a + p_b + p_c + p_d)^{2x} - (p_b + p_c + p_d)^{2x} \\ P_x(ab|abcd) &= (p_a + p_b + p_c + p_d)^{2x} - (p_b + p_c + p_d)^{2x} - (p_a + p_c + p_d)^{2x} \\ &\quad + (p_c + p_d)^{2x} \\ P_x(abc|abcd) &= (p_a + p_b + p_c + p_d)^{2x} - (p_b + p_c + p_d)^{2x} - (p_a + p_c + p_d)^{2x} \\ &\quad - (p_a + p_b + p_d)^{2x} + (p_c + p_d)^{2x} + (p_b + p_d)^{2x} \\ &\quad + (p_a + p_d)^{2x} - p_d^{2x}, x > 1 \\ P_x(abcd|abcd) &= (p_a + p_b + p_c + p_d)^{2x} - (p_b + p_c + p_d)^{2x} - (p_a + p_c + p_d)^{2x} \\ &\quad - (p_a + p_b + p_d)^{2x} - (p_a + p_b + p_c)^{2x} + (p_c + p_d)^{2x} \\ &\quad + (p_b + p_d)^{2x} + (p_b + p_c)^{2x} + (p_a + p_d)^{2x} + (p_a + p_c)^{2x} \\ &\quad + (p_a + p_b)^{2x} - p_a^{2x} - p_b^{2x} - p_c^{2x} - p_d^{2x}, x > 1 \end{aligned}$$

Examples

For example, suppose a rape sample has alleles $\{e\} = abcd$. The victim is a known contributor and has alleles ab. The prosecution explanation C may be that a suspect, with alleles cd, is also a known contributor and the set $\{k\}$ is therefore $abcd$. There are no unknown people from the prosecution viewpoint, and no unexplained alleles, and the required probability is $P_\phi = 1$. The defense explanation $\bar{C}$ may be that the victim was a contributor to the sample, but alleles cd are from an unknown contributor, and the required probability is $P_1(cd|abcd)$. The likelihood ratio for assessing the strength of the evidence is

$$\begin{aligned} L &= \frac{P_\phi}{P_1(cd|abcd)} \\ &= \frac{1}{2p_c p_d} \end{aligned}$$

which is the same result as if the victim's profile had not been included in the sample.

Alternatively, the defense explanation may be that neither contributor to the evidentiary sample is known, as could happen if the sample was taken from a location other than the victim's person. In that case, the probability for the defense explanation is $P_2(abcd|abcd)$, and

$$\begin{aligned} L &= \frac{P_\phi}{P_2(abcd|abcd)} \\ &= \frac{1}{24p_a p_b p_c p_d} \end{aligned}$$

It is not known whether the two unknowns are ab, cd or ac, bd or ad, bc.

Situations can arise where the numerator of the likelihood ratio is not one, such as in a crime where it is known that there are two perpetrators who contributed to the evidentiary sample, but only one suspect is at hand. If the sample has alleles $abcd$ and the suspect is of type ab, then the prosecution explanation is that the contributors were the suspect and some other person, whereas the defense explanation may be that the contributors were two unknown people. The likelihood ratio is

$$\begin{aligned} L &= \frac{P_1(cd|abcd)}{P_2(abcd|abcd)} \\ &= \frac{2p_c p_d}{24p_a p_b p_c p_d} \\ &= \frac{1}{12p_c p_d} \end{aligned}$$

showing that the strength of the evidence against the suspect is diminished by a factor of six over what it would have been had the stain shown only that person's profile.

A more complex example is when sample E has only three alleles abc and includes contributions from a victim and a perpetrator. If the victim has genotype ab and a suspect has profile aa, then the explanation C including the suspect still requires another contributor for allele c. The alternative explanation $\bar{C}$ may still include the victim but excludes the suspect, and so includes some unknown contributors. For example, if $\bar{C}$ supposes two unknown contributors

$$\begin{aligned} L &= \frac{P_1(c|abc)}{P_2(c|abc)} \\ &= \frac{1}{(p_a + p_b + p_c)^2 + (p_a + p_b)^2} \end{aligned}$$

However, if $\bar{C}$ had only one unknown contributor to account for allele c then the likelihood ratio is 1. The suspect is not excluded as a contributor, but the mixed sample is equally likely to be of type abc if came from the suspect and the victim or if it came from an unknown person and the victim.

Further discussion on the interpretation of mixed samples, including the effects of unseen alleles and the numbers of contributors and their ethnicities, is given by Weir et al. (1996).

ERROR RATES

There are three possible explanations for why a suspect may be declared to have the same genetic profile as an evidentiary sample:

- The suspect provided the sample.
- The suspect did not provide the sample, and has the profile by chance.
- The suspect did not provide the sample, and the matching is a false positive.

It is the second of these explanations that depends on population genetic principles and has been discussed in this chapter. The third explanation falls outside the province of this book.

The possibility of laboratory mistakes needs to be reflected in laboratory protocols, chain of custody rules, and so on, that are designed to prevent errors. Proficiency tests are necessary to identify sources of error, but they do not provide false positive error rates. At most, proficiency test error rates provide prior probabilities of error that need to be modified in light of the procedures used in a particular case. The suggestion of Lempert (1991) that the mistake rate and estimated profile frequency be combined does not seem to be appropriate since the specific conditions surrounding the analysis of a particular forensic case argue against the concept or application of an average error rate. The issue was discussed by Roeder (1994).

EXERCISES

Exercise 6.1

Determine the coancestry coefficients for uncle and niece, parent and offspring, grandparent and grand-offspring, and first cousins. Assume all individuals are noninbred.

Exercise 6.2

Find the probability with which an uncle and nephew both have Aa genotypes, if they belong to a population in Hardy-Weinberg equilibrium with frequencies p_A, p_a.

Exercise 6.3

The seven-locus profiles of three people, OS, NB and RG, and a sample taken from a sock, are as follows:

Locus	OS	NB	RG	Sock
LDLR	A, B	A, B	A, B	A, B
GYPA	B, B	A, B	A, A	A, B
HBGG	B, C	A, B	A, A	A, B
D7S8	A, B	A, B	B, B	A, B
GC	B, C	A, C	A, A	A, C
DQα	1.1,1.2	1.1,1.1	1.3,4	1.1,1.1
D1S80	24,25	18,18	24,24	18,18

What is the conditional probability of the sock profile if that profile was from some person, unrelated to NB, belonging to either of two populations (African American or Caucasian) for which the following sets of allele frequencies are appropriate? How would these probabilities change if the sock profile and the NB profile both came from (different) people in a population for which $\theta = 0.03$ and which had the Caucasian allele frequencies? Ignore the possibility of error.

	Database	Allele			
LDLR		A	B		
	Afr. Am.	0.224	0.776		
	Caucasian	0.453	0.547		
GYPA		A	B		
	Afr. Am.	0.479	0.521		
	Caucasian	0.584	0.416		
HBGG		A	B	C	
	Afr. Am.	0.506	0.197	0.297	
	Caucasian	0.470	0.527	0.003	
D7S8		A	B		
	Afr. Am.	0.614	0.386		
	Caucasian	0.615	0.385		
GC		A	B	C	
	Afr. Am.	0.103	0.707	0.190	
	Caucasian	0.257	0.172	0.571	
DQα*		1.1	1.2	1.3	4
	Afr. Am.	0.117	0.314	0.055	0.338
	Caucasian	0.122	0.176	0.041	0.328
D1S80*		18	24	25	
	Afr. Am.	0.072	0.228	0.048	
	Caucasian	0.230	0.341	0.061	

Exercise 6.4

Calculate the paternity indices for the case where the true father is an uncle, or a first cousin, of the alleged father of type A_iA_j when the obligate paternal allele is A_i.

Exercise 6.5

The victim of a rape has genotype *ab* and the suspect has genotype *ac*. If the mixed sample from the victim and the perpetrator has profile *abc*, what is the likelihood ratio for the hypotheses that the suspect was the perpetrator versus some person unrelated to the suspect being the perpetrator? The analysis could include the possibility of an unseen fourth allele in the mixed profile.

Chapter 7

Linkage

INTRODUCTION

When data are available from successive generations there is the opportunity to study the degree of linkage between pairs of loci. Estimates of the recombination fraction enable loci to be ordered and placed onto a genetic map. With loci that are genetic markers, a genetic map can be established, and then an estimate of the strength of linkage between markers and traits will be used to indicate the relative positions of markers and genes affecting those traits. Plant and animal breeders can carry out specified crosses to simplify the detection of linkage and the estimation of its strength. For example, methods have been developed to make use of F_2 populations between parental lines and backcrosses to those lines. Linkage studies in humans make use of family pedigrees where genes of interest are segregating. In both situations, heavy use is made of likelihood procedures. It may also be possible to infer the strength of linkage by estimating the amount of association between markers and traits in populations without having direct observations on successive generations.

The general means for constructing genetic maps, depending on the joint inheritance patterns of pairs of markers, has been established for some time. The challenges presented by molecular data have to do with the need to map many loci at the same time, and the need to map loci that are close together.

Good sources for linkage studies are books by Mather (1951) and Bailey (1961), while comprehensive accounts of procedures for human populations are contained in the books by Ott (1991) and Terwilliger and Ott (1994). This chapter will consider estimation of recombination fractions between loci considered two at a time, rather than more general multipoint methods (Ott 1991).

DISTANCES BETWEEN GENES

Suppose loci A and B, with alleles A, a and B, b are located on the same arm of a chromosome and they are not transmitted independently between generations. The ***recombination fraction*** c_{AB} between them is the probability that the gamete transmitted by an individual is recombinant, having been derived by crossing over between the two parental gametes received by that individual. In the absence of any segregation distortion, a double heterozygote AB/ab that received gametes AB and ab is expected to transmit the four possible gametic types in the proportions

$$\frac{1-c_{AB}}{2}AB + \frac{c_{AB}}{2}Ab + \frac{c_{AB}}{2}aB + \frac{1-c_{AB}}{2}ab$$

The recombination fraction is expected to lie in the interval $[0, \frac{1}{2}]$, and is sometimes transformed to a ***linkage parameter*** λ by

$$\lambda_{AB} = 1 - 2c_{AB}$$

which ranges from 0 for unlinked loci (free recombination) to 1 for completely linked loci (no recombination).

Two related quantities need to be introduced. There is a ***physical distance*** d_{AB} between any two regions on a chromosome. Although this may be difficult to define for a structural gene with many exons and introns, once some feature is specified there is a definite distance, usually measured in thousands of bases, kilobases (kb). A complete DNA sequence between two restriction sites, for example, allows this distance to be given precisely. In other cases, lengths can be estimated from migration distances of DNA fragments on an electrophoretic gel. Maps based on physical distances can be constructed by radiation hybrid mapping (Cox et al. 1990), whereby hybrid cell lines containing chromosomal fragments produced by radiation breakage are screened to identify the hybrids that have retained a given locus. Similar retention patterns imply proximity, and have been used as part of the process of generating detailed physical maps (e.g. Hudson et al. 1995). The human genome, the complete set of DNA passed from one parent to a child, has a length of about 3.3×10^6 kb (Vogel and Motulsky 1986).

As a result of linkage and other studies it is possible to specify the relative positions of genes on the chromosomes, and to assign them a position on a genetic map. The distance between two genes can therefore also be described by their ***genetic map distance*** x_{AB}. As a unit of map distance, the ***Morgan*** (M) is defined as that distance along which one crossing over is expected to occur per gamete per generation. The total length of the (sex-averaged) human map is about 33 M, or about 3.3×10^3 centiMorgans, cM (Ott 1991).

The rule of thumb, therefore, is that in humans 1 cM $\approx$ 1000 kb. Although map distance increases with physical distance, there is no simple relation between them. In a review of the kb/cM ratios for various species, Meagher et al. (1988) stress that the ratio can vary considerably within a genome to the point at which it may be quite misleading to work with single values.

Now the recombination fraction is a dimensionless quantity. To estimate it from the kinds of data discussed below, it is necessary to identify recombinant gametes, which in turn result from an odd number of crossovers. The recombination fraction increases with map distance, so the tendency for crossing over can be quantified by map distance, but once again there is no universal or simple relation between them. When crossovers occur independently, Haldane (1919) showed that the recombination fraction c could be expressed in terms of the map distance x as

$$c_{AB} = \frac{1}{2}\left(1 - e^{-2x_{AB}}\right)$$

Note that c_{AB} increases from 0 to 0.5 as x_{AB} increases from 0. Rewriting this expression as

$$x_{AB} = -\frac{1}{2}\ln(1 - 2c_{AB})$$

allows a comparison to other mapping functions. Kosambi (1944) allowed for ***interference***, whereby one crossover tends to prevent other crossovers in the same region, to derive

$$x_{AB} = \frac{1}{4}\ln\left(\frac{1 + 2c_{AB}}{1 - 2c_{AB}}\right)$$

The amount of interference allowed in the Kosambi map function decreases as the loci get further apart, and is zero for unlinked loci ($c = 0.5$). For both the Haldane and the Kosambi functions, $x_{AB} \approx c_{AB}$ when c_{AB} is small. More complex functions have been found empirically (e.g., Rao et al. 1977).

ESTIMATION OF RECOMBINATION

Backcross Method

From two lines homozygous for different alleles at each locus, say $AABB$ and $aabb$, the F_1 cross population is entirely doubly heterozygous AB/ab and the backcross $AB/ab \times aabb$ yields four classes of offspring. In the first place, these four categories allow a check to be made on the segregations at each locus and the joint segregation at the two loci. The notation employed

Table 7.1 Two-locus counts for a backcross. Each locus has two alleles, with one dominant to the other.

	Offspring				
Genotype	AB/ab	Ab/ab	aB/ab	ab/ab	
Phenotype	AB	Ab	aB	ab	Total
Observed	a	b	c	d	n
Expected	$\frac{n}{4}$	$\frac{n}{4}$	$\frac{n}{4}$	$\frac{n}{4}$	n

by Bailey (1961) is shown in Table 7.1. The table also shows the phenotypes of the four offspring types when alleles A, B are dominant to a, b.

At locus A there are $a+b$ individuals that received allele A and $c+d$ that received a, so that a chi-square test statistic for independent segregation of alleles at this locus is given by

$$X_A^2 = \frac{(a+b-n/2)^2}{n/2} + \frac{(c+d-n/2)^2}{n/2} = \frac{(a+b-c-d)^2}{n} \tag{7.1}$$

with a similar statistic for locus B:

$$X_B^2 = \frac{(a+c-b-d)^2}{n} \tag{7.2}$$

For individuals that receive allele B, segregation at locus A is measured by $a-c$, whereas for individuals receiving b it is measured by $b-d$. Linkage between the loci causes alleles not to be transmitted independently, so the segregation at one locus is affected by the allele present at the other. For locus A this effect can be quantified by $(a-c)-(b-d)$. The same quantity would result from looking at the effect of A on segregation at locus B. Linkage is therefore detected by comparing $a+d$ to $b+c$. These terms should be the same for independent loci. An appropriate chi-square, by analogy to Equations 7.1 and 7.2, is

$$X_{AB}^2 = \frac{(a+d-b-c)^2}{n} \tag{7.3}$$

As Bailey (1961) points out, these three chi-squares, each with 1 d.f., add up to the goodness-of-fit chi-square with 3 d.f. found by comparing the four counts to one-fourth of the sample size

$$\begin{aligned} X^2 &= \frac{(a-\frac{1}{4}n)^2}{\frac{1}{4}n} + \frac{(b-\frac{1}{4}n)^2}{\frac{1}{4}n} + \frac{(c-\frac{1}{4}n)^2}{\frac{1}{4}n} + \frac{(d-\frac{1}{4}n)^2}{\frac{1}{4}n} \\ &= X_A^2 + X_B^2 + X_{AB}^2 \end{aligned}$$

As an example, consider the maize data reported by Goodman et al. (1980) for enzymes *Mdh5* with alleles *E12, E15*, and *Got3* with alleles *U4, U8* in maize. The following testcross was set up

$$\frac{E12,U4}{E15,U8} \times \frac{E12,U4}{E12,U4}$$

and the resulting offspring array was

$\frac{E12,U4}{E12,U4}$	$\frac{E12,U8}{E12,U4}$	$\frac{E15,U4}{E12,U4}$	$\frac{E15,U8}{E12,U4}$
53	13	9	55

The chi-square test statistics for these data are

$$X^2_{Mdh5} = \frac{(66 - 64)^2}{130} = 0.03$$

$$X^2_{Got3} = \frac{(62 - 68)^2}{130} = 0.28$$

$$X^2_{MDH5,Got3} = \frac{(108 - 22)^2}{130} = 56.89$$

$$X^2 = \frac{20.5^2 + 19.5^2 + 23.5^2 + 22.5^2}{32.5} = 57.20$$

Both loci are segregating normally, but there is strong evidence for linkage between the two loci.

With two alleles per locus and reasonable sample sizes, the chi-square tests perform well for detecting linkage. It would be possible to use an exact test to determine if the observed offspring array (essentially a gametic array), conditional on the observed allele arrays, had an unusually low probability. The procedure described in Chapter 3 for testing for gametic linkage disequilibrium would be used.

The chi-square test approach obscures one feature of testing for linkage: it is usually assumed that there is a biological constraint that the number of recombinants should not be more than the number of nonrecombinants. For unlinked loci ($c = 0.5$), each of the four gametic types AB, Ab, aB, ab from a double heterozygote AB/ab is expected to be equally frequent, and for completely linked loci ($c = 0$) there are expected to be no recombinants. As the amount of recombination increases, the two parental gametes AB, ab are expected to decrease in frequency and the two recombinant gametes Ab, aB to increase. However, it is not expected that the frequency of recombinant

Table 7.2 Two-locus counts for a backcross for linked loci.

	Offspring				
Genotype	AB/ab	Ab/ab	aB/ab	ab/ab	
Phenotype	AB	Ab	aB	ab	Total
Observed	a	b	c	d	n
Expected	$\frac{n}{2}(1-c_{AB})$	$\frac{n}{2}c_{AB}$	$\frac{n}{2}c_{AB}$	$\frac{n}{2}(1-c_{AB})$	n

gametes should exceed the frequency of parental gametes, so $c \leq 0.5$. But the test statistic in Equation 7.3 can be large if there are more *or* less observed recombinant gametes $(b+c)$ than nonrecombinant gametes $(a+d)$. A significantly large number of recombinant gametes would indicate a problem with the underlying genetic model. If a chi-square test is performed for the case where it is known which offspring are parental and which are recombinant (i.e. "phase known" in the treatment of Doerge 1995), the test is one-sided and the significance level from the chi-square table should be double the required value. For cases in which there are two classes of offspring with proportions $\tilde{p}$ and $1-\tilde{p}$, one parental and one recombinant, but it is not known which is which, the estimated recombination fraction c is taken to be the proportion of the least frequent class.

$$\hat{c} = \min(\tilde{p}, 1-\tilde{p})$$

The alternative hypothesis in this "phase-unknown" case is that the proportion p is not 0.5 and the test is two-sided. The chi-square tables may be used without adjustment.

For the backcross $AB/ab \times ab/ab$, the expected values of the four offspring types when the loci are linked are shown in Table 7.2 as functions of the unknown recombination fraction c_{AB}.

The likelihood of c_{AB} for these counts is

$$L(c_{AB}) \propto (1-c_{AB})^{a+d}(c_{AB})^{b+c}$$

so that the score and expected information (Chapter 2) are

$$S_{c_{AB}} = -\frac{a+d}{1-c_{AB}} + \frac{b+c}{c_{AB}}$$

$$I_{c_{AB}} = \frac{n}{c_{AB}(1-c_{AB})}$$

Table 7.3 Two-locus counts for F_2 population. Each locus has two alleles with one dominant to the other.

	Offspring				
Genotype	AB/ab	Ab/ab	aB/ab	ab/ab	
Phenotype	AB	Ab	aB	ab	Total
Observed	a	b	c	d	n
Expected	$\frac{9n}{16}$	$\frac{3n}{16}$	$\frac{3n}{16}$	$\frac{n}{16}$	n

The MLE and its variance are therefore given by

$$\hat{c}_{AB} = \frac{b+c}{n}$$

$$\text{Var}(\hat{c}_{AB}) = \frac{c_{AB}(1-c_{AB})}{n}$$

There is a restriction to be placed on the estimate. Maximum likelihood estimates must lie in the range of validity of the parameter, so that $\hat{c}$ must lie in the interval [0, 0.5]. If the ratio $(b+c)/n$ is greater than 0.5, the estimate is set to 0.5, and the correct expression for the MLE is

$$\hat{c} = \min\left(\frac{b+c}{n}, 0.5\right)$$

F_2 Population Method

The other common situation in plant or animal breeding is where F_2 data are available. When the F_1 double heterozygotes AB/ab are crossed there are 10 possible genotypes among the offspring, and these fall into four recognizable classes when there is dominance at both loci. For independent segregation within and between loci the data now have the expectations shown in Table 7.3.

Repeating the argument made above for the double backcross leads to the following three test statistics:

$$X_A^2 = \frac{(a+b-\frac{3}{4}n)^2}{\frac{3}{4}n} + \frac{(c+d-\frac{1}{4}n)^2}{\frac{1}{4}n}$$

$$= \frac{(a+b-3c-3d)^2}{3n}$$

Table 7.4 Two-locus counts for F_2 population for linked loci.

	Offspring				
Genotype	AB/ab	Ab/ab	aB/ab	ab/ab	
Phenotype	AB	Ab	aB	ab	Total
Observed	a	b	c	d	n
Expected	$\frac{n}{4}(2+\theta)$	$\frac{n}{4}(1-\theta)$	$\frac{n}{4}(1-\theta)$	$\frac{n}{4}\theta$	n

$$X_B^2 = \frac{(a+c-3b-3d)^2}{3n}$$

$$X_{AB}^2 = \frac{(a+9d-3b-3c)^2}{9n}$$

The expected segregation at each locus is now 3:1, instead of 1:1 as for the backcross.

To estimate the recombination fraction from F_2 data, the expectations are displayed in Table 7.4 using Mather's (1951) simplifying notation of $\theta = (1-c_{AB})^2$. The likelihood is

$$L(\theta) \propto (2+\theta)^a(1-\theta)^{b+c}(\theta)^d$$

The first two derivatives with respect to θ give the score and information

$$S(\theta) = -\frac{n\theta^2 - (a-2b-2c-d)\theta - 2d}{(2+\theta)\theta(1-\theta)}$$

$$I(\theta) = \frac{n(1+2\theta)}{2\theta(1-\theta)(2+\theta)}$$

Recall that the inverse of the information provides the variance of a maximum likelihood estimate. Estimation of θ therefore requires the solution of a quadratic equation. Invoking the properties of MLE's allows the MLE of c_{AB} to then be found as

$$\hat{c}_{AB} = 1 - \sqrt{\hat{\theta}}$$

and the variance of this estimate is

$$\begin{aligned}\text{Var}(\hat{c}_{AB}) &= \left(\frac{d\hat{c}_{AB}}{d\hat{\theta}}\right)^2 \text{Var}(\hat{\theta}) \\ &= \frac{(1-\theta)(2+\theta)}{2n(1+2\theta)}\end{aligned}$$

A method of moments estimator has been suggested for the F_2 population situation (Fisher and Balmakund 1928, Immer 1930). The ratio $K = ad/bc$, with each term replaced by its expectation, has a simple form:

$$\frac{\mathcal{E}(a)\mathcal{E}(d)}{\mathcal{E}(b)\mathcal{E}(c)} = \frac{\theta(2+\theta)}{(1-\theta)^2}$$

suggesting that $\hat{\theta}$ be found as a solution to

$$(K-1)\theta^2 - 2(K+1)\theta + K = 0$$

Immer (1930) constructed tables of $\hat{\theta}$ for values of $K = ad/bc$. Since the estimator is a homogeneous equation in multinomial counts, Fisher's variance formula can be used:

$$\begin{aligned}
\text{Var}(K) &= \left(\frac{\partial T}{\partial a}\right)^2 (2+\theta) + \left(\frac{\partial T}{\partial b}\right)^2 (1-\theta) + \left(\frac{\partial T}{\partial c}\right)^2 (1-\theta) \\
&\quad + \left(\frac{\partial T}{\partial d}\right)^2 \theta - \left(\frac{\partial T}{\partial n}\right)^2 \\
&= \frac{2\theta(2+\theta)(1+2\theta)}{n(1-\theta)^5}
\end{aligned}$$

However,

$$\text{Var}(K) = \left(\frac{dK}{d\hat{\theta}}\right)^2 \text{Var}(\hat{\theta})$$

and

$$\frac{dK}{d\theta} = \frac{2(1+2\theta)}{(1-\theta)^3}$$

so that

$$\text{Var}(\hat{\theta}) = \frac{2\theta(1-\theta)(2+\theta)}{n(1+2\theta)}$$

Now this result is the same as that obtained for the MLE of θ. This finding is unusual in that MLE's are known to be the most ***efficient estimators***, meaning that they have the smallest variances. It is rare for another estimator to be equally efficient.

Mather (1951) allowed a more general treatment of the F_2 method in that different recombination fractions can be assigned to males and females. If the fractions are c_{AB_m}, c_{AB_f} in males and females, then the above analysis holds with

$$\theta = (1 - c_{AB_m})(1 - c_{AB_f})$$

Direct Method

Inferences about recombination in humans need to be made from families, since crosses between homozygous lines cannot be arranged. In his review of methodology, Ott (1991) covered the ***direct method*** in which the recombination fraction is estimated as the observed number of recombinant gametes in a set of offspring. His example is shown in Figure 7.1. Individuals in this three-generation pedigree have been classified by *ABO* blood type and by status of the autosomal dominant trait *CMT1* (hereditary motor and sensory neuropathy type 1). Normal individuals (empty symbols in Figure 7.1) are homozygous for the normal allele t. Affected individuals (symbols marked with *Aff.*) from normal parents are heterozygous Tt. The blood groups of children often allow the *ABO* genotype of a parent to be inferred: individual 3.1 is known to be *AA* or *AO* on the basis of his blood group, but the occurrence of his *OO* children rules out *AA*. Following through the pedigree allows the two-locus genotype of 3.1 to be determined as *TA/tO*, and those of his four offspring to be *TA/tO*, *TO/tO*, *TO/tO*, and *TO/tO*, respectively. Evidently individual 3.1 produced one nonrecombinant gamete *TA* and three recombinant *TO* gametes or ***haplotypes***. No estimate is possible from the other parent, 3.2, since she is double homozygous *tO/tO*. The estimated recombination fraction between the *CMT1* and *ABO* loci is therefore 3/4, and this analysis has been called the ***three-generation method***. (The estimate would be reported as 1/2.)

Lod Scores with Phase Known

The only other method to be treated here is the likelihood method, introduced for human studies by Haldane and Smith (1947) in their estimation of the recombination fraction between the X-linked genes for color blindness and hemophilia. These authors sought to maximize the likelihood $L(c)$ for the recombination fraction, but did this by a variant of the grid-search method (Chapter 2) instead of using numerical iterations. To demonstrate that two loci were linked, they compared the likelihood of some c value with the likelihood of $c = 0.5$ (unlinked loci). The logarithm to base 10 of this ratio is termed the ***lod score*** (for log-odds) $Z(c)$ for that c value

$$Z(c) \quad = \quad \log\left[\frac{L(c)}{L(0.5)}\right]$$

For the pedigree in Figure 7.1, the likelihood calculated for the four children is that of obtaining three recombinants, each with probability c, and one nonrecombinant, with probability $1 - c$, among four independent

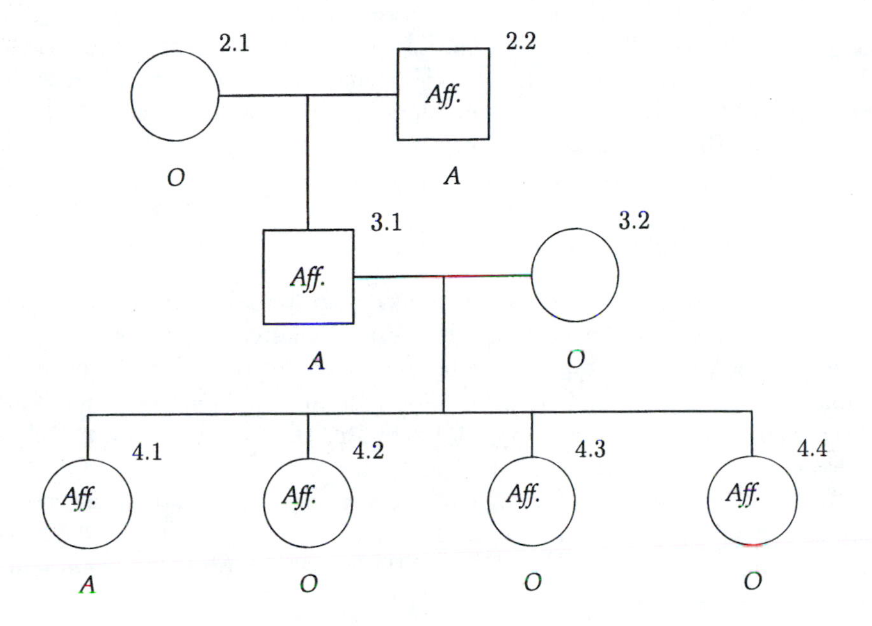

Figure 7.1 Three-generation pedigree. Individuals marked *Aff.* have the trait CMT1. Letters under each individual show *ABO* blood type. Source: Ott 1991.

gametes. The multinomial distribution is appropriate

$$L(c) \propto (1-c)c^3$$

and the lod score is

$$Z(c) = \log[16(1-c)c^3]$$

The maximum likelihood estimate of the recombination fraction is that value for which the lod score is maximized. When the MLE $\hat{c} < 0.5$ is more likely than $c = 0.5$, the lod score will be positive, and a conventional rule is to conclude that loci are linked whenever the lod score exceeds 3 ("odds of 1000 to 1"). Conversely, if the MLE $\hat{c}$ <0.5 is less likely than c =0.5, the lod score is negative, and the convention is to conclude a lack of linkage when $Z(c)$ is less than -2. For $\hat{c}$ values giving a lod score between these limits, Morton (1955) suggests the collection of more data. For the pedigree in Figure 7.1, the lod score is maximized, in the allowable range of c values, for $\hat{c}$ =0.5. As $Z(0.5)$=0, however, no definite conclusions about linkage can be drawn from such a small amount of data.

A key feature of linkage analysis from human families is that information can be combined from independent families. The joint likelihood calculated from a set of independent families is just the product of the separate likelihoods, and the overall lod score is therefore the sum of the separate scores. The same lod score from the family with four children in Figure 7.1 could have been obtained from two independent families, of the same structure, one family with evidence of two recombination events and one family with one recombinant and one nonrecombinant.

Lod Scores with Phase Unknown

The probabilities of offspring combinations cannot be calculated simply from the binomial theorem when the parents are double heterozygotes of unknown phase. Suppose one parent is a double homozygote ab/ab but the other is a double heterozygote and no information in the available data indicates whether this parent has genotype AB/ab or Ab/aB. It is generally assumed in this situation that each phase is equally likely.

A child of genotype AB/ab or ab/ab is a nonrecombinant if the second parent was in the AB/ab phase, and is a recombinant if the parent was Ab/aB. Such a child is said by Ott (1991) to be of type 1. For dominant loci, this may be phrased as a child of phenotype AB or ab being of type 1. Other children are of type 2. The probability of obtaining a type 1 child is the probability of an AB/ab parent and no recombination plus the probability of an Ab/aB parent and recombination

$$\Pr(\text{Type 1 child}) = \frac{1}{2}(1-c) + \frac{1}{2}c = \frac{1}{2}$$

Similarly the probability of a type 2 child is 0.5. However, the probability of two type 1 children is

$$\Pr(\text{Two type 1 children}) = \frac{1}{2}(1-c)^2 + \frac{1}{2}c^2 \neq \frac{1}{4}$$

demonstrating that, although the recombination events in distinct gametes are independent, the genotypes of two children in the same family are not independent. A family with two children of type 2 (i.e., those that are recombinant if parental phase AB/ab is assumed and nonrecombinant if phase Ab/aB is assumed) gives a lod score of

$$Z(c) = \log[2(1-2c+2c^2)]$$

The lod score for two independent families, one with two type 1 children and one with one of each type of child, is the sum of the two separate lods

$$Z(c) = \log[2(1-2c+2c^2)] + \log[4c(1-c)]$$

Calculating likelihoods for pedigrees with more than two generations can be very difficult, but general algorithms have been developed. Reference should be made to Elston and Stewart (1971), Ott (1974), and Lange and Elston (1975).

GENETIC MAP CONSTRUCTION

If the recombination fractions are known between each pair of markers, it is possible in principle to order the markers. The actual construction of a genetic map from estimated recombination fractions, however, can be quite difficult. As Thompson (1987) pointed out, loci cannot be ordered until there has been at least one recombinant observed between every pair of loci. A good review of the locus-ordering problem was given by Weeks (1991).

A preliminary notion is based on just three loci, *A, B, C*. Suppose the true order is alphabetical. There will be a detectable recombinant between *A* and *C* if there is recombination between *A* and *B* but not between *B* and *C*, and vice versa. If there is no interference,

$$c_{AC} = c_{AB}(1 - c_{BC}) + (1 - c_{AB})c_{BC}$$

The sum of adjacent recombination coefficients, ***sar***, for the order *A, B, C* is $c_{AB} + c_{BC}$.

If the order of the loci was mistakenly believed to be *B, A, C*, then the ***sar*** would be calculated as $c_{AB} + c_{AC}$. From the relation derived under the true order,

$$\begin{aligned} c_{AB} + c_{AC} &= c_{AB} + c_{AB}(1 - c_{BC}) + (1 - c_{AB})c_{BC} \\ &= c_{AB} + c_{BC} + c_{AB}(1 - 2c_{BC}) \\ &\geq c_{AB} + c_{BC} \end{aligned}$$

since $(1 - 2c_{BC}) \geq 0$. The sar under the wrong order is not less than that under the correct order. The same holds if the order is mistakenly believed to be *A, C, B*, showing that the true order is the one with the minimum sar. This property holds for any number of loci.

There are two problems with using the sar as a criterion for determining the correct order for a set of loci. Only the estimated recombination fractions are known, and the best order for this set may not be the true order simply because of sampling variation. The other problem is that there are too many orders to examine once the number of markers gets large. The number of ways of ordering m markers is $m!/2$, and for $m = 10$ this is 1,814,400. The same problems face the use of other optimizing criteria. The sum of adjacent lods ***sal*** uses (the logarithm of) the maximum likelihoods defined on the gametic counts for the intervals defined by each pair of markers assumed to be adjacent. Alternative strategies are needed.

Seriation

Buetow and Chakravarti (1987a,b) provided an algorithm that copes with the problem of ordering a large number of markers. Their method of ***seriation*** was based on earlier work by Gelfand (1971). For a set of m loci and estimated recombination coefficients $\hat{c}_{uv}$ between loci u, v, repeat the following procedure for each locus i:

1. Write the locus L_i.

2. Find the locus L_j from the remaining $(m-1)$ loci that has the smallest value of $\hat{c}_{ij}$. Place this to the right of L_i.

3. For each of the remaining $(m-2)$ loci, choose the locus L_k from the remaining unplaced loci with the smallest value of $\hat{c}_{ik}$. Place it on the end of the cluster of placed loci with which it has the smallest $\hat{c}$ value.

This algorithm requires only m orders to be examined, and that with the lowest sar can be taken as the best. Buetow and Chakravarti (1987a) give some rules for handling cases of orders with equal values of the criterion. Although the method works quite well in practice, there is no assurance that the correct order will be found.

Simulated Annealing

A method which examines a much larger portion of the set of all possible orders is termed ***simulated annealing*** (Kirkpatrick et al. 1983) and was set up with language depending on the observation that when liquids are cooled very slowly, they will crystallize in a state of minimum energy. With this language, suppose the energy of a locus order is written as E_1, meaning the value of a criterion such as sal or sar. The order is changed by some random permutation of loci and the new energy, E_2, is determined.

If $E_2 < E_1$, the new order is adopted. However, if $E_2 > E_1$, the new order is adopted with probability P, where

$$P = \frac{1}{k_b T} e^{-(E_2 - E_1)}$$

The intention of allowing a worse order to be chosen is to prevent the algorithm from settling down at a local optimum order and ignoring a global optimum somewhere else in the space of orders. The new order is permuted and the process continues.

The quantity T corresponds to temperature, and k_b is a physical constant known as the Boltzman constant. Initially k_bT is given a low value, meaning that worse orders will be allowed with a high probability. There is virtually no restriction on which order may be chosen next. As the process continues, P is lowered, meaning that it becomes increasingly difficult for a worse order to be examined further. In practice, it is desirable to keep an updated list of the best set of orders examined to date, in case no further improvement is found.

Evidently, several decisions need to be made to implement the algorithm. There have been suggestions that, for m loci, $100m$ new orders or $10m$ improved orders be examined for each value of T, and that T be changed to $0.95T$ at that point. The algorithm stops after 200 values of T have been used, or after 100 permutations have failed to provide a better order. Once again, the method has been used successfully, but there is no guarantee of obtaining the true order.

Branch and Bound

There is an algorithm that is guaranteed to find the best order. Because this ***branch and bound*** method (Thompson 1987) has such a guarantee, there will be occasions when it would require an examination of all orders but generally it requires only a small set of all possible orders to be examined.

Thompson took the implied total number of crossover events for a data set when a particular order is assumed as the criterion by which to judge orders. This is related to, but is not the same as, the sar. The key observation is that this criterion is additive over gametes: it cannot decrease when more gametes are observed or when more loci are added. The number of crossover events is computed by noting the grandparental origins of the alleles at each locus. For each gamete, each locus is assigned a score of 1 if it carries a grandpaternal allele, a score of 2 if it carries a grandmaternal allele, and a score of 0 if the grandparental origin cannot be determined. Reading the scores along the gamete causes the number of crossover events to increase by one each time the score changes from 1 to 2 or from 2 to 1.

The branch and bound method is an example of a tree-structured search. The nodes, or branch points, of the tree correspond to loci and the alternative branches to alternative points of insertion of loci. From Figure 3 in Thompson (1987), suppose the first two loci considered are *C, D*. The next locus *B* can be inserted in one of three positions, corresponding to orders *BCD*, *CBD*, and *CDB*. The fourth locus, *F*, can be inserted at four different positions for each of those three orders, and so on. At level i, when the ith locus is to be inserted into the growing order, there are i branch points. The implied number of recombinants in the data set is calculated for each branch. Because this number cannot decrease as more loci are added later, any subtree can be removed from further consideration if it has a number of recombinants greater than that (the comparison value) for another subtree. If the number of recombinants for order *BCD* is less than those for *CBD* and *CDB*, for example, there is no need to consider any subsequent ordering which has loci *B,C,D* in either of the orders *CBD* or *CDB*.

The algorithm gives a very efficient search through the space of orders if an order with a low criterion value is found early in the search to give a low comparison value, and if high increases in the criterion are acquired at low levels in the tree to reduce the chances of having to search many subtrees. The method works well when the locus chosen to be inserted next gives the largest average immediate increase in implied number of recombinants over all alternative positions of insertion. Unfortunately, there is no guarantee that all orders will not need to be examined.

Error Detection

Although sampling variation is expected in recombination estimates, mistakes in typing or recording genotypes can cause problems in using genetic maps. The problems increase as the maps become more dense. Methods for detecting these errors have been discussed by Lincoln and Lander (1992), Ott (1993) and Ehm et al. (1996). Errors tend to introduce spurious recombination events, and so inflate the genetic map. As the map becomes denser, and the recombination rate between adjacent markers decreases to the error rate, thought to be about $0.5\% \sim 3\%$, a significant proportion of observed recombination events will be spurious.

The simplest approach to detecting errors is to search for apparent double crossovers in short intervals. Details of a likelihood approach are given by Lincoln and Lander (1992) and by Ehm et al. (1996).

LINKAGE DISEQUILIBRIUM MAPPING

The growing abundance of molecular markers, especially microsatellites, is enabling genetic maps of the entire human genome to be constructed (e.g. Gyapay et al. 1994). The methods of this chapter can be employed to detect linkage between markers on the map and disease genes from family pedigree studies, or among the offspring of specific mating types. The number of informative meioses in pedigrees is generally such that map distances below

Table 7.5 Two-way table of haplotype frequencies for showing marker-disease associations.

	Disease Status		
Marker Status	Normal	Disease	Total
Allele M_1	n_{N1}	n_{D1}	n_1
Allele M_2	n_{N2}	n_{D2}	n_2
Total	n_N	n_D	n

about 1 cM cannot be estimated with any precision. For this reason, along with the possibility of there not being any family data, it may be desired to detect proximity of marker and disease loci by population associations. In essence, this approach of using linkage disequilibrium to locate disease genes increases the sample size by using all the recombination events between the two loci since the origin of the disease mutation, instead of only those events in the past few generations of a family. The disadvantage of the approach is that the associations between marker and disease phenotypes are also affected by evolutionary forces such as mutation, drift, selection, and admixture.

Random Samples

Genotype Data The simplest situation is when data are collected from a sample taken randomly from the whole population. If the genotypes at disease and marker loci can be determined, then composite coefficients of linkage disequilibrium can be estimated and the hypothesis of no disequilibrium can be tested by the methods in Chapter 3. It is expected that those markers found to be in disequilibrium will be close to the disease locus.

Haplotype Data Now suppose gamete, or haplotype, data are available from the whole population and it is known whether the gamete has the normal (N) or defective (D) form of the disease gene. Associations between markers and disease genes are revealed by detection of linkage disequilibrium between marker and disease gene, with the data being set out as in Table 7.5. Marker alleles M_1, M_2 may correspond to the presence or absence of a restriction site, or to different alleles for a microsatellite locus.

Table 7.6 Associations of RFLP's with sickle cell anemia in three separate samples. At the disease locus, N and S are the normal and defective alleles.

Sample	Disease	Marker Genotype				Number of
	Genotype	7,7	7,13	13,13	n	M_1 Alleles
1	*NN*	39	4	0	$n_1 = 43$	$x_1 = 82$
2	*NS*	32	53	1	$n_2 = 86$	$x_2 = 117$
3	*SS*	4	24	30	$n_3 = 58$	$x_3 = 32$

Source: Kan and Dozy (1978).

A significant association is indicated by large values of the test statistic

$$X^2 = \frac{n(n_{N1}n_{D2} - n_{N2}n_{D1})^2}{n_1 n_2 n_N n_D} \tag{7.4}$$

Linkage disequilibrium $\mathcal{D}$ between marker and disease loci has the usual estimate of

$$\hat{\mathcal{D}} = \frac{n_{N1}}{n} - \frac{n_N}{n} \times \frac{n_1}{n}$$

Conditional Data

For rare diseases, a sample taken randomly from the population may not contain any individuals with the disease. Associations are therefore sought from samples taken from the various disease categories in the population. A procedure for the estimation of linkage disequilibrium in such conditional samples was given by Maiste and Weir (1992), as is now shown.

Genotype Data One of the first reports of an association between RFLP's and a disease was that between *Hpa*I fragments and sickle cell anemia. Data published by Kan and Dozy (1978) are shown in Table 7.6. These data are from three separate samples, one for each of the disease genotypes.

Simply detecting an association between disease and marker genotypes may be accomplished with a contingency table chi-square test for the data in Table 7.6. This tests for heterogeneity of marker genotypes in the three disease categories. If the value of linkage disequilibrium between disease and marker loci in the whole population is wanted, however, a more elaborate analysis is needed.

For a general discussion, suppose a disease locus has alleles N, D and a marker locus has two alleles M_1, M_2. In the whole population, the allele frequencies are p_N, p_D at the disease locus and m_1, m_2 at the marker locus. Allowing for linkage disequilibrium $\mathcal{D}$ between the loci, the four gametes have frequencies

$$\begin{aligned} NM_1: y_1 &= p_N m_1 + \mathcal{D} \\ NM_2: y_2 &= p_N m_2 - \mathcal{D} \\ DM_1: y_3 &= p_D m_1 - \mathcal{D} \\ DM_2: y_4 &= p_D m_2 + \mathcal{D} \end{aligned}$$

For a random-mating population, the conditional frequencies α of marker allele M_1 in each of the three disease genotypic categories (i.e., in each of the three separate samples) are

$$\begin{aligned} a_1 = \Pr(M_1|NN) &= \frac{y_1^2 + y_1 y_2}{p_N^2} \\ &= m_1 + \frac{\mathcal{D}}{p_N} \\ a_3 = \Pr(M_1|DD) &= \frac{y_3^2 + y_3 y_4}{p_D^2} \\ &= m_1 - \frac{\mathcal{D}}{p_D} \\ a_2 = \Pr(M_1|ND) &= \frac{y_1 y_4 + y_2 y_3 + 2y_1 y_3}{2p_N p_D} \\ &= m_1 + \frac{\mathcal{D}}{2p_N} - \frac{\mathcal{D}}{2p_D} \\ &= \frac{1}{2}(a_1 + a_3) \end{aligned}$$

As random mating is assumed, the samples with homozygous disease genotypes are in Hardy-Weinberg equilibrium and the marker allele numbers follow a binomial distribution. In the NN class, marker homozygotes M_1M_1 have frequency $(y_1/p_N)^2 = \alpha_1^2$ for example. For the heterozygous disease class, however, the marker genotypes M_1M_1, M_1M_2, M_2M_2 have frequencies $a_1a_3, a_1(1-a_3) + a_3(1-a_1), (1-a_1)(1-a_3)$ (Chakravarti et al. 1984) and HWE cannot be assumed unless there is no linkage disequilibrium.

If the population frequencies p_N, p_D of the disease alleles are known, the two unknowns $m_1, \mathcal{D}$ can be estimated from the three separate samples. The two parameters a_1, a_3 are transformations of the parameters $m_1, \mathcal{D}$. There are 6 d.f. in the data, ruling out use of Bailey's method for estimating the parameters. The likelihoods for the three samples are

$$\begin{aligned} L(a_1, a_3)_1 &\propto (a_1)^{x_1}(1-a_1)^{2n_1-x_1} \\ L(a_1, a_3)_2 &\propto (a_1a_3)^{n_{21}}(a_1+a_3-2a_1a_3)^{n_{22}}(1-a_1-a_3+a_1a_3)^{n_{23}} \\ L(a_1, a_3)_3 &\propto (a_3)^{x_3}(1-a_3)^{2n_3-x_3} \end{aligned}$$

where x_i is the number of M_1 alleles among the $2n_i$ genes in sample $i, i = 1, 3$, and $n_{2j}, j = 1, 2, 3$ are the three marker genotypic counts in sample 2. The total likelihood for the three independent samples is the product of the three separate likelihoods. Simple algebraic expressions for the maximum likelihood estimates of a_1, a_3 are not available, and the numerical methods discussed in Chapter 2 must be used. The maximum likelihood approach provides estimates of the variances of the parameter estimates. By inverting the information matrix

$$\begin{aligned} \text{Var}(\hat{a}_1) &= \delta_1\beta_1\beta_{13}/\gamma \\ \text{Var}(\hat{a}_3) &= \delta_3\beta_3\beta_{13}/\gamma \\ \text{Cov}(\hat{a}_1, \hat{a}_3) &= -n_2\beta_1\beta_3\beta_{13}/\gamma \end{aligned}$$

where

$$\begin{aligned} \beta_1 &= a_1(1-a_1), \ \ \beta_3 = a_3(1-a_3), \ \ \beta_{13} = a_1+a_3-2a_1a_3 \\ \delta_1 &= \beta_{13}(2n_3+n_2) - n_2\beta_1, \ \ \delta_3 = \beta_{13}(2n_1+n_2) - n_2\beta_3 \\ \gamma &= \delta_1\delta_3 - n_2^2\beta_1\beta_3 \end{aligned}$$

It may be simpler to perform a grid search. The likelihood could be evaluated at all combinations of, say, 100 values for each of a_1 and a_3. The maximum value of L, if unique, would provide the MLE's of a_1, a_3 to two decimal place accuracy.

Once the solutions are found, the estimates of interest and their variances can be obtained from

$$\begin{aligned} \hat{m}_1 &= p_N\hat{a}_1 + p_D\hat{a}_3 \\ \hat{\mathcal{D}} &= p_Np_D(\hat{a}_1 - \hat{a}_3) \\ \text{Var}(\hat{m}_1) &= p_N^2\text{Var}(\hat{a}_1) + p_D^2\text{Var}(\hat{a}_3) + 2p_Np_D\text{Cov}(\hat{a}_1, \hat{a}_3) \\ \text{Var}(\hat{\mathcal{D}}) &= p_N^2p_D^2\left[\text{Var}(\hat{a}_1) + \text{Var}(\hat{a}_3) - 2\text{Cov}(\hat{a}_1, \hat{a}_3)\right] \end{aligned}$$

For the data in Table 7.6 the estimates are $\hat{m}_1 = 0.8804, \hat{\mathcal{D}} = 0.0237$ if it is assumed that $p_N = 0.95, p_D = 0.05$. The standard deviation for $\hat{m}_1$ is estimated as 0.0429, and for $\hat{D}$ as 0.0040.

Yule's coefficient of association, ϕ, has been defined in the conditional data context (Nei and Li 1980) as a function of only the conditional allele frequencies

$$\phi = \frac{a_1 - a_3}{a_1(1-a_3) + a_3(1-a_1)}$$

It will be zero when the marker and disease loci are independent.

Approximate estimates of the conditional allele frequencies are obtained by ignoring the disease heterozygotes

$$\begin{aligned}\hat{a}_1 &= \frac{1}{2n_1}(2n_{11} + n_{12}) \\ \hat{a}_3 &= \frac{1}{2n_3}(2n_{31} + n_{32})\end{aligned}$$

and an approximate test statistic for the hypothesis of no disequilibrium is

$$X^2 = \frac{(\hat{a}_1 - \hat{a}_3)^2}{\frac{\hat{a}_1(1-\hat{a}_1)}{2n_1} + \frac{\hat{a}_3(1-\hat{a}_3)}{2n_3}}$$

Haplotype Data With pedigree information available, it may be possible to determine phase of double heterozygotes and therefore construct haplotype (gamete) frequencies. If a sample of n_D disease allele bearing haplotypes is sampled independently of a set of n_N normal allele bearing haplotypes, the data may be set out as in Table 7.5. The hypothesis of independence of marker and disease loci is that marker allele frequencies are the same on normal and disease haplotypes. In other words, a test of homogeneity of the two samples is appropriate, and the test statistic is that given in Equation 7.4.

With haplotype data there are two conditional frequencies for marker alleles:

$$\begin{aligned}a_1 = \Pr(M_1|N) &= \frac{\Pr(NM_1)}{\Pr(N)} \\ &= m_1 + \frac{\mathcal{D}}{p_N} \\ a_3 = \Pr(M_1|D) &= \frac{\Pr(DM_1)}{\Pr(D)} \\ &= m_1 - \frac{\mathcal{D}}{p_D}\end{aligned}$$

Maximum likelihood estimates of linkage disequilibrium and marker allele frequencies require knowledge of the disease locus frequencies

$$\begin{aligned}\hat{m}_1 &= p_N \frac{n_{N1}}{n_N} + p_D \frac{n_{D1}}{n_D} \\ &= p_N \hat{a}_1 + p_D \hat{a}_3\end{aligned}$$

$$\begin{aligned}\hat{\mathcal{D}} &= p_N p_D \left(\frac{n_{N1}}{n_N} - \frac{n_{D1}}{n_D}\right) \\ &= p_N p_D (\hat{a}_1 - \hat{a}_3)\end{aligned}$$

Heterozygote Data Pedigrees studied for human diseases are often ascertained through an affected child. For recessive diseases, this implies that each parent is heterozygous at the disease locus and each family can contribute these two (independent) heterozygotes for the analysis of association. In other words, only sample number 2 (with sample size n_2) of the situation illustrated in Table 7.6 is available (Chakraborty 1986). Maximizing the likelihood leads to the gene-counting (EM algorithm) estimate of a_2

$$\hat{a}_2 = \frac{2n_{21} + n_{22}}{2n_2}$$

since Bailey's method provides

$$\widehat{a_1 a_3} = \frac{n_{21}}{n_2} \quad , \quad \widehat{a_1 + a_3} = \frac{2n_{21} + n_{22}}{n_2}$$

Recovering estimates of m_1 and D from these requires the solution of quadratic equations.

Family data often allow the phase of double heterozygotes to be determined, and haplotypic data to be recovered. From the sample of heterozygotes at the disease locus, the four haplotypes have the observed and expected frequencies shown in Table 7.7. The n_{22} double heterozygotes have been split into n_{22}^c of NM_1/DM_2 and n_{22}^r of NM_2/DM_1. The estimates of conditional allele frequency a_2 and conditional linkage disequilibrium $\mathcal{D}_c$ are

$$\hat{a}_2 = \frac{2n_{21} + n_{22}}{2n_2} \quad , \quad \hat{\mathcal{D}}_c = \frac{n_{22}^c - n_{22}^r}{2n_2}$$

and it is such conditional values that were reported for some cystic fibrosis families by Weir (1989b). With knowledge of the population frequencies p_N, p_D of the disease alleles, the population estimates are

$$\hat{m}_1 = \hat{a}_2 - (p_D - p_N)\hat{\mathcal{D}}_c \quad , \quad \hat{\mathcal{D}} = 2p_N p_D \hat{\mathcal{D}}_c$$

Other Tests

Case-Control Tests The comparison of marker allele frequencies in samples of disease and normal genotypes or haplotypes is called a case-control test. Ideally, individuals in the two samples are paired in such a way that they are expected to differ only in their disease status. If the two samples, cases and controls, are taken from a large population there is the danger that marker allele frequencies may differ between the samples because of population stratification rather than proximity to the disease locus. Accordingly, comparisons are often made within families to remove these population effects. Two such tests are known as the mean-haplotype sharing test and the transmission disequilibrium test.

Mean Haplotype-sharing Test For a family with more than one affected child, there is a test for linkage that requires knowledge of which marker alleles parents transmit to each child (Ott 1991). The ***affected sib-pair test*** considers the marker alleles received by two affected children. It can often be determined whether the two sibs receive copies of the same parental allele, and such pairs of alleles are identical by descent, ibd. If there is no linkage between disease and marker loci, the distribution of ibd alleles should have nothing to do with disease status. In particular, the proportions of families in which sib pairs have 0, 1 or 2 ibd pairs of alleles should be 1:2:1. Linkage will cause higher fractions of ibd alleles.

Table 7.7 Observed and expected frequencies for four haplotypes when only heterozygotes at a disease locus are available.

	M_1		M_2		
	Exp.	Obs.	Exp.	Obs.	Total
N	$\frac{1}{2}a_1$	$\frac{1}{2n_2}(n_{21}+n_{22}^c)$	$\frac{1}{2}(1-a_1)$	$\frac{1}{2n_2}(n_{23}+n_{22}^r)$	$\frac{1}{2}$
D	$\frac{1}{2}a_3$	$\frac{1}{2n_2}(n_{21}+n_{22}^r)$	$\frac{1}{2}(1-a_3)$	$\frac{1}{2n_2}(n_{23}+n_{22}^c)$	$\frac{1}{2}$
Total	a_2	$\frac{1}{2n_2}(2n_{21}+n_{22})$	$1-a_2$	$\frac{1}{2n_2}(2n_{23}+n_{22})$	1

Table 7.8 Genotypes and transmitted gametes for heterozygous markers.

Genotype	Frequency	Probability of				
		D	$T:M_1D$ $NT:M_1$	$T:M_1D$ $NT:M_2$	$T:M_2D$ $NT:M_1$	$T:M_2D$ $NT:M_2$
M_1D/M_1D	$\Pr(M_1D)^2$	1	1	0	0	0
M_1D/M_1N	$2\Pr(M_1D)\Pr(M_1N)$	1/2	1/2	0	0	0
M_1N/M_1N	$\Pr(M_1N)^2$	0	0	0	0	0
M_1D/M_2D	$2\Pr(M_1D)\Pr(M_2D)$	1	0	1/2	1/2	0
M_1D/M_2N	$2\Pr(M_1D)\Pr(M_2N)$	1/2	0	$(1-c)/2$	$c/2$	0
M_1N/M_2D	$2\Pr(M_1N)\Pr(M_2D)$	1/2	0	$c/2$	$(1-c)/2$	0
M_1N/M_2N	$2\Pr(M_1N)\Pr(M_2N)$	0	0	0	0	0
M_2D/M_2D	$\Pr(M_2D)^2$	1	0	0	0	1
M_2D/M_2N	$2\Pr(M_2D)\Pr(M_2N)$	1/2	0	0	0	1/2
M_2N/M_2N	$\Pr(M_2N)^2$	0	0	0	0	0

Transmission Disequilibrium Test When a parent heterozygous at the mark locus transmits a marker allele along with a disease-locus allele, there another marker allele that is not transmitted. The numbers of times an alle is either transmitted or not transmitted are expected to be the same when th disease and marker loci are neither linked nor exhibit linkage disequilibriun The version of this ***TDT*** test proposed by Spielman et al. (1993) is no described.

Consider a recessive disease where affected individuals must receive alle D from each parent. What is needed is the probabilities with which an ind vidual transmits one marker allele and not the other given that it transmi a D allele. The parent has an unknown genotype at the disease locus. Table 7.8, the possible parental genotypes are listed, along with the prob bilities that such genotypes transmit (T) or do not transmit (NT) either two marker alleles M_1, M_2 and transmit D alleles. These probabilities ref to families in which there is a single affected child. In the table, $\Pr(M_1L$ is the frequency of M_1D haplotypes. If m, p are the frequencies of allel M_1, D and $\mathcal{D}$ is the population linkage disequilibrium

$$\Pr(M_1D) = pm + \mathcal{D}$$

Putting together all the terms in the table, the conditional transmissic probabilities are

$$\begin{aligned}
\Pr(T: M_1, NT: M_1|T: D) &= m^2 + m\mathcal{D}/p \\
\Pr(T: M_1, NT: M_2|T: D) &= m(1-m) + (1-c-m)\mathcal{D}/p \\
\Pr(T: M_2, NT: M_1|T: D) &= m(1-m) + (c-m)\mathcal{D}/p \\
\Pr(T: M_2, NT: M_2|T: D) &= (1-m)^2 - (1-m)\mathcal{D}/p
\end{aligned}$$

as given by Spielman et al. (1993).

The recombination fraction c affects only the probabilities from heterozygous M_1M_2 parents, and conditioning on marker heterozygosity:

$$\begin{aligned}
\Pr(T: M_1, NT: M_2|M_1M_2, T: D) &= \frac{1}{2} + \frac{(1-2c)\mathcal{D}}{2[2m(1-m)p + \mathcal{D}(1-2m)]} \\
\Pr(T: M_2, NT: M_2|M_1M_2, T: D) &= \frac{1}{2} - \frac{(1-2c)\mathcal{D}}{2[2m(1-m)p + \mathcal{D}(1-2m)]}
\end{aligned}$$

If there is no linkage disequilibrium between the two loci, or if the loci are not linked, then each marker allele is equally likely to be transmitted along with the disease allele. The joint hypothesis of no linkage *or* no linkage disequilibrium can be tested for by comparing the proportion of marker heterozygotes passing a specific one of their marker alleles to affected children to the value of one half. Evidently this test will detect linkage only when there is linkage disequilibrium.

The treatment given here has been for recessive diseases. More generally, each disease genotype has a certain probability of giving an affected phenotype. Under this more general model, the calculations set out in Table 7.8 were for each parent in families with an affected child, and require that the marker locus has only two alleles. For markers with more alleles, the contributions of the two parents are not independent because there are several different heterozygous marker genotypes (Bickeböller and Clerget-Darpoux 1995). It is necessary to consider transmitted "genotypes" (i.e. the alleles transmitted by both parents to affected offspring) rather than transmitted gametes.

Population admixture The linkage disequilibrium upon which the TDT test depends may reflect the relatively young age of the disease, or it may reflect population stratification of the sort that can accompany population admixture. Suppose a population contains two subpopulations, I and II, in proportions α and $1-\alpha$. Even if each subpopulation is in linkage equilibrium for disease allele D and marker allele M,

$$\begin{aligned}
p_{DM_I} &= p_{D_I}p_{M_I} \\
p_{DM_{II}} &= p_{D_{II}}p_{M_{II}}
\end{aligned}$$

the whole population will show linkage disequilibrium. If the whole population frequencies are p_{DM}, p_D, p_M, then

$$\begin{aligned} p_{DM} &= \alpha p_{DM_I} + (1-\alpha)p_{DM_{II}} \\ &= [\alpha p_{D_I} + (1-\alpha)p_{D_{II}}][\alpha p_{M_I} + (1-\alpha)p_{M_{II}}] \\ &\quad + \alpha(1-\alpha)(p_{D_I} - p_{D_{II}})(p_{M_I} - p_{M_{II}}) \\ &= p_D p_M + \alpha(1-\alpha)(p_{D_I} - p_{D_{II}})(p_{M_I} - p_{M_{II}}) \end{aligned}$$

showing that the linkage disequilibrium, in amount $\alpha(1-\alpha)(p_{D_I}-p_{D_{II}})(p_{M_I}-p_{M_{II}})$, reflects the differences in allele frequencies at both loci in the two subpopulations. This is the stratification that may cause case-control studies to be invalid, but can allow TDT tests to work.

The disequilibrium caused by initial admixture will decline over time because of recombination between the loci, although this can be slowed by continuing genetic contact with the founding populations as discussed by Stephens et al. (1994). These authors considered using detectable linkage disequilibrium between disease and marker loci in admixed populations to locate disease genes. They noted that the disequilibrium between unlinked loci will halve each generation, and so quickly become non-detectable. Only for linked loci will disequilibrium remain longer, such as over the time period corresponding to admixture within the US African-American and Hispanic populations.

Estimating Marker-Disease Distances

The previous section was concerned with testing for associations between disease and marker loci. It is expected that the closest marker loci will show the strongest associations with the disease, but it is not simple to quantify this relationship. Two different scenarios lead to theory predicting the amount of disequilibrium as a function of the recombination fraction: either the disease is so old that an equilibrium has been established in the population subject to evolutionary forces such as drift, recombination, and mutation, or the disease is so young that recombination is the force of most importance.

Old Diseases When the disease mutation has been present in the population for a long time, the amount of linkage disequilibrium even to nearby markers is expected to be zero. There may be disequilibrum, but the sign is arbitrary and does not indicate which marker alleles were present on the ancestral chromosome where the disease mutation first occurred. For this reason, it is appropriate to work with the squared disequilibrium and theory is available for the squared correlation r^2 of allele frequencies

$$\begin{aligned} r^2 &= \frac{\hat{\mathcal{D}}^2}{\hat{p}_N \hat{p}_D \hat{m}_1 \hat{m}_2} \\ &= X^2/n \end{aligned}$$

The X^2 statistic is the one given in Equation 7.4. If only drift and recombination are affecting frequencies at the disease and marker loci, population genetic theory predicts that r^2 has an approximate expected value of

$$\mathcal{E}(r^2) = \frac{1}{1+4Nc}$$

when N is the population size and c the recombination fraction between disease and marker loci. If there is also a low amount of mutation, this result becomes (Weir and Hill 1986)

$$\mathcal{E}(r^2) = \frac{10+4Nc}{(2+4Nc)(11+4Nc)}$$

Although these results predict greater associations for more tightly linked loci, they are not useful for estimating c (or Nc) because of the large sampling variance of r^2. The variance due to statistical sampling can be minimized by increasing sample size, but that due to genetic sampling cannot be controlled and can be quite substantial. Simply equating observed and expected values of r^2 is like replacing a single observation by its mean and ignores these variances. Hill and Weir (1994) used simulations to generate the distribution of gametic frequencies, and hence of r^2, so that the likelihood for Nc could be given for a specific set of gamete counts. Maximum likelihood methods then provide an estimate and a confidence interval for Nc. This was for random samples. Kaplan and Weir (1992) gave an analogous analysis for conditional sampling. When a likelihood curve is constructed, both the maximum likelihood estimate of Nc and a confidence interval are provided. The interval is found as the range over which the log-likelihood is no more than 2 less than its maximum value, and it corresponds approximately to a 95% confidence interval.

The expected value of linkage disequilibrium in equilibrium populations is zero, even though the expected value of r^2 is not zero. This means that such populations do not have the enrichment in linkage disequilibrium needed for the TDT test.

Young Diseases For young diseases, chromosomes carrying the disease allele are also likely to carry the marker allele that was present on the chromosome(s) on which the disease mutation occurred. Over time, recombination introduces other marker alleles onto disease chromosomes. The population frequencies of disease haplotypes DM_i, carrying the disease allele D and the marker allele M_i, will depend on the recombination fraction c, and if these frequencies were known then the likelihood for c could be constructed for any set of observed disease haplotype counts. Kaplan et al. (1995) used simulations to generate growing disease populations. These simulated populations provide the population frequencies needed for likelihood calculation, with the set of simulations conducted for a particular c value leading to one point on the likelihood curve. This likelihood-based method makes use of the whole distribution of the relevant statistics, rather than just the first moment.

The moment method of Hästbacka et al. (1994) estimates the recombination fraction by equating observed and expected marker-allele frequency "excesses." This depends on the frequency p_{Di} of ancestral marker allele M_i on disease chromosomes being greater than the frequency p_{Ni} of that allele on normal chromosomes. If α is the probability that a disease-allele chromosome descends from the ancestral chromosome on which the disease mutation arose, and Pr(no recombination)$= (1-c)^G$ is the probability that such chromosomes have not undergone mutation between disease and marker loci in the G generations since that time, then p_{Di} and p_{Ni} are related by

$$p_{Di} = \alpha \Pr(\text{no recombination}) + [1 - \alpha \Pr(\text{no recombination})] p_{Ni}$$

or

$$p_{\text{excess}_i} = \frac{p_{Di} - p_{Ni}}{1 - p_{Ni}} = \alpha \Pr(\text{no recombination})$$

to define p_{excess}. If $\alpha = 1$, this last relation suggests that c can be found as

$$c = 1 - (p_{\text{excess}_i})^{1/G}$$

Comparisons of the moment and maximum likelihood estimates were given by Kaplan and Weir (1995), who showed that the MLE and its associated upper confidence limit are larger than the corresponding moment method results.

EXERCISES

Exercise 7.1

To estimate the recombination fraction between a series of esterase loci in barley, Allard et al. (1972) used a series of crosses between inbred lines. The F_1 progeny were crossed to produce an intercross F_2 generation. The crosses involved different alleles at the various loci. For loci A and B, they used 3 sets of crosses, and observed F_2 progeny in 1, 2, and 6 families respectively. Use the whole data set to estimate the recombination fraction between loci A and B. Set up a likelihood that is the product of the likelihoods for all 9 families.

			Family					
Parents	F_1	Intercross Progeny	1	2	3	4	5	6
$A_2A_2B_1B_1$	$A_2A_3B_1B_2$	$A_2A_2B_1B_1$	12					
$A_3A_3B_2B_2$		$A_3A_3B_2B_2$	15					
		$A_2A_3B_1B_2$	35					
$A_2A_2B_1B_1$	$A_2A_3B_1B_3$	$A_2A_2B_1B_1$	16	46				
$A_3A_3B_3B_3$		$A_3A_3B_3B_3$	33	32				
		$A_2A_3B_1B_3$	44	105				
		$A_2A_2B_1B_3$	3	0				
		$A_2A_3B_1B_1$	2	0				
		$A_2A_3B_3B_3$	2	0				
$A_2A_2B_1B_1$	$A_2A_3B_1B_2$	$A_2A_2B_1B_1$	71	90	92	81	80	95
$A_3A_3B_2B_2$		$A_3A_3B_2B_2$	69	99	108	60	57	80
		$A_2A_3B_1B_2$	138	158	164	142	118	137
		$A_2A_3B_1B_1$	1	0	0	0	0	0
		$A_3A_3B_1B_2$	1	0	0	0	0	0
		$A_2A_3B_2B_2$	1	0	0	0	0	0

Exercise 7.2

Suppose two parental inbred lines are $AABB$ and $aabb$ with A dominant to a and B dominant to b. The F_2 population has x, y, z, w counts of phenotypes AB, Ab, aB, ab. Test for linkage between loci A and B when $x = 30, y = 15, z = 14, w = 1$. What about $x = 30, y = 17, z = 13, w = 0$?

Exercise 7.3

Suppose two parental inbred lines are $AAbb$ and $aaBB$ with A dominant to a and B dominant to b. The F_2 population has x, y, z, w counts of phenotypes AB, Ab, aB, ab. Test for linkage between loci A and B when $x = 30, y = 15, z = 14, w = 1$. What about $x = 30, y = 17, z = 13, w = 0$?

Exercise 7.4

Suppose disease-locus genotypes DD, DN and NN have probabilities f_{DD}, f_{DN} and f_{NN} of giving the disease phenotype. What is the probability that a parent with the M_1M_2 marker genotype transmits M_1D or M_1N? What, therefore, is the probability that an M_1M_2 parent transmits M_1 to an affected child? How does this impose a dependence on the transmitted haplotypes of the two parents?

Exercise 7.5

Two completely inbred lines P_1, P_2 are fixed for alternative alleles of five marker loci, and the F_1 is backcrossed to P_1. The five-locus marker genotypes for 100 members of the backcross population are shown below (1 means the P_1 genotype and 2 means the F_1 genotype). The physical order of the markers is not known. Construct a genetic map for the markers. [Ignore the problem of unknown phase in these data, and take the proportion of "12" or "21" genotypes as a measure of distance for each pair of loci.]

M_1	M_2	M_3	M_4	M_5	M_1	M_2	M_3	M_4	M_5	M_1	M_2	M_3	M_4	M_5
1	1	1	1	1	2	2	1	1	2	2	2	1	1	1
2	2	2	1	1	2	2	1	1	1	2	2	2	1	2
2	2	2	1	2	1	2	1	1	2	2	2	1	1	1
2	1	1	2	1	2	2	1	1	1	2	2	1	2	1
2	1	1	2	1	2	1	1	1	2	2	1	2	1	1
2	1	1	2	1	1	1	2	2	2	1	1	2	2	2
2	2	2	2	2	1	2	2	1	2	2	2	2	2	1
2	2	2	2	2	1	1	2	2	2	2	1	1	2	1
2	1	1	2	1	2	2	1	1	1	2	2	1	1	1
2	2	2	1	1	1	1	1	1	1	1	1	2	2	2
1	1	2	1	2	1	1	1	2	2	2	2	1	2	1
1	1	1	2	1	2	2	1	1	1	1	1	1	2	2
1	1	2	2	2	1	2	1	1	2	1	2	2	1	2
1	1	2	2	2	1	2	2	1	2	2	2	1	1	1
2	2	1	2	1	1	1	1	2	2	2	2	1	1	1
1	1	2	2	2	2	1	1	2	1	1	1	2	2	2
2	1	1	2	1	1	2	1	2	2	2	2	1	1	1
2	2	1	1	1	1	1	2	2	2	1	1	1	2	2
1	2	2	2	2	1	1	2	2	2	2	2	1	2	1
2	1	1	2	1	1	1	2	2	1	2	2	2	1	1
2	1	2	2	2	2	2	1	1	2	1	1	2	1	2
2	2	1	1	1	2	2	2	1	1	2	2	1	1	1
1	1	2	1	2	1	1	1	1	1	2	1	2	2	2
2	1	1	2	1	2	2	1	1	1	2	2	1	1	2
2	1	1	1	1	2	2	1	1	1	1	1	1	2	2
2	1	1	2	1	2	2	1	1	1	2	2	1	1	1
2	2	1	1	1	1	1	1	1	1	2	2	1	1	2
1	1	2	2	2	1	2	2	1	2	1	1	1	2	1
1	2	1	1	2	1	1	2	2	2	2	2	1	2	1
2	2	1	1	1	1	1	1	1	2	1	2	2	1	2
2	2	2	1	2	2	2	2	1	1	1	2	2	1	2
1	2	2	1	2	1	1	2	2	2	2	2	1	1	1
1	1	2	2	2	1	1	2	2	2	2	2	1	2	1
					2	2	1	1	1					

Chapter 8

Outcrossing and Selection

When genetic data are collected from individuals in successive generations, inferences can be made on the processes concerned with the transmission of genetic material from parent to offspring. Attention is paid in this chapter to two such processes: outcrossing versus selfing, and selection. Reference should be made to Endler (1986) and Manly (1985) for further discussion of the estimation of selection coefficients.

ESTIMATION OF OUTCROSSING

Plant data are often obtained from seed collected from a set of maternal parent plants, leaving open the question of paternal parentage. In the simplest approach, estimates of the proportion of offspring produced by outcrossing as opposed to selfing are based on the proportion of heterozygous offspring carrying one allele that could not be of maternal origin. Various methods have been used to estimate outcrossing depending on whether the maternal genotype is known or not, whether single or multiple loci are used, and whether one or more paternal parents are possible for offspring from a single female parent. Maximum likelihood estimation is generally used, although this may require numerical methods.

Estimation from Homozygous Female Parents

Outcrossing rates can be estimated from progeny arrays derived from female parents used because they are known to be homozygous. Suppose there is a constant probability, t, that each offspring of any maternal plant A_uA_u is an outcross, and probability $1-t$ that it is a self. An offspring is heterozygous only if it is both an outcross and it received an allele different from that carried by the mother. This probability is

$$\sum_{v \neq u} tp_v = t(1 - p_u)$$

where p_u is the frequency of A_u alleles in the pollen pool. The heterozygote numbers $\tilde{H}_u$ in a sample of size N_u from A_uA_u mothers are binomially distributed

$$\tilde{H}_u \sim B[N_u, t(1 - p_u)]$$

Combining over maternal genotypes gives the likelihood of t

$$L(t) \propto \prod_u [t(1 - p_u)]^{\tilde{H}_u} [1 - t(1 - p_u)]^{N_u - \tilde{H}_u}$$

so that the score (Chapter 2) for t is

$$\begin{aligned} S(t) &= \frac{\partial \ln L(t)}{\partial t} \\ &= \frac{1}{t} \sum_u \tilde{H}_u - \sum_u \left[\frac{(N_u - \tilde{H}_u)(1 - p_u)}{1 - t(1 - p_u)} \right] \end{aligned}$$

Setting the score to zero shows that the MLE $\hat{t}$ for t must satisfy an equation that requires numerical solution

$$\frac{1}{\hat{t}} \sum_u H_u = \sum_u \frac{(N_u - \tilde{H}_u)(1 - p_u)}{1 - \hat{t}(1 - p_u)} \tag{8.1}$$

Note that this equation assumes the allele frequencies p_u are known.

When only one maternal genotype, say AA, is used, the estimate can be found explicitly as

$$\hat{t} = \frac{\tilde{H}_A}{N_A(1 - p_A)}$$

if the N_A offspring contain $\tilde{H}_A$ heterozygous offspring.

The data shown in Table 8.1 are for esterase frequencies used by Allard et al. (1972) to estimate the outcrossing rate in a barley population. From Equation 8.1, $\hat{t}$ must satisfy

$$\frac{10}{\hat{t}} = \frac{277.6608}{1 - 0.8928\hat{t}} + \frac{357.8382}{1 - 0.6518\hat{t}} + \frac{446.5235}{1 - 0.4723\hat{t}} + \frac{58.0029}{1 - 0.9831\hat{t}}$$

The easiest approach to solving such equations is to use iterations. An initial guess is used on the right-hand side of the equation, and the value found from

Table 8.1 Outcrossing data for an esterase locus in a barley population.

Maternal Class i	No. of Offspring N_u	Heterozygotes H_u	Gene Frequencies p_u
1	315	4	0.1072
2	554	5	0.3482
3	946	1	0.5277
4	59	0	0.0169
Totals	1874	10	1.0000

Source: R. W. Allard (personal communication).

the left-hand side is regarded as the next iterate. In this particular example, solving the quartic equation directly produces only one valid root in the range between 0 and 1, and that is $\hat{t} = 0.0087$, reinforcing the impression from Table 8.1 that there is very little outcrossing in the population.

For such small outcrossing values, the products $\hat{t}(1 - p_u)$ can be ignored as being much smaller than 1, and Equation 8.1 simplifies to

$$\hat{t} = \frac{\sum_u \tilde{H}_u}{\sum_u (N_u - \tilde{H}_u)(1 - p_u)} \tag{8.2}$$

To find the variance of the MLE $\hat{t}$, notice that it is a function of the counts $\tilde{H}_u$ and the pollen frequencies p_u. In the simple case considered by Allard et al. (1972) the allele frequencies were estimated from different samples, and so were independent of the heterozygote counts. Sample values $\tilde{p}_u$ replaced parameters p_u in Equation 8.2 and the delta method was used to calculate the variance

$$\begin{aligned}\text{Var}(\hat{t}) &= \sum_u \left(\frac{\partial \hat{t}}{\partial \tilde{H}_u}\right)^2 \text{Var}(\tilde{H}_u) + \sum_u \left(\frac{\partial \hat{t}}{\partial \tilde{p}_u}\right)^2 \text{Var}(\tilde{p}_u) \\ &\quad + \sum_u \sum_{v \neq u} \left[\frac{\partial \hat{t}}{\partial \tilde{p}_u} \frac{\partial \hat{t}}{\partial \tilde{p}_v} \text{Cov}(\tilde{p}_u, \tilde{p}_v)\right]\end{aligned}$$

Derivatives of $\hat{t}$ can be found from Equations 8.1 or 8.2.

Estimation from Equilibrium Population

A population with a constant proportion t of outcrossing and a proportion $1 - t$ of selfing eventually reaches an equilibrium in which genotypic frequencies remain constant over time for neutral loci. For a locus with alleles A_u these frequencies are

$$\begin{aligned} P_{uu} &= p_u^2 + p_u(1-p_u)\frac{1-t}{1+t}, \quad \text{for} A_uA_u \text{ homozygotes} \\ P_{uv} &= 2p_up_v\frac{2t}{1+t}, \quad \text{for} A_uA_v \text{ heterozygotes} \end{aligned}$$

which suggested to Fyfe and Bailey (1951) that observed genotypic frequencies could provide estimates of both gene frequencies p_u and outcrossing rate t. The procedures are the same as were considered in Chapter 2, where gene frequencies and the inbreeding coefficient f were estimated. Outcrossing and inbreeding coefficients are related by

$$f = \frac{1-t}{1+t} \tag{8.3}$$

It is a simple matter to invert Equation 8.3 to provide an estimate and its variance for t

$$\begin{aligned} \hat{t} &= \frac{1-\hat{f}}{1+\hat{f}} \\ \text{Var}(\hat{t}) &= \left(\frac{d\hat{t}}{d\hat{f}}\right)^2 \text{Var}(\hat{f}) \\ &= \frac{4}{(1+f)^4}\text{Var}(\hat{f}) \end{aligned}$$

with the variance of $\hat{f}$ following from Exercise 2.4.

Estimation from Offspring of Arbitrary Female Parents

It is not necessary to assume that the population is in equilibrium to estimate outcrossing rates and allele frequencies. The approach of Brown et al. (1975) assumes that sufficient offspring are available per parent that the maternal genotype can be inferred, and here the EM algorithm method given by Cheliak et al. (1983) will be treated.

The EM algorithm accommodates the fact that some offspring genotypes may arise from either selfing or outcrossing, and it is not known which event occurred. The E step consists of separating offspring numbers into these two classes. Cheliak et al. (1983) required maternal genotypes to be known, and appropriate notation is shown in Table 8.2. In this table, ${}_{ab}x_{cd}$ indicates the observed count of A_cA_d genotypes from A_aA_b female parents; ${}_{ab}s_{cd}$ is for selfed and ${}_{ab}c_{cd}$ is for crossed offspring, respectively. An asterisk denotes

Table 8.2 Notation for estimating selfed and crossed offspring frequencies with the EM algorithm. Asterisks denote maternal alleles.

Maternal Genotype	Offspring Genotype	Observed Count	Probability	Expected Count
$A_u^*A_u^*$	$A_u^*A_u^*$	${}_{uu}x_{uu}$	$1-t$	${}_{uu}s_{uu}$
	$A_u^*A_u$		tp_u	${}_{uu}c_{uu}$
	$A_u^*A_v$	${}_{uu}x_{uv}$	tp_v	${}_{uu}c_{uv}$
$A_u^*A_v^*$	$A_u^*A_u^*$	${}_{uv}x_{uu}$	$(1-t)/4$	${}_{uv}s_{uu}$
	$A_u^*A_u$		$tp_u/2$	${}_{uv}c_{uu}$
	$A_v^*A_v^*$	${}_{uv}x_{vv}$	$(1-t)/4$	${}_{uv}s_{vv}$
	$A_v^*A_v$		$tp_v/2$	${}_{uv}c_{vv}$
	$A_u^*A_v^*$	${}_{uv}x_{uv}$	$(1-t)/2$	${}_{uv}s_{uv}$
	$A_u^*A_v$		$tp_v/2$	${}_{uv}c_{uv}$
	$A_v^*A_u$		$tp_u/2$	${}_{uv}c_{vu}$
	$A_u^*A_w$	${}_{uv}x_{uw}$	$tp_w/2$	${}_{uv}c_{uw}$
	$A_v^*A_w$	${}_{uv}x_{vw}$	$tp_w/2$	${}_{uv}c_{vw}$

a maternal gene. The E step of the EM algorithm starts by assuming initial values of the unknowns t and p_u and then the numbers of crossed and outcrossed offspring can be estimated:

$$ {}_{uu}s_{uu} = \left[\frac{1-t}{(1-t)+tp_u}\right] {}_{uu}x_{uu} \quad , $$

$$
\begin{aligned}
{}_{uu}c_{uu} &= \left[\frac{tp_u}{(1-t)+tp_u}\right] {}_{uu}x_{uu} \\
{}_{uv}s_{uu} &= \left[\frac{1-t}{(1-t)+2tp_u}\right] {}_{uv}x_{uu} \quad , \\
{}_{uv}c_{uu} &= \left[\frac{2tp_u}{(1-t)+2tp_u}\right] {}_{uv}x_{uu} \\
{}_{uv}s_{vv} &= \left[\frac{1-t}{(1-t)+2tp_v}\right] {}_{uv}x_{vv} \quad , \\
{}_{uv}c_{vv} &= \left[\frac{2tp_v}{(1-t)+2tp_v}\right] {}_{uv}x_{vv} \\
{}_{uv}s_{uv} &= \left[\frac{1-t}{(1-t)+t(p_u+p_v)}\right] {}_{uv}x_{uv} \quad , \\
{}_{uv}c_{uv} &= \left[\frac{tp_v}{(1-t)+t(p_u+p_v)}\right] {}_{uv}x_{uv}
\end{aligned}
$$

With all offspring genotypic counts estimated, gene frequencies in the pollen pool can be estimated straightforwardly as the proportion of outcrossed offspring that receives pollen carrying allele A_u

$$\hat{p}_u = \frac{{}_{..}c_{.u}}{{}_{..}c_{..}}$$

A dot indicates summation. Similarly, the outcrossing rate is estimated as the proportion of crossed offspring among all offspring

$$\hat{t} = \frac{{}_{..}c_{..}}{{}_{..}s_{..} + {}_{..}c_{..}}$$

These estimates are used in another E step, and the EM algorithm continues until the estimates stabilize.

Cheliak et al. (1983) applied the EM method to data on three isozyme loci in the white lupin, *Lupinus albus*, collected by Green at al. (1980). The appropriate version of Table 8.2 for the *Pgm*-1 locus is given as Table 8.3, and the data are shown in Table 8.4. Progress of the iterations for each of the three loci is shown in Table 8.5. The same end-points were obtained for other initial values of t and p_F, unless $t = 1$, in which case it remained at 1. The three estimates of t are not significantly different, although the allele frequencies estimated for the pollen pool are different from those found in the maternal population (Green et al. 1980).

It may be that the actual numbers of offspring per parent are not large enough to infer the maternal genotype. It is then necessary to assign probabilities to each possible maternal genotype, as explained by Clegg et al. (1978), and their methodology now follows. Offspring data are collected

Table 8.3 Notation for applying the EM algorithm to *Pgm*-1 in *Lupinus albus* data of Green et al. (1980). Asterisks denote maternal alleles.

Maternal Genotype	Offspring Genotype	Observed Count	Probability	Expected Count
F^*F^*	F^*F^*	120	$1-t$	$FF^{s}FF$
	F^*F		tp_F	$FF^{c}FF$
	F^*S	5	tp_S	$FF^{c}FS$
S^*S^*	S^*S^*	104	$1-t$	$SS^{s}SS$
	S^*S		tp_S	$SS^{c}SS$
	S^*F	9	tp_F	$SS^{c}SF$
S^*F^*	S^*S^*	16	$(1-t)/4$	$SF^{s}SS$
	S^*S		$tp_S/2$	$SF^{c}SS$
	F^*F^*	11	$(1-t)/4$	$SF^{s}FF$
	F^*F		$tp_F/2$	$SF^{c}FF$
	S^*F^*		$(1-t)/2$	$SF^{s}SF$
	S^*F	28	$tp_F/2$	$SF^{c}SF$
	F^*S		$tp_S/2$	$SF^{c}FS$

Table 8.4 Parent-offspring genotype combinations for *Lupinus albus* data of Green et al. (1980).

Maternal Genotype	Offspring Genotype	Locus *Pgm*-1	*6Pgd*-2	*Aat*-2
FF	*FF*	120	108	80
	FS	5	4	7
FS	*FF*	11	17	14
	FS	28	35	30
	SS	16	14	16
SS	*FS*	9	16	4
	SS	104	263	283

from a series of maternal plants, and the probability π_{my} that the yth plant has genotype m needs to be estimated. In the population, the probability

Table 8.5 Outcrossing estimates for *Lupinus albus* data of Green et al. (1980). Initial iterates were 0.5 for t and p_F.

Iterate	*Pgm*-1 t	*Pgm*-1 p_F	*6Pgd*-2 t	*6Pgd*-2 p_F	*Aat*-2 t	*Aat*-2 p_F
1	0.118	0.704	0.089	0.702	0.068	0.468
2	0.117	0.678	0.088	0.674	0.065	0.300
3	0.118	0.663	0.089	0.658	0.069	0.228
4	0.118	0.655	0.089	0.648	0.073	0.197
5	0.118	0.650	0.090	0.641	0.077	0.182
6	0.118	0.648	0.091	0.636	0.081	0.173
7	0.118	0.646	0.091	0.633	0.083	0.166
8	0.118	0.646	0.092	0.630	0.086	0.162
9	0.118	0.645	0.092	0.628	0.087	0.159
10	0.118	0.645	0.092	0.627	0.089	0.156
20	0.118	0.644	0.093	0.623	0.094	0.148
30	0.118	0.644	0.093	0.623	0.094	0.148

Table 8.6 Offspring genotypes and probabilities for partial outcrossing for a single locus with arbitrary number of alleles.

Maternal Genotype m	Offspring Genotype o	Probability θ_{mo}	
A_uA_u	A_uA_u	$(1-t)+tp_u$	
	A_uA_v	tp_v	
A_uA_v	A_uA_u	$(1-t)/4+tp_u/2$	
$(u \neq v)$	A_vA_v	$(1-t)/4+tp_v/2$	
	A_uA_v	$(1-t)/2+t(p_u+p_v)/2$	
	A_uA_w	$tp_w/2$	$w \neq u, v$
	A_vA_w	$tp_w/2$	$w \neq u, v$

that an individual has genotype m is P_m.

Let θ_{mo} be the probability an offspring from female parent of type m is of type o, as shown in Table 8.6 for a single locus. The multinomial probability of observing offspring array $\{n_{yo}\}$, where n_{yo} is the number of type o from parent y, assuming parent y is of genotype m, is

$$\Pr[\{n_{yo}\}|m] = \left(\sum_o n_{yo}\right)! \prod_o \frac{(\theta_{mo})^{n_{yo}}}{n_{yo}!} \tag{8.4}$$

The joint probability of maternal type m and offspring array $\{n_{yo}\}$ is

$$\Pr(\{n_{yo}\}, m) = P_m \Pr(\{n_{yo}\}|m)$$

Then the probability of maternal type m given the offspring array follows from Bayes' theorem as

$$\begin{aligned} \pi_{my} &= \Pr(m|\{n_{yo}\}) \\ &= \frac{\Pr(\{n_{yo}\}, m)}{\Pr(\{n_{yo}\})} \\ &= \frac{P_m \Pr(\{n_{yo}\}|m)}{\sum_m P_m \Pr(\{n_{yo}\}|m)} \end{aligned}$$

The estimation process is begun by assigning values to t, p_u, and P_m and using them to calculate the π_{my} values. These, in turn, lead to revised

Table 8.7 Identification of outcross events with multiple loci in hypothetical data.

Offspring Number	Loci A	B	C	D	E
			Maternal genotypes		
	11	22	12	13	23
			Offspring genotypes		
1	11	22	12	13	13*
2	11	22	22	11	33
3	11	12*	12	23*	13*

* A discriminatory offspring genotype.

estimates of the genotypic frequencies from

$$P_m = \frac{1}{M}\sum_{y=1}^{M}\pi_{my}$$

and iteration continues until the P_m values stabilize. Summation over y values means summing over all the M parents available. Equation 8.4 is then used to provide maximum likelihood estimates of the quantities t and p_u, with numerical methods being needed. At this point the whole process needs to be repeated. Maternal genotypic frequencies are estimated and then allele frequencies and the outcrossing rate is reestimated until all estimates have stabilized.

Multilocus Estimates

As data are usually collected at several loci for the same individual, and as all loci are subject to the same outcrossing process, it seems appropriate to base estimation procedures on all these loci. The immediate benefit of multiple loci is that it is easier to identify outcross individuals, as shown by Shaw et al. (1981). Their hypothetical data are shown in Table 8.7, illustrating that, with more loci, it is easier to detect outcross events. For offspring 1 in that table, locus E shows that the offspring is an outcross even though the other four loci were nondiscriminatory.

Shaw et al. (1981) distinguish between outcrosses and ambiguous outcrosses that cannot be identified as such on the basis of their genotypes. The probability t of outcrossing can be modified to $t(1-\alpha)$, where α is the probability that an outcross will not be discerned. The number n of discernible outcrosses in a sample of size N can be taken to be binomially distributed

$$n \sim B[N, t(1-\alpha)]$$

so that t can be estimated as

$$\hat{t} = \frac{n}{N(1-\alpha)}$$

if a value for α is available.

A discernible outcross is one in which at least one locus carries an allele that is different from both maternal alleles at that locus. The probability of an outcross being discernible is 1 minus the probability that outcrosses cannot be detected at all loci. If, at locus ℓ, the probability that outcrossing cannot be detected is β_ℓ

$$1-\alpha = 1-\prod_\ell \beta_\ell$$

or

$$\alpha = \prod_\ell \beta_\ell$$

Note that β_ℓ refers to an outcross that has occurred but is not discernible. Shaw et al. (1981) give β_ℓ as

$$\beta_\ell = \sum_u P_{\ell u,\ell u} p_{\ell u} + \sum_u \sum_{v \neq u} P_{\ell u,\ell v}(p_{\ell u} + p_{\ell v})$$

and they weight the maternal genotype frequencies $P_{\ell u,\ell u}$ and $P_{\ell u,\ell v}$ by the number of offspring contributed by that type. The allele frequencies $p_{\ell u}$ refer to the pollen pool. This equation points out a potential problem. The probability of not detecting an outcross varies with maternal genotype, but Shaw et al. (1981) apply the same average value α to all offspring in the sample, regardless of maternal genotype. For small values of t this may not be too much of a problem.

Yeh and Morgan (1987) extended the EM approach to the multilocus case. Based on all loci, offspring are classified as discernible outcrosses or as ambiguous. Table 8.2 needs additional lines for the discernible outcrosses. The numbers of selfs and outcrosses among the ambiguous types are estimated, with the estimates now needing multilocus gametic frequencies instead of single-locus gene frequencies. The possibility of gametic linkage disequilibrium among pollen gametes needs to be accommodated.

Table 8.8 Probabilities of mother-father pairs and offspring arrays when there is a single father for a family.

Mother	Father	Pr(Parents)	Pr(Offspring)		
			AA	Aa	aa
AA	AA	p_A^4	1	0	0
	Aa	$2p_A^3(1-p_A)$	0.5	0.5	0
	aa	$p_A^2(1-p_A)^2$	0	1	0
Aa	AA	$2p_A^3(1-p_A)$	0.5	0.5	0
	Aa	$4p_A^2(1-p_A)^2$	0.25	0.5	0.25
	aa	$p_A(1-p_A)^3$	0	0.5	0.5
aa	AA	$p_A^2(1-p_A)^2$	0	1	0
	Aa	$2p_A(1-p_A)^3$	0	0.5	0.5
	aa	$(1-p_A)^4$	0	0	1

Estimating Number of Paternal Parents

As well as estimating the proportion of individuals that arises from outcrossing in otherwise selfing species, there is the related question of estimating the number of paternal parents for a family when outcrossing has occurred. Multiple paternity is found in natural populations of many plant and animal species, with a recent review being given by Williams and Evarts (1989). Their approach to distinguishing between one and two paternal parents will now be discussed.

Although a multiple-allele treatment is possible, Williams and Evarts(1989) worked with a two-allele model, obtained by focusing attention on one allele at a locus and combining all others into a single allelic class. Observations are made on a set of n offspring from a single maternal parent. If there is a single father for all the offspring, then the probabilities of observing a family with counts n_{AA}, n_{Aa}, n_{aa} for genotypes AA, Aa, aa are given in Table 8.8. The father is not observed, and the genotype of the mother is supposed not to be available, so Table 8.8 lists all nine possible mother-father combinations. Each combination has a probability that can be expressed in terms of the allele frequency p_A, assuming Hardy-Weinberg equilibrium. For the first row, the probability of seeing counts n_{AA}, n_{Aa}, n_{aa} is

$$p_A^4 \frac{n!}{n_{AA}!n_{Aa}!n_{aa}!}(1)^{n_{AA}}(0)^{n_{Aa}}(0)^{n_{aa}}$$

Similar expressions hold for the other rows, and summing over all nine rows in the table gives the overall probability $P_1(n|p_A)$ of observing the offspring array $n = n_{AA}, n_{Aa}, n_{aa}$ for a specific value of p_A.

If offspring in a family have two father between them, there are 27 possible combinations of parental genotypes. These are listed in Table 8.9 along with the corresponding probabilities. If random mixing of male gametes takes place, so that each of the two males is equally likely to be the father of any of the offspring, the probabilities of each offspring genotype for the 27 parental trios are also displayed in the table. The overall probability $P_2(n|p_A)$ of an offspring array given the allele frequency is now the sum of 27 terms.

Now the data allow the estimation of the unknown allele frequency as well as the probability ϕ that an offspring comes from a family with multiple paternity. Allowing for both one (P_1) or two (P_2) fathers, the total probability of an offspring array is

$$\Pr(n|p_A, \phi) = (1-\phi)P_1(n|p_A) + \phi P_2(n|p_A)$$

In practice, data will be available from several families, and if these are independent their probabilities may be multiplied. Maximum likelihood can be used to estimate the two parameters, p_A, ϕ, from the support function

$$\ln L(p_A, \phi) = \sum_n \log[(1-\phi)P_1(n|p_A) + \phi P_2(n|p_A)]$$

Numerical methods will generally be needed to find the MLE's.

Most Likely Paternal Plants

Paternity analyses have been extended to plant populations. An examination of the genotypes of maternal plant, offspring, and all available paternal plants may lead to a determination of which male is the most likely father. Such work is generally based strictly on the genotypes of mother, offspring, and male, without regard to the geographic distance between male and mother. Pollen dispersal distance effects are ignored. If R_1 is the event that male A is the father F of offspring C from female parent M, and R_2 the event that

Table 8.9 Probabilities of mother-male trios and offspring arrays when two males inseminate the mother.

Mother	Male 1	Male 2	Pr(Parents)	Pr(Offspring)		
				AA	*Aa*	*aa*
AA	*AA*	*AA*	p_A^6	1	0	0
		Aa	$2p_A^5(1-p_A)$	0.75	0.25	0
		aa	$p_A^4(1-p_A)^2$	0.5	0.5	0
	Aa	*AA*	$2p_A^5(1-p_A)$	0.75	0.25	0
		Aa	$4p_A^4(1-p_A)^2$	0.5	0.5	0
		aa	$2p_A^3(1-p_A)^3$	0.25	0.75	0
	aa	*AA*	$p_A^4(1-p_A)^2$	0.5	0.5	0
		Aa	$2p_A^3(1-p_A)^3$	0.25	0.75	0
		aa	$p_A^2(1-p_A)^4$	0	1	0
Aa	*AA*	*AA*	$2p_A^5(1-p_A)$	0.5	0.5	0
		Aa	$4p_A^4(1-p_A)^2$	0.375	0.5	0.125
		aa	$2p_A^3(1-p_A)^3$	0.25	0.5	0.25
	Aa	*AA*	$4p_A^4(1-p_A)^2$	0.375	0.5	0.125
		Aa	$8p_A^3(1-p_A)^3$	0.25	0.5	0.25
		aa	$4p_A^2(1-p_A)^4$	0.125	0.5	0.375
	aa	*AA*	$2p_A^3(1-p_A)^3$	0.25	0.5	0.25
		Aa	$4p_A^2(1-p_A)^4$	0.125	0.5	0.375
		aa	$2p_A(1-p_A)^5$	0	0.5	0.5
aa	*AA*	*AA*	$p_A^4(1-p_A)^2$	0	1	0
		Aa	$2p_A^3(1-p_A)^3$	0	0.75	0.25
		aa	$p_A^2(1-p_A)^4$	0	0.5	0.5
	Aa	*AA*	$2p_A^3(1-p_A)^3$	0	0.75	0.25
		Aa	$4p_A^2(1-p_A)^4$	0	0.5	0.5
		aa	$2p_A(1-p_A)^5$	0	0.25	0.75
	aa	*AA*	$p_A^2(1-p_A)^4$	0	0.5	0.5
		Aa	$2p_A(1-p_A)^5$	0	0.25	0.75
		aa	$(1-p_A)^6$	0	0	1

Table 8.10 One-locus, two-allele lod scores for possible paternal plants for one locus with two alleles.

Mother M	Offspring C	Male A	$\Pr(C\|M)$	$\Pr(C\|M,A)$	Ratio
Aa	AA	AA^*	p_A	1	$1/p_A$
		Aa	p_A	1/2	$1/(2p_a)$
		aa	p_a	0	0

*Most likely father.

the male is simply a random male from the population and not related to C, then the likelihoods

$$\begin{aligned} L(R_1) &= \Pr(C|M,A)\Pr(M)\Pr(A) \\ L(R_2) &= \Pr(C|M)\Pr(M)\Pr(A) \end{aligned}$$

are both functions of the genotypes of M and A and the probabilities of C conditional on the known parent(s). There is a direct similarity to the work on disputed paternity in humans. The logarithm of the ratio of these likelihoods, the lod score, is a function of quantities that can be determined by transmission probabilities for the loci scored

$$\text{lod}(R_1 : R_2) = \log\left[\frac{\Pr(C|M,A)}{\Pr(C|M)}\right]$$

These lods are calculated for each possible male A, and the male with the highest score, provided that score exceeds some preset value, is said to be the father of the offspring. Some simple cases are set out in Table 8.10 under the assumption of random mating. This methodology has been developed and used by Meagher (1986).

ESTIMATION OF SELECTION

There are several methods available for detecting the presence of natural selection operating within a population. Many of these depend on estimating genotypic frequencies in successive generations, and comparing the results to those expected on the basis of no selection. When specific selection models are formulated, the appropriate parameters may be estimated from the relation between successive frequencies.

Goodness-of-Fit Test for Selection

Lewontin and Cockerham (1959) showed that selection may produce genotypic frequencies that are indistinguishable from those that would be obtained in the absence of selection. They were concerned with attempts to demonstrate the presence of selection from genotypic frequencies within a single population.

For a locus A with alleles A, a, and allele frequencies p_A, p_a among adults, the frequencies of genotypes in the offspring generation before the action of any selection are

AA	Aa	aa
p_A^2	$2p_Ap_a$	p_a^2

If the relative viabilities of the AA and aa homozygotes with respect to the heterozygote Aa are w_{AA} and w_{aa}, then the genotypic frequencies after selection will be

AA	Aa	aa
$w_{AA}p_A^2/\bar{w}$	$2p_Ap_a/\bar{w}$	$w_{aa}p_a^2/\bar{w}$

where the mean fitness $\bar{w}$ is $\bar{w} = w_{AA}p_A^2 + 2p_Ap_a + w_{aa}p_a^2$ so that the allele frequencies among adult offspring are

$$p'_A = \frac{w_{AA}p_A^2 + p_Ap_a}{\bar{w}}$$

$$p'_a = \frac{w_{aa}p_a^2 + p_Ap_a}{\bar{w}}$$

If a sample is taken from the adult offspring generation, will the action of selection result in evidence for departures from Hardy-Weinberg equilibrium? If a sample of size n has genotypic numbers n_{AA}, n_{Aa}, and n_{aa}, the sample allele frequencies

$$\tilde{p}_A = \frac{2n_{AA} + n_{Aa}}{2n} \quad , \quad \tilde{p}_a = \frac{2n_{aa} + n_{Aa}}{2n}$$

provide expected values for the test of HWE:

AA	Aa	aa
$n\tilde{p}_A^2$	$2n\tilde{p}_A\tilde{p}_a$	$n\tilde{p}_a^2$

Recall that the sample allele frequencies $\tilde{p}_A, \tilde{p}_a$ estimate the offspring adult frequencies p'_A, p'_a. Hardy-Weinberg genotypic frequencies exist in the offspring population if

$$\begin{aligned}(p'_A)^2 &= \frac{w_{AA}p_A^2}{\bar{w}} \\ 2(p'_A)(p'_a) &= \frac{2p_Ap_a}{\bar{w}} \\ (p'_a)^2 &= \frac{w_{aa}p_a^2}{\bar{w}}\end{aligned}$$

Lewontin and Cockerham (1959) point out that this requires

$$w_{AA}w_{aa} = 1$$

so that selection schemes with this relation will produce populations that have Hardy-Weinberg proportions. Evidently then, neutrality is one of the sufficient conditions for HWE, but not a necessary condition. Li (1988) showed that HWE proportions can also be found for various nonrandom mating situations.

If the population is in gene-frequency equilibrium (but not Hardy-Weinberg equilibrium), meaning that frequencies do not change over time,

$$p'_A = p_A \quad , \quad p'_a = p_a$$

then it is possible to estimate the two relative viabilities w_{AA} and w_{aa}. In this situation

$$\bar{w} = w_{AA}p_A + p_a = w_{aa}p_a + p_A$$

so that there are only two independent parameters, p_A and w_{AA}. Applying Bailey's rule (Chapter 2) gives maximum likelihood estimates of

$$\begin{aligned}\hat{p}_A &= \frac{2n_{AA} + n_{Aa}}{2(n_{AA} + n_{Aa} + n_{aa})} \\ \hat{w}_{AA} &= \frac{2n_{AA}(2n_{aa} + n_{Aa})}{n_{Aa}(2n_{AA} + n_{Aa})}\end{aligned}$$

With gene-frequency equilibrium, Hardy-Weinberg deviations can be written in terms of the disequilibrium coefficient D_A

$$\begin{aligned}D_A &= \frac{w_{AA}p_A^2}{\bar{w}} - \left(\frac{w_{AA}p_A^2 + p_Ap_a}{\bar{w}}\right)^2 \\ &= \frac{p_A^2p_a^2(w_{AA}w_{aa} - 1)}{\bar{w}^2}\end{aligned}$$

When the disequilibrium is not zero, the chi-square goodness-of-fit statistic has noncentrality parameter ν

$$\begin{aligned}\nu &= n\left[\frac{p_A p_a(w_{AA}w_{aa}-1)}{\bar{w}^2}\right]^2 \\ &= n\left[\frac{p_A p_a(w_{AA}w_{aa}-1)}{(w_{AA}p_A+p_a)(w_{aa}p_a+p_A)}\right]^2\end{aligned}$$

to correct the expressions given in Lewontin and Cockerham (1959) and in Weir and Cockerham (1978). As in Chapter 3, this allows determination of sample size to detect specified departures of $w_{AA}w_{aa}$ from 1.

Estimation within One Generation

Another way of detecting selection is to compare frequencies of one class of individual with those of a standard type that has an equal expected frequency in the absence of selection. Suppose A, B are phenotypes with observed numbers a, b in a sample of size n. If the viability of B relative to that of A is v, the expected proportions of A and B are

$$A : \frac{1}{v+1} \quad , \quad B : \frac{v}{v+1}$$

A naïve estimate of v is b/a, but this can be quite biased. A better estimate (Haldane 1956) is

$$\hat{v} = \frac{b}{a+1}$$

and this has been used in several experiments on *Drosophila* by Mukai (e.g., Mukai et al. 1974). Properties of this estimate follow from the binomial distribution $b \sim B(n, \frac{v}{v+1})$. The expected value of the estimate is

$$\begin{aligned}\mathcal{E}(\hat{v}) = \sum_{b=0}^{n} \hat{v}\,\Pr(b) &= \sum_{b=0}^{n} \frac{b}{a+1}\frac{n!}{a!b!}\left(\frac{v}{v+1}\right)^b\left(\frac{1}{v+1}\right)^a \\ &= v\sum_{b=1}^{n} \frac{n!}{(b-1)![n-(b-1)]!}\frac{v^{(b-1)}}{(v+1)^n} \\ &= v\left[1-\left(\frac{v}{v+1}\right)^n\right] \approx v \text{ for large } n\end{aligned}$$

By contrast, the naïve estimate does not have a finite expectation since there is the possibility that a will be zero in a sample. This problem could be removed by using the distribution of b conditional on b being less than n (and a being greater than zero):

$$\Pr(b|b < n) = \frac{\Pr(b)}{1 - \Pr(b = n)}$$

To find the variance of $\hat{v}$ note that it is not a function to which Fisher's approximate formula (Chapter 2) may be applied. Instead, $\hat{v}$ can be expressed in terms of b and the delta method used:

$$\hat{v} = \frac{b}{n + 1 - b}$$

$$\begin{aligned} \text{Var}(\hat{v}) &\approx \left(\frac{\partial \hat{v}}{\partial b}\right)^2 \text{Var}(b) \\ &= \frac{v(v+1)^2}{n} \end{aligned}$$

Components of Selection

Detecting selection from an analysis of frequencies in successive generations is complicated by the existence of several types of selection that act at different stages of the life cycle. A typical representation is

$$\text{Mating} \longrightarrow \text{Zygote}_t \overset{\text{early}}{\longrightarrow} \text{Adult}_t \overset{\text{late}}{\longrightarrow} \text{Mating} \longrightarrow \text{Zygote}_{t+1} \overset{\text{early}}{\longrightarrow} \text{Adult}_{t+1}$$

This representation shows viability (early) selection between zygotes and adults in each generation, and fertility (late) selection between adults in one generation and zygotes in the next.

A sample may be taken from the population at the same stage in successive generations with the aim of making inferences about selection. This procedure was discussed by Prout (1965). He considered a two-allele locus acted on by "early" (viability) and "late" (fertility) selection. Observations are taken between the early and late stages. Assuming random mating, the genotypic frequencies are given in Table 8.11. Divisors for mean fitnesses have been omitted in Table 8.11 as they drop out of the following ratios. Within each of the two generations, the ratios of the AA homozygote frequencies to the heterozygote frequencies are

$$\begin{aligned} R_{AA_t} &= \frac{P_{AA_t}}{P_{Aa_t}} \\ &= \frac{p_A^2 E_{AA}}{2p_A p_a} \end{aligned}$$

Table 8.11 Genotypic frequencies under Prout's model of early and late selection.

	AA	Aa	aa
Generation t			
Preselection	p_A^2	$2p_Ap_a$	p_a^2
Early selection	$p_A^2E_{AA}$	$2p_Ap_a$	$p_a^2E_{aa}$
Late selection	$p_A^2E_{AA}L_{AA}$	$2p_Ap_a$	$p_a^2E_{aa}L_{aa}$
Generation $t+1$			
Preselection	$X_A^{2\,*}$	$2X_AX_a$	X_a^2
Early selection	$X_A^2E_{AA}$	$2X_AX_a$	$X_a^2E_{aa}$

$^*X_A = p_A^2E_{AA}L_{AA} + p_Ap_a,\ \ X_a = p_a^2E_{aa}L_{aa} + p_ap_A$

with a similar expression for R_{aa_t}, and

$$\begin{aligned} R_{AA_{t+1}} &= E_{AA}\frac{p_A^2E_{AA}L_{AA} + p_Ap_a}{2(p_a^2E_{aa}L_{aa} + p_Ap_a)} \\ &= \frac{E_{AA}}{2}\frac{2R_{AA_t}L_{AA} + 1}{2R_{aa_t}L_{aa} + 1} \end{aligned}$$

and similarly for $R_{aa_{t+1}}$. It is the ratios R that are the observable quantities.

The total fitness of a genotype may be defined as the product of early and late selection coefficients

$$w = EL$$

When there is no late selection, the expressions for the ratios in the second generation can be rearranged to provide estimators of fitness

$$\begin{aligned} w_{AA} = E_{AA} &\hat{=} 2R_{AA_{t+1}}\frac{2R_{aa_t} + 1}{2R_{AA_t} + 1} \\ w_{aa} = E_{aa} &\hat{=} 2R_{aa_{t+1}}\frac{2R_{AA_t} + 1}{2R_{aa_t} + 1} \end{aligned} \qquad (8.5)$$

If late selection is present, however, these last expressions will not provide estimates of either early or total fitnesses. Even though observations are being made at the same point of the life cycle, Equations 8.5 do not lead to estimates of total fitness, as can be seen by including late fitness in those equations.

$$\hat{w}_{AA} = 2R_{AA_{t+1}}\frac{2R_{aa_t} + 1}{2R_{AA_t} + 1}$$

$$\begin{aligned}
&= E_{AA}\frac{(2R_{AA_t}L_{AA}+1)(2R_{aa_t}+1)}{(2R_{aa_t}L_{aa}+1)(2R_{AA_t}+1)} \\
\hat{w}_{aa} &= 2R_{aa_{t+1}}\frac{2R_{AA_t}+1}{2R_{aa_t}+1} \\
&= E_{aa}\frac{(2R_{aa_t}L_{aa}+1)(2R_{AA_t}+1)}{(2R_{AA_t}L_{AA}+1)(2R_{aa_t}+1)}
\end{aligned}$$

and in order for

$$\begin{aligned}
\hat{w}_{AA} &= E_{AA}L_{AA} \\
\hat{w}_{aa} &= E_{aa}L_{aa}
\end{aligned}$$

it is necessary that

$$L_{AA}L_{aa} = 1$$

and

$$(4R_{AA_t}R_{aa_t}-1)(L_{AA}-1) = 0$$

This last expression requires either that $L_{AA}=1$, implying the absence of late selection, or that $4R_{AA_t}R_{aa_t}=1$, implying Hardy-Weinberg frequencies among the observed individuals. Prout (1965) explained that when these conditions are not met, the use of $\hat{w}_{AA}$ as an estimate of total fitness will give spurious indications of frequency-dependent selection. If selection has not been completed at the time observations are made, the estimated viabilities may not represent net fitness values. For samples taken before selection operates ($E_{AA}=E_{aa}=1$), then $\hat{w}_{AA}\hat{w}_{aa}=1$, giving a misleading suggestion of the heterozygote having fitness intermediate between those of the two homozygotes. Estimation of selection coefficients must take into account the mode of selection and the stage of the life cycle at which observations are made.

In a classical set of experiments on the fitnesses associated with *Drosophila* chromosomes, Dobzhansky and Levene (1951) used information on successive pairs of observations, but pointed out that more than one pair is needed. Their notation is given in Table 8.12, showing that they wrote x for the ratio $p_A/(1-p_A)$ in one generation.

In the next generation

$$x' = x\frac{w_{AA}x+1}{x+w_{aa}}$$

Table 8.12 Dobzhansky and Levene's selection model with one mode of selection.*

	AA	Aa	aa
Before Selection	$x^2(1-p_A)^2$	$2x(1-p_A)^2$	$x(1-p_A)^2$
After Selection	$w_{AA}x^2$	$2x$	w_{aa}

$^*x = p_A/(1-p_A)$

and this can be arranged to give

$$w_{aa} = \left(\frac{x}{x'} - x\right) + x\left(\frac{x}{x'}\right) w_{AA}$$

In other words, the data provide only a linear relation between the two fitnesses, rather than separate estimates of each. At least two such data sets are needed to determine the two selection coefficients separately. The coefficients are found as the point of intersection of the two linear relations, one for each data set. Should the lines not intersect for positive w's, it may be concluded either that the model is wrong, or that there is a large sampling error.

Maximum Likelihood Estimation of Viability Selection

A general maximum likelihood method for estimating viability selection in experimental populations was given by DuMouchel and Anderson (1968). The method depends on setting up transition equations for allele frequencies, under the assumption of random mating, and then comparing observed and predicted frequencies in generations subsequent to an initial generation that has known frequencies.

Consider a locus with alleles A_u that have frequencies p_{u_t} at the beginning of generation t. Under the random mating assumption, genotypic frequencies are $p_{u_t}^2$ for A_uA_u homozygotes and $2p_{u_t}p_{v_t}$ for A_uA_v heterozygotes. Selection acts during the generation to modify these genotypic frequencies to $w_{uu}p_{u_t}^2/\bar{w}_t$ and $2w_{uv}p_{u_t}p_{v_t}/\bar{w}_t$, respectively, where the mean fitness is

$$\bar{w}_t = \sum_u \sum_v w_{uv} p_{u_t} p_{v_t}$$

Allele frequencies at the beginning of generation $t+1$ are just those among the parents of generation t:

$$\begin{aligned} p_{u_{t+1}} &= \frac{w_{uu}p_{u_t}^2 + \sum_{v \neq u} w_{uv}p_{u_t}p_{v_t}}{\bar{w}_t} \\ &= p_{u_t}\frac{\sum_v w_{uv}p_{v_t}}{\bar{w}_t} \end{aligned}$$

DuMouchel and Anderson (1968) were able to use these recurrence equations by taking known allele frequencies at the beginning of an experiment and predicting what the frequencies should be in a series of subsequent generations. At the beginning of generation t, still under the assumption of random mating, the numbers n_{u_t} of alleles in a sample are multinomially distributed. Therefore the likelihood of a set of selection coefficients is

$$L(\{w_{uv}\})_t \propto \prod_u (p_{u_t})^{n_{u_t}}$$

When data are available from several generations

$$L(\{w_{uv}\}) = \prod_t L(\{w_{uv}\})_t$$

Once again, numerical methods are needed to find the set of w_{uv} values that maximizes the likelihood.

Selection for Partial Selfers

Workman and Jain (1966) considered the estimation of selection coefficients in plant species that practice self-mating and random outcrossing. With the addition of selfing, the transition equations for allele frequencies are not as simple as in the previous section. Workman and Jain repeated Prout's (1965) warning that estimation must take account of the stages of the life cycle at which selection acts and at which observations are made. They phrased their development in terms of the arrangement

$$\begin{array}{cccccc} \text{Stage} & \text{zygotes} & \xrightarrow{1} & \text{mature adults} & \xrightarrow{2} & \text{zygotes} \\ \text{Generation} & t & & t & & t+1 \end{array}$$

In their model I, both selection and observations occur at stage 1 with selection having occurred prior to scoring. This model seems to be appropriate for seed characters that are determined by the maternal genotype. Model II allows selection at either stages 1 or 2 but requires that scoring be done before the action of any selection. This model is used for allozyme loci scored on very young plants. Model III has only fecundity selection, meaning that selection is at stage 2, and scoring is prior to any selection. Each model therefore has selection acting during a prescribed stage and scoring not taking place at a partially selected stage.

Workman and Jain (1966) gave estimates for loci with two alleles under each of their three models. Generalizations to multiple alleles and two loci were given by Allard et al. (1972) and Weir et al. (1974). Under Model II, the genotypic transition equations for the frequencies P_{uu} of A_uA_u homozygotes and P_{uv} of A_uA_v heterozygotes are

$$\begin{aligned} P_{uu}^{t+1} &= s\left(w_{uu}P_{uu}^t + \frac{1}{4}\sum_{v\neq u} w_{uv}P_{uv}^t\right)/\bar{w}_t \\ &\quad + (1-s)\left(w_{uu}P_{uu}^t + \frac{1}{2}\sum_{v\neq u} w_{uv}P_{uv}^t\right)^2/\bar{w}_t^2 \\ P_{uv}^{t+1} &= s\left(\frac{1}{2}w_{uv}P_{uv}^t\right)/\bar{w}_t + 2(1-s)\left(w_{uu}P_{uu}^t + \frac{1}{2}\sum_{x\neq u} w_{ux}P_{ux}\right) \\ &\quad \times \left(w_{vv}P_{vv}^t + \frac{1}{2}\sum_{x\neq v} w_{vx}P_{vx}\right)/\bar{w}_t^2, \ \ v\neq u \end{aligned}$$

where s and $1-s$ are the selfing and outcrossing rates, respectively, and $\bar{w}_t$ is the mean fitness in generation t:

$$\bar{w}_t = \sum_u w_{uu}P_{uu}^t + \sum_u \sum_{v\neq u} w_{uv}P_{uv}^t$$

To simplify things, use is made of the fact that the allele frequencies in generation $t+1$ are given by

$$p_u^{t+1} = (w_{uu}P_{uu}^t + \sum_{v\neq u} w_{uv}P_{uv}^t)/\bar{w}_t$$

This allows the transition equations to be rewritten as

$$\begin{aligned} P_{uu}^{t+1} - (p_u^{t+1})^2 &= s(w_{uu}P_{uu}^t + \frac{1}{4}\sum_{v\neq u} w_{uv}P_{uv}^t)/\bar{w}_t \\ P_{uv}^{t+1} - 2p_u^{t+1}p_v^{t+1} &= s(\frac{1}{2}w_{uv}P_{uv}^t)/\bar{w}_t \end{aligned}$$

As only *relative* viabilities can be estimated from genotypic proportions, some condition needs to be placed on the coefficients w_{uv}. For example, genotype A_1A_1 could be assigned unit fitness and then the other w_{uv}'s will measure fitness relative to this genotype. With this convention, the method of moments applied to the transition equations leads to the following estimates as functions of the observed genotypic frequencies in generations t and $t+1$, and observed allele frequencies in generation $t+1$:

$$\hat{w}_{uu} = \frac{\tilde{P}_{11}^t}{\tilde{P}_{uu}^t} \left\{ \frac{2\tilde{P}_{uu}^{t+1} - \tilde{p}_u^{t+1}[s + 2(1-s)\tilde{p}_u^{t+1}]}{2\tilde{P}_{11}^{t+1} - \tilde{p}_1^{t+1}[s + 2(1-s)\tilde{p}_1^{t+1}]} \right\}$$

$$\hat{w}_{uv} = \frac{2\tilde{P}_{11}^t}{\tilde{P}_{uv}^t} \left\{ \frac{\tilde{P}_{uv}^{t+1} - (1-s)\tilde{p}_u^{t+1}\tilde{p}_v^{t+1}}{\tilde{P}_{11}^{t+1} - \tilde{p}_1^{t+1}[s + 2(1-s)\tilde{p}_1^{t+1}]} \right\}, \; v \neq u$$

Variances of these estimates can be found by the delta method, assuming independent multinomial sampling of genotypes within each of the two successive generations.

Under Workman and Jain's Model I, the transition equations are much simpler, since the selection coefficients enter linearly:

$$P_{uu}^{t+1} = \frac{w_{uu}}{\bar{w}_t} \left[s(P_{uu}^t + \frac{1}{4}\sum_{v \neq u} P_{uv}^t) + (1-s)(P_{uu}^t + \frac{1}{2}\sum_{v \neq u} P_{uv}^t)^2 \right]$$

$$P_{uv}^{t+1} = \frac{w_{uv}}{\bar{w}_t} \left[s(\frac{1}{2}P_{uv}^t) + 2(1-s)(P_{uu}^t + \frac{1}{2}\sum_{x \neq u} P_{ux})(P_{vv}^t + \frac{1}{2}\sum_{x \neq v} P_{vx}) \right]$$

With the same convention of unit fitness for A_1A_1, the method of moments estimators are

$$\hat{w}_{uu} = \frac{\tilde{P}_{11}^{t+1}}{\tilde{P}_{uu}^{t+1}} \left[\frac{s(\tilde{P}_{uu}^t + \frac{1}{4}\sum_{v \neq u} \tilde{P}_{uv}^t)}{s(\tilde{P}_{11}^t + \frac{1}{4}\sum_{v \neq 1} \tilde{P}_{1v}^t)} \right]$$

$$\hat{w}_{uv} = \frac{\tilde{P}_{uv}^{t+1}}{\tilde{P}_{uu}^{t+1}} \left[\frac{s(\frac{1}{2}\tilde{P}_{uv}^t) + 2(1-s)(\tilde{P}_{uu}^t + \frac{1}{2}\sum_{x \neq u} \tilde{P}_{ux})(\tilde{P}_{vv}^t + \frac{1}{2}\sum_{x \neq v} \tilde{P}_{vx})}{s(\tilde{P}_{11}^t + \frac{1}{4}\sum_{v \neq 1} \tilde{P}_{1v}^t)} \right]$$

$v \neq u$

For both Model I and II estimators it is necessary that independent estimates of the selfing rate s are available.

Use of Mother-Offspring Data

Another treatment of components of selection, with special reference to animal populations, was given by Christiansen and Frydenberg (1973). These authors considered the case in which data are collected on mothers and offspring, as may be the case in fish populations. They identified four selection components:

Zygotic selection. Differential survival of genotypes from zygote to adult.
Sexual selection. Differential success of genotypes at mating.
Fecundity selection. Differential zygote production of matings.
Gametic selection. Distorted segregation in heterozygotes.

Disentangling these components requires observations at different stages of the life cycle. An ideal data set would consist of

1. Sample of the population of zygotes.

2. Sample of population of adults (males and females).

3. Sample of breeding population (males, females, and mate pairs).

4. Numbers of zygotes produced by mate pairs.

5. Sample of new population of zygotes.

Data of types 1 and 2 would allow estimation of the zygotic selection coefficients, types 2 and 3 estimation of sexual selection coefficients, type 4 estimation of fecundity selection coefficients, and types 3, 4, and 5 estimation of gametic selection coefficients. For species of plants or animals in which zygotes produced by individual females can be identified, all these data types may be available. In the fish populations studied by Christiansen and Frydenberg, the adult population was sampled when females were pregnant, so that samples consisted of males, and breeding and nonbreeding females. Such data differ from the ideal in that information about males is limited to male gametes transmitted to offspring. Gametic and sexual selection on males cannot be identified separately with such data.

The counts of each of the three offspring genotypes for a locus with two alleles from each of the maternal genotypes are shown in Table 8.13, in the notation of Christiansen and Frydenberg. The authors assumed genotypes were available for one randomly chosen fetal offspring per pregnant female. Counts for nonpregnant females and adult males are also shown. It can be seen from the table that there are seven offspring counts, three nonbreeding female counts and three adult male counts, giving a total of $6 + 2 + 2 = 10$ d.f. for estimating parameters and testing hypotheses.

With no assumptions about the structure of the data, the expected proportions could be defined as in Table 8.14. Within each of the three types, MLE's for the parameters follow simply as the observed proportions:

$$\hat{\gamma}_{uv} = \frac{C_{uv}}{C_{..}}, \quad \hat{\sigma}_u = \frac{S_u}{S_.}, \quad \hat{\alpha}_u = \frac{A_u}{A_.}$$

Table 8.13 Notation for mother-offspring counts for the model of Christiansen and Frydenberg (1973).

		Offspring				Nonbreeding	Adult
		A_1A_1	A_1A_2	A_2A_2	Sums	Females	Males
Mother	A_1A_1	C_{11}	C_{12}	–	$F_1 = C_{11} + C_{12}$	S_1	M_1
	A_1A_2	C_{21}	C_{22}	C_{23}	$F_2 = C_{21} + C_{22} + C_{23}$	S_2	M_2
	A_2A_2	–	C_{32}	C_{33}	$F_3 = C_{32} + C_{33}$	S_3	M_3
Male gametes : $G_1 = C_{11} + C_{21} + C_{32},\ G_2 = C_{12} + C_{23} + C_{33}$							

There is an overall test available for random mating and no selection, when the genotypic counts would be in Hardy-Weinberg proportions. If the allele frequency of A_1 is written as π_1, the various genotypic classes have the proportions shown in Table 8.15. Writing $\pi_2 = 1 - \pi_1$, the likelihood for the 10 counts under the Hardy-Weinberg hypothesis is

$$\begin{aligned} L(\pi_1) \;\propto\; & \pi_1^{3C_{11}}[\pi_1^2(1-\pi_1)]^{(C_{12}+C_{21})}[\pi_1(1-\pi_1)]^{C_{22}}[\pi_1(1-\pi_1)^2]^{(C_{23}+C_{32})} \\ & \times (1-\pi_1)^{3C_{23}}\pi_1^{2(S_1+M_1)}[2\pi_1(1-\pi_1)]^{(S_2+M_2)}(1-\pi_1)^{2(S_3+M_3)} \end{aligned}$$

so that

$$\ln L(\pi_1) \;\propto\; X \ln \pi_1 + Y \ln(1-\pi_1)$$

Table 8.14 Expected proportions for mother-offspring data for the model of Christiansen and Frydenberg (1973).

Category	Expected Proportion	Observed Count
Offspring u, Mother v	γ_{uv}	C_{uv}
Mother u	ϕ_u	F_u
Nonbreeding female u	σ_u	S_u
Adult male u	α_u	M_u

Table 8.15 Hardy-Weinberg proportions for mother-offspring data for the model of Christiansen and Frydenberg (1973).

		Offspring			Nonbreeding Females and Adult Males
		A_1A_1	A_1A_2	A_2A_2	
Mother	A_1A_1	π_1^3	$\pi_1^2\pi_2$	–	π_1^2
	A_1A_2	$\pi_1^2\pi_2$	$\pi_1\pi_2$	$\pi_1\pi_2^2$	$2\pi_1\pi_2$
	A_2A_2	–	$\pi_1\pi_2^2$	π_2^3	π_2^2

and

$$\hat{\pi}_1 = \frac{X}{X+Y}$$

where X is the number of A_1 alleles and $X+Y$ is the total number of alleles in the sample

$$\begin{aligned} X &= 3C_{11} + 2(C_{12} + C_{21}) + C_{22} + C_{23} \\ &\quad + C_{32} + 2(S_1 + M_1) + S_2 + M_2 \\ X+Y &= 3C_{..} - C_{22} + 2S. + 2M. \end{aligned}$$

Dots in this notation indicate summation over those subscripts. The estimation of π_1 accounts for 1 d.f., and the remaining 9 can be accounted for with a goodness-of-fit chi-square on the 10 categories:

$$X^2 = \frac{(C_{11} - \hat{\pi}_1^3)^2}{\hat{\pi}_1^3} + \cdots + \frac{[M_3 - (1-\hat{\pi}_1)^2]^2}{(1-\hat{\pi}_1)^2}$$

An alternative procedure is to look at the three types of counts separately and set up a series of tests for various components of selection. Christiansen and Frydenberg identify six hypotheses, as listed in Table 8.16.

Only hypothesis H_1 will be treated in detail here. The hypothesis is of no selection among female gametes, and it implies that heterozygous females contribute equal proportions of each of their two alleles to offspring and

$$\gamma_{22} = \frac{1}{2}\phi_2$$

This leads to

$$\gamma_{22} = \gamma_{21} + \gamma_{23}$$

Table 8.16 Hypotheses for selection components with mother-offspring data. From Christiansen and Frydenberg (1973).

	Hypothesis	Component if Hypothesis Rejected
H_1	Half offspring from A_1A_2 mothers are heterozygotes.	Gametic selection in A_1A_2 females.
H_2	Frequency of transmitted male gametes independent of the genotype of the mother.	Nonrandom mating in the breeding population, and female specific selection of male gametes.
H_3	Frequency of transmitted male gametes equals the gene frequency in adult males.	Differential male mating success, and gametic selection in males.
H_4	Equal genotypic frequencies among mothers and nonbreeding females.	Differential female mating success.
H_5	Equal genotypic frequencies among adult females and adult males.	Unequal zygotic selection in males and females.
H_6	Adult population equals estimated zygote population.	Zygotic selection.

The likelihood for the counts of mothers and offspring is

$$L \propto \prod_{u,v} \gamma_{uv}^{C_{uv}}$$

With no constraints, the MLE's are just the observed frequencies, as given above. When the likelihood is constrained by the hypothesis, γ_{22} is replaced by $(\gamma_{21} + \gamma_{23})$

$$L \propto \gamma_{11}^{C_{11}} \gamma_{12}^{C_{12}} \gamma_{21}^{C_{21}} (\gamma_{21} + \gamma_{23})^{C_{22}} \gamma_{23}^{C_{23}} \gamma_{32}^{C_{32}} \times (1 - \gamma_{11} - \gamma_{12} - 2\gamma_{21} - 2\gamma_{23} - \gamma_{32})^{C_{33}}$$

the maximum likelihood estimates follow from equating the scores to zero:

$$\frac{\partial L}{\partial \gamma_{uv}} = \frac{C_{uv}}{\gamma_{uv}} - K, \quad (u,v) \neq (2,1), (2,3)$$

$$\frac{\partial L}{\partial \gamma_{21}} = \frac{C_{21}}{\gamma_{21}} + \frac{C_{22}}{\gamma_{21} + \gamma_{23}} - K$$

$$\frac{\partial L}{\partial \gamma_{23}} = \frac{C_{23}}{\gamma_{23}} + \frac{C_{22}}{\gamma_{21} + \gamma_{23}} - K$$

where

$$K = \frac{C_{33}}{1 - \gamma_{11} - \gamma_{12} - 2\gamma_{21} - 2\gamma_{23} - \gamma_{32}}$$

The solutions for heterozygous offspring are

$$\hat{\gamma}_{21} = \frac{C_{21}(C_{21} + C_{22} + C_{23})}{2C_{..}(C_{21} + C_{23})}$$

$$\hat{\gamma}_{23} = \frac{C_{23}(C_{21} + C_{22} + C_{23})}{2C_{..}(C_{21} + C_{23})}$$

which imply that

$$\hat{\gamma}_{22} = \frac{C_{21} + C_{22} + C_{23}}{2C_{..}}$$

The hypothesis can be tested with a chi-square test on heterozygous offspring

$$X^2 = \sum_{j=1}^{3} \frac{(C_{2j} - C_{..}\hat{\gamma}_{2j})^2}{C_{..}\hat{\gamma}_{2j}}$$

A more general approach to testing nested sets of hypotheses, including those of Christiansen and Frydenberg, is described in Williams et al. (1990).

SUMMARY

Data from successive generations allow many different genetic processes to be studied, from recombination between loci, to mating system, to selection. In each case a genetic model needs to be specified, and statements made about the expected behavior of statistics under such models. For each process considered in this chapter, inferences are generally based on likelihood methods. Although the underlying principles are straightforward, computational procedures can quickly become cumbersome.

EXERCISES

Exercise 8.1

Estimate the outcrossing parameter for the following data, arranged as in Table 8.1:

Maternal Class i	No. of Offspring N_u	Heterozygotes H_u	Gene Frequencies p_u
1	937	2	0.4401
2	164	2	0.1568
3	51	1	0.0027
4	144	1	0.4004
Totals	1296	6	1.0000

Exercise 8.2

Christiansen and Frydenberg (1973) gave the following data, set out in the format of Table 8.15. Test the hypothesis called H_1 in Table 8.16.

	Mother – offspring			Sterile females	Males
	A_1A_1	A_1A_2	A_2A_2		
A_1A_1	41	70	–	8	54
A_1A_2	65	173	119	32	200
A_2A_2	–	127	187	29	177

Chapter 9

Sequence Data

INTRODUCTION

Extensive data on restriction site, repeat copy number and sequence variation are now available for population genetic studies. Many of the analyses can be performed with the tools already developed in this book, but there are additional aspects that will be considered in this chapter.

The presence or absence of a recognition sequence for a restriction enzyme may be treated formally as two codominant alleles at a locus. In many studies, however, the restriction sites surveyed are quite close together so that assumptions of independence are unlikely to be valid. Once suitable models are formulated, restriction site variation can be used to infer the amount of underlying DNA sequence variation or to indicate the presence of a gene of interest. Data on copy number variants, from VNTR or STR loci, can be regarded as multi-allele data. For loci with very many alleles, there is the issue of distinguishing alleles on the basis of migration distances of the corresponding fragments on electrophoretic gels. The most appropriate analyses in this case may be continuous, but discrete methods are simpler and are treated in this book.

DNA sequence data require several new analyses, if only because of their sheer magnitude. Analyses of several copies of a sequence of 1000 bases or more are heavily dependent on computer algorithms. Many of the preliminary characterizations of sequence data are concerned with base composition, patterns of bases, and regions of similarity between sequences, and may be performed without reference to the biological processes that give rise to such features. A full analysis must also incorporate the effects of evolutionary forces such as drift, mutation, and recombination.

RESTRICTION SITE DATA

Estimating Fragment Lengths

When genomic DNA is cut by a restriction enzyme, and the resulting fragments separated by size on an electrophoretic gel, two types of inference are possible. In the first place, the numbers and sizes of fragments allow conclusions to be drawn about the presence or absence of restriction sites. These ***restriction fragment length polymorphism*** (RFLP) data address variation for the restriction sites. In other systems, the ***variable number of tandem repeat*** (VNTR) or ***short tandem repeat*** (STR) loci, it is not the restriction sites that are of interest but the variation in fragment length due to varying number of repeat units between restriction sites. For either type of inference, information is needed on fragment length, and lengths are inferred from migration distances.

Migration distance on a gel is inversely proportional to fragment length, and a set of standard fragments of known length run on a gel provides a means of calibrating that gel. There is a physical limit to the resolution of bands on a gel that prevents a band from appearing as a line. Agard et al. (1981) describe a convolution of an "ideal profile" and a "smearing function" that results in observed bands. They establish a methodology for deconvoluting these two factors, by estimating the smearing function empirically. It is important to recognize that measurement error, beyond the control of the observer, limits the accuracy with which the positions of bands on a gel can be estimated. There is also the problem that very small bands may be lost off the gel.

The relation between migration distance and fragment length (or the logarithm of fragment length) is not linear, except over quite small ranges, so that simply drawing a calibration curve for the standard fragments will not be satisfactory. Nonlinear regression (Parker et al. 1977) of length on distance for the standard fragments may be used to predict the length of another fragment whose migration distance is measured. Another method makes use of a physical model that suggests migration distance m and fragment length L are related by

$$(m - m_0)(L - L_0) = c$$

where m_0, L_0, and c are constants to be estimated from the set of standard fragments.

Schaffer (1983) suggested the following least-squares method for such estimation. Suppose the set of n standard fragments has lengths and migration distances $\{L_i, m_i\}$. Schaffer minimized the sum of squares

$$Q = \sum_i (c_i - \bar{c})^2$$

where the average value $\bar{c}$ follows from the model as

$$\begin{aligned} \bar{c} &= \frac{1}{n}\sum_i c_i \\ &= \frac{1}{n}\sum_i m_i L_i - m_0 \bar{L} - L_0 \bar{m} + m_0 L_0 \end{aligned}$$

with $\bar{m}$ and $\bar{L}$ being the means of the distances and lengths of the standard fragments. This allows Q to be written as

$$Q = \sum_i \left[m_i L_i - \frac{1}{n}\sum_j m_j L_j - m_0(L_i - \bar{L}) - L_0(m_i - \bar{m}) \right]^2$$

Differentiating Q with respect to m_0 and L_0, and writing

$$\begin{aligned} S_{LL} &= \sum_i (L_i - \bar{L})^2 \\ S_{mL} &= \sum_i (m_i - \bar{m})(L_i - \bar{L}) \\ S_{mm} &= \sum_i (m_i - \bar{m})^2 \\ U_{mL} &= \sum_i m_i L_i (L_i - \bar{L}) \\ V_{mL} &= \sum_i m_i L_i (m_i - \bar{m}) \end{aligned}$$

provides the least-squares estimates

$$\begin{aligned} \hat{m}_0 &= \frac{S_{mm} U_{mL} - S_{mL} V_{mL}}{S_{LL} S_{mm} - S_{mL}^2} \\ \hat{L}_0 &= \frac{S_{LL} V_{mL} - S_{mL} U_{mL}}{S_{LL} S_{mm} - S_{mL}^2} \end{aligned}$$

and

$$\begin{aligned} \hat{c}_i &= (m_i - \hat{m}_0)(L_i - \hat{L}_0) \\ \bar{\hat{c}} &= \frac{1}{n}\sum_i (m_i - \hat{m}_0)(L_i - \hat{L}_0) \end{aligned}$$

Table 9.1 Estimating the sizes of *Hind*III restriction fragments for lambda.*

Distance (mm)		Length (kb)	
m_i	$(m_i - \bar{m})$	L_i	$(L_i - L)$
11.5	−22.04	23.1	15.14
17.2	−16.29	9.4	1.43
23.5	−10.04	6.6	−1.37
34.0	0.46	4.4	−3.57
55.0	21.46	2.3	−5.67
60.0	26.46	2.0	−5.97

Data supplied by A. H. D. Brown.

*$\bar{m} = 33.54$ $\bar{L} = 7.97$
$S_{mm} = 2012.273$ $S_{mL} = -624.1879$ $S_{LL} = 313.5568$
$U_{mL} = 2074.8015$ $V_{mL} = -4096.0475$
$\hat{m}_0 = 6.7054$ $\hat{L}_0 = 0.0443$ $\bar{\hat{c}} = 108.5359$

For another fragment with migration distance, or mobility, m, the estimated length $\hat{L}$ is

$$\hat{L} = \hat{L}_0 + \frac{\bar{\hat{c}}}{m - \hat{m}_0}$$

An example of the use of this procedure is shown in Table 9.1 for data supplied by A. H. D. Brown (personal communication). The standard fragments result from cutting the DNA of bacteriophage lambda with *Hind*III. Using the estimates $\hat{m}_0, \hat{L}_0, \bar{\hat{c}}$ from the footnote to Table 9.1

$$\hat{L} = 0.0443 + \frac{108.5359}{m - 6.7054}$$

so that a fragment with migration distance 30.5 mm on that gel has an estimated length of 4.61 kb.

Elder and Southern (1987) describe this method of using the whole set of standard fragment migration distances and lengths as a ***global*** analysis, and they contrast it with their ***local*** analysis. The same reciprocal relationship $(m - m_0)(L - L_0) = c$ as before is used, but the three unknowns m_0, L_0, c are estimated from three standard fragments close to the fragment whose length is to be estimated. If the three standards are indicated by subscripts 1, 2 or 3, then Elder and Southern write

$$A = \frac{(m_3 - m_2)(L_2 - L_3)}{(m_2 - m_1)(L_3 - L_2)}$$

and propose using

$$m_0 = \frac{m_3 - m_1 A}{1 - A}$$

$$L_0 = \frac{(m_3 - m_0)L_3 - (m_1 - m_0)L_1}{m_3 - m_1}$$

$$c = (m_1 - m_0)(L_1 - L_0)$$

This procedure is used twice: once with the two closest standards with higher mobility and the closest standard with lower mobility, and once with the closest standard of higher mobility and the two closest standards with lower mobility. The average of the two resulting length estimates is used. For the data in Table 9.1, to estimate the length of a fragment with mobility 30.5, the two sets of standards have mobilities 17.2, 23.5, 34 and 23.5, 34, 55. The two length estimates are 4.95 and 5.00, with an average of 4.97. The global estimate was 4.61.

If there is only one standard with either lower or higher mobility than the unknown, then there is only one set of three standards to use. If the unknown is beyond the mobility range of the standards, then the prudent course is not to attempt a length estimation. Elder and Southern report more accurate results with the local than the global procedure, since the reciprocal relationship does not hold with the same parameter values over the whole range of migration distances. However, they use both procedures as a check. The set of standard fragments can also be examined by estimating the length of each of them from the mobilities of the remainder and comparing these estimates to the true values.

Ordering Fragments

Once the sizes of a series of fragments have been determined, the problem of restriction mapping is to infer the locations of the restriction sites that bound these fragments. The task is made easier by applying different restriction enzymes singly and in pairs. Ordering can often be done by a trial and error process by hand, but there have been attempts to construct algorithms to automate the process. The algorithm of Pearson (1982) is now illustrated.

Suppose that applying the restriction enzyme *Eco*RI to a 10-kb piece of DNA produces two fragments of sizes 9 and 1 kb, whereas two fragments of sizes 6 and 4 kb result from applying enzyme *Bam*HI. The relative positions of the single recognition sites for each enzyme are not known. As shown in Figure 9.1, there are two possible orders, with order A suggesting that restriction by both enzymes will produce fragments of lengths 6, 3, and 1 kb. The alternative order will produce lengths of 4, 5, and 1 kb, which is the order found in an actual double digest. Order B is taken to be the actual order.

Pearson's algorithm begins by considering the fragment sets found from each of the first two enzymes applied singly. For each possible order of the recognition sites for the first enzyme, every possible ordering of sites for the second enzyme is investigated. Each arrangement is evaluated by comparing the fragment sizes from hypothetical and actual double digests. Arrangements are compared on the basis of the sums of squares of differences of fragment lengths between hypothetical and actual double digests. These differences are unlikely to be zero because of migration distance measurement error, and uncertainty in translating migration distances into fragment lengths. There are also problems such as fragments of similar lengths not being distinguished on a gel, or small fragments not being seen on the gel.

Pearson's algorithm considers all possible orders of restrictions sites, and this is not feasible for large numbers of sites. Lander (1989) reviews other approaches.

Estimating Location of Restriction Sites

Once the order of a set of restriction sites has been determined, there may still be a problem in determining the relative locations of the sites, since experimental error will prevent (estimated) fragment sizes adding exactly to the total length between the two outside sites. For a collection of r sites, there is a need to estimate $r-2$ coordinates for the interior sites.

If the coordinate of the ith site is written as X_i, and if the length of the fragment between sites i and j has been estimated to be d_{ij}, Schroeder and Blattner (1978) minimized the sum of squared relative errors to estimate coordinates. They chose X_i and X_j to minimize

$$Q = \sum_i \sum_j \left[\frac{d_{ij} - (X_j - X_i)}{d_{ij}} \right]^2$$

Longer lengths receive smaller weights in this expression. In general, minimization of Q will require numerical methods. Negative coordinates may indicate an incorrect ordering.

Inferring Nucleotide Variation

Although a complete picture of variation at the nucleotide level depends on DNA sequencing, estimates of this variation can be provided by restriction

1. Ordering of fragments.

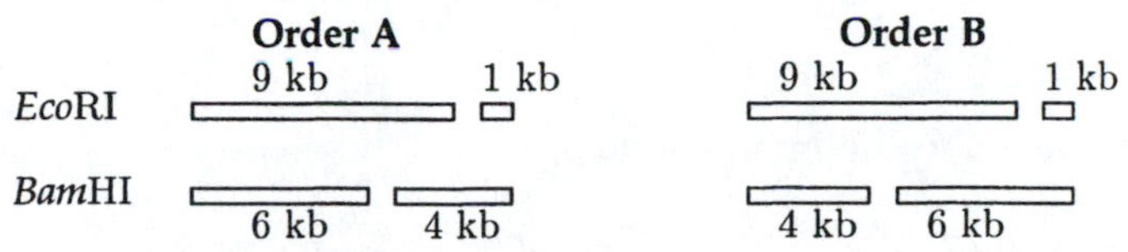

2. Calculation of hypothetical double digest fragments.

Order A

6 kb 3 kb 1 kb

Order B

4 kb 5 kb 1 kb

3. Comparison of hypothetical and actual double digest fragments.

Order A

6 kb

3 kb

1 kb

Actual Order

5 kb

4 kb

1 kb

Order B

5 kb

4 kb

1 kb

4. Error calculation

Order A: $(6-5)^2 + (3-4)^2 + (1-1)^2 = 2$

Order B: $(5-5)^2 + (4-4)^2 + (1-1)^2 = 0$

Figure 9.1 Pearson's restriction site ordering algorithm illustrated for hypothetical data.

site variation providing a suitable model is used.

A restriction site is a subsequence of particular bases, usually of length 4, 5, or 6. Mutation of any one of the bases to another type will cause the subsequence to cease being a recognition site, and the restriction enzyme will not cut the sequence at that point. Evidently the chance of mutation introducing a new recognition site will be much less than that of destroying a present site. Using standard population genetic arguments, Ewens (1983) considered a population of size N and restriction sites of length j that can be lost or gained by mutation or lost by genetic sampling (drift). For a small mutation rate μ it may be supposed that at most one mutation will occur in a site per generation. From standard population genetic theory, the probability of a single nucleotide site being polymorphic in the population is

$$P = 4N\mu \ln 2N$$

and of being polymorphic in a sample of n

$$p = 4N\mu \ln 2n$$

Under the assumption of independence of adjacent nucleotides, the probability of a restriction site being polymorphic in the sample is approximately jp so, if k of the m observed restriction sites are found to be polymorphic, a naïve estimate of p would be

$$\hat{p} = \frac{k}{jm}$$

Ewens pointed out, however, that a restriction site must be observed at least once in the sample to be included in these calculations. A restriction site at any location is known to exist in a sample only if it is seen at least once in that sample. The probability of a restriction site appearing on x of n gametes in a sample is given by

$$\begin{aligned} \Pr(x) &= \Pr(x|x > 0) \\ &= \frac{\frac{n!}{x!(n-x)!}p^x(1-p)^{n-x}}{1-(1-p)^n} \end{aligned}$$

Using such arguments, Ewens found that the conditional probability of polymorphism, given at least one occurrence, is $2jp$ and the correct estimate of p is

$$\hat{p} = \frac{k}{2jm}$$

with corresponding population value

$$\hat{P} = \frac{k \ln 2N}{2jm \ln 2n}$$

For the data on four restriction sites in Table 1.6, the sample size was $n = 17$. Eight enzymes of length $j = 6$ were used, and a total of $m = 24$ restriction sites found. Of these only $k = 4$ were polymorphic, meaning that 20 were present in every sample member. These figures provide an estimate of $\hat{p}$ =0.014. In other words, a level of polymorphism of (4/24=0.167) at the restriction site level translates into a level of polymorphism of 0.014 at the DNA level. The population size N is unknown.

Table 9.2 Base compositions for nine complete DNA sequences, to illustrate different compositions between sequences.

Sequence	Locus Name*	Base A	C	G	T	Total
Bacteriophages						
Lambda	*LAMCG*	0.25	0.24	0.25	0.26	48502
T7	*PT7*	0.27	0.23	0.24	0.26	39936
ϕX174	*PX1CG*	0.24	0.22	0.31	0.23	5386
Viruses						
Cauliflower mosaic	*MCACGDH*	0.37	0.21	0.23	0.19	8016
Human papovirus BK	*PVBMM*	0.30	0.20	0.30	0.20	4963
Hepatitis B	*HPBAYW*	0.28	0.22	0.23	0.27	3182
Mitochondria						
Human	*HUMMT*	0.31	0.31	0.25	0.13	16569
Bovine	*BOVMT*	0.33	0.26	0.27	0.14	16338
Mouse	*MUSMT*	0.35	0.24	0.29	0.12	16295

* Names in GenBank database.

DNA SEQUENCE DATA

Base Composition

An obvious first summary of a DNA sequence is just the distribution of the four base types. Although it would be convenient for mathematical modeling if the four bases were equally frequent, almost all empirical studies show an unequal distribution. The examples in Tables 9.2 and 9.3 show that bases have different frequencies within and between sequences. Table 9.2 contains data from the sequences of nine entire DNA molecules, and Table 9.3 is for two fetal globin genes, ${}^{G}\gamma$, ${}^{A}\gamma$, each of which has three exons and two introns (Shen et al. 1981). An arbitrary 500-bp region on each side of each gene is called "flanking." The intergenic region is the remaining region between the two genes.

Table 9.3 Base compositions for different regions of human fetal globin genes ${}^{G}\gamma$ and ${}^{A}\gamma$, to illustrate differences in composition within one sequence.

Region (number)	Length	A	C	G	T
5′ Flanking (2)	1000	0.33	0.23	0.22	0.22
3′ Flanking (2)	1000	0.29	0.15	0.26	0.30
Introns (4)	1996	0.27	0.17	0.27	0.29
Exons (6)	882	0.24	0.25	0.28	0.22
Intergenic (1)	2487	0.32	0.19	0.18	0.31

Source: Locus *HSGLBN* in EMBL database.

Dinucleotide Frequencies

One of the principal difficulties in analyzing DNA sequences is that the frequencies of neighboring bases are not independent. In particular, the frequencies of adjacent bases are generally different from the products of the frequencies of single bases. Algebraically, if p_u is the frequency of base type u in the sequence, and p_{uv} is the frequency with which successive bases are of types u and v, then

$$p_{uv} \neq p_u p_v$$

Dinucleotide frequencies have been studied empirically by Nussinov (1984), and some of her findings are shown in Table 9.4. These data are from 166 vertebrate sequences, totaling 136,731 bases. The ratios in the table are of the 16 dinucleotide frequencies divided by the products of the corresponding single base frequencies. Relative abundances for dinucleotides in mitochondrial, papovirus and retrovirus genomes were given by Karlin and Cardon (1994) and showed similar ranges of values.

As a specific example, consider the sequence for the chicken hemoglobin β chain mRNA shown in Figure 9.2. This figure, reading row by row, gives the 438 bases in that coding region. The four base and the 16 dinucleotide counts are displayed in Table 9.5. Regarding this table as a 4×4 contingency table and calculating the chi-square statistic for independence of rows and columns gives $X^2 = 59.3$, showing strong evidence for associations between successive nucleotide pairs.

In coding regions, there will be constraints on the DNA sequence to preserve the sequence of amino acids, and these constraints at the codon level are related to constraints at the dinucleotide level (Santibánez-Koref

Table 9.4 Examples of dinucleotide frequencies in some vertebrate sequences. The frequency of each of the 16 dinucleotides is compared to the product of the corresponding pair of base frequencies.

Pair	Observed/Expected
TG	1.29
CT	1.26
CC	1.18
AG	1.16
AA	1.15
CA	1.15
GG	1.14
TT	1.07
GA	1.04
TC	1.00
GC	0.99
AT	0.85
AC	0.84
GT	0.82
TA	0.65
CG	0.42

Source: Nussinov (1984).

and Reich 1986, Nussinov 1987). In Table 9.6 the nuclear genetic code is shown, along with the numbers of the various codons in the sequence

```
  1 GTGCACTGGA CTGCTGAGGA GAAGCAGCTC ATCACCGGCC TCTGGGGCAA GGTCAATGTG
 61 GCCGAATGTG GGGCCGAAGC CCTGGCCAGG CTGCTGATCG TCTACCCCTG GACCCAGAGG
121 TTCTTTGCGT CCTTTGGGAA CCTCTCCAGC CCCACTGCCA TCCTTGGCAA CCCCATGGTC
181 CGCGCCCACG GCAAGAAAGT GCTCACCTCC TTTGGGGATG CTGTGAAGAA CCTGGACAAC
241 ATCAAGAACA CCTTCTCCCA ACTGTCCGAA CTGCATTGTG ACAAGCTGCA TGTGGACCCC
301 GAGAACTTCA GGCTCCTGGG TGACATCCTC ATCATTGTCC TGGCCGCCCA CTTCAGCAAG
361 GACTTCACTC CTGAATGCCA GGCTGCCTGG CAGAAGCTGG TCCGCGTGGT GGCCCATGCC
421 CTGGCTCGCA AGTACCAC
```

Figure 9.2 DNA sequence for the coding region of the chicken β-globin gene (GenBank database Locus *CHKHBBM*, Accession Number J00860).

Table 9.5 Dinucleotide counts for the chicken β-globin sequence in Figure 9.2.

		Second Base				
		A	C	G	T	Total
First base	A	23	26	23	15	87
	C	37	51	14	41	143
	G	25	38	36	19	118
	T	2	29	44	14	89
Total		87	144	117	89	437

in Figure 9.2. Although the numbers are quite small, and strong statistical conclusions cannot be drawn, it does appear that the different (synonymous) codons that encode the same amino acid are not represented equally in the chicken β-globin sequence. Such codon bias is bound to be reflected in associations at the dinucleotide level. Table 9.6 also makes it clear that there is much less constraint on the base present in the third position of a codon than in the second since changing the third base often does not change the amino acid.

There are several other biological reasons why successive base frequencies may not be independent. The average rate of recombination between adjacent bases is extremely low, of the order of 10^{-8} per generation, and may be less in some regions. Associations established by new base substitutions are therefore expected to remain in a population for a very long time. There are also features such as possible avoidance of the pair CG because it is a methylation site in mammals and so prevents cleavage by restriction enzymes (Lewin 1994). Other authors (Smith et al. 1983) have related dinucleotide frequencies to functional features of sequences. Dependencies may also follow from the tendency of some regions to be AT-rich or GC-rich. Karlin and Burge (1995) present evidence that dinucleotide relative abundances constitute a "signature" of a genome that may allow discrimination between sequences from different organisms. Whatever the reasons, or implications, it is of interest to characterize the extent of associations between adjacent nucleotides.

Table 9.6 The genetic code. All 64 possible base triplets (codons) and the corresponding amino acids are shown. The numbers are those for the sequence in Figure 9.2.

UUU Phe 3	UCU Ser 0	UAU Tyr 0	UGU Cys 2
UUC Phe 5	UCC Ser 5	UAC Tyr 2	UGC Cys 1
UUA Leu 0	UCA Ser 0	UAA Stop 0	UGA Stop 0
UUG Leu 0	UCG Ser 0	UAG Stop 0	UGG Trp 4
CUU Leu 1	CCU Pro 1	CAU His 3	CGU Arg 0
CUC Leu 6	CCC Pro 4	CAC His 4	CGC Arg 3
CUA Leu 0	CCA Pro 0	CAA Gln 1	CGA Arg 0
CUG Leu 11	CCG Pro 0	CAG Gln 4	CGG Arg 0
AUU Ile 1	ACU Thr 3	AAU Asn 1	AGU Ser 0
AUC Ile 6	ACC Thr 4	AAC Asn 6	AGC Ser 2
AUA Ile 0	ACA Thr 0	AAA Lys 1	AGA Arg 0
AUG Met 1	ACG Thr 0	AAG Lys 9	AGG Arg 3
GUU Val 0	GCU Ala 4	GAU Asp 1	GGU Gly 1
GUC Val 5	GCC Ala 11	GAC Asp 5	GGC Gly 4
GUA Val 0	GCA Ala 0	GAA Glu 4	GGA Gly 0
GUG Val 7	GCG Ala 1	GAG Glu 3	GGG Gly 3

SINGLE SEQUENCE ANALYSES

Markov Chain Analysis

Associations between adjacent bases will lead to associations between more distant bases, and an estimate of how far the relations extend may be found from Markov chain theory (Tavaré and Giddings 1989). Without invoking any biological mechanism, a ***Markov chain of order*** k supposes that the base present at a certain position in a sequence depends only on the bases present at the previous k positions. A chain of order 1 supposes that the probability of a particular base being present at position i depends only on the probabilities of the four bases present at position $i-1$. A sequence composed of independent bases will correspond to a Markov chain of order 0. The order can be estimated by likelihood methods.

Markov chain analyses are of use at the genome level, rather than at the level of an individual gene. If the genome of *Escherichia coli* is being studied, for example, there would be little point in seeking the order of a Markov chain describing the sequence of a gene since this involves only about 1 kb, which is not sufficient to demonstrate the presence of high-order chains (Tavaré and Giddings 1989). Although such a short sequence may indicate only a second-order chain, it is unlikely that the entire *E. coli* genome can be described in terms of single base and dinucleotide frequencies. On the other hand, it is unlikely that the same Markov chain could describe the whole genome. If a Markov chain model has been fitted to a genome, no biological mechanism is implied, but useful questions can be answered. The frequency of particular subsequences can be predicted, so that the expected number of fragments produced when a specific restriction enzyme is applied to the genome can be estimated (Bishop et al. 1983). Arnold et al. (1988) looked at the frequency of sequences up to six bases in length in 392 kb of yeast DNA, but their methods also apply to the longer synthetic oligonucleotides being used as PCR primers for DNA amplification.

As a notational device, suppose that X_i is the nucleotide at position i in the sequence, and that it is of base type a_i. If the sequence can be regarded as the result of a Markov chain of order k, the nth base depends only on the previous k bases:

$$\begin{aligned} &\Pr(X_n = a_n | X_{n-1} = a_{n-1}, X_{n-2} = a_{n-2}, \ldots, X_0 = a_0) \\ &= \Pr(X_n = a_n | X_{n-1} = a_{n-1}, X_{n-2} = a_{n-2}, \ldots, X_{n-k} = a_{n-k}) \end{aligned}$$

The probability of a transition from a run of, say, three particular bases to a fourth, $a_1 \rightarrow a_2 \rightarrow a_3 \rightarrow a_4$, is estimated as the observed number of such tetranucleotides divided by the observed number of trinucleotides, $a_1 \rightarrow a_2 \rightarrow a_3$. Symbolically

$$\Pr(a_4|a_1, a_2, a_3) \;\hat{=}\; \frac{n_{a_1,a_2,a_3,a_4}}{n_{a_1,a_2,a_3,\bullet}}$$

so that, for CTATAATAG, the probability that an ATA is followed by an A is

$$\begin{aligned} \Pr(a_i = \mathtt{A}|a_{i-3} = \mathtt{A}, a_{i-2} = \mathtt{T}, a_{i-1} = \mathtt{A}) &= \Pr(\mathtt{A}\,|\,\mathtt{ATA}) \\ &\hat{=} \frac{n_{\mathtt{ATAA}}}{n_{\mathtt{ATA}\bullet}} = \frac{1}{2} \end{aligned}$$

The dot in the trinucleotide notation is to emphasize that all sets of four nucleotides are being examined, but only those beginning with specific bases a_1, a_2, a_3 are being counted.

Under the hypothesis of an order k chain, the likelihood of a sequence is found as the product of the probability of the first k bases and the probabilities of each successive group of $k+1$ bases. The likelihood of an order two chain describing CTATAATAG, for example, is

$$\begin{aligned} L(2) &= \Pr(\texttt{CT})\Pr(\texttt{A}|\texttt{CT})\Pr(\texttt{T}|\texttt{TA})\Pr(\texttt{A}|\texttt{AT}) \\ &\quad\times \Pr(\texttt{A}|\texttt{TA})\Pr(\texttt{T}|\texttt{AA})\Pr(\texttt{A}|\texttt{AT})\Pr(\texttt{G}|\texttt{TA}) \end{aligned}$$

where Pr(A|CT) means the probability that an A follows a CT pair. This probability is estimated as the number of CTA triples divided by the number of CT pairs. The likelihood of an order three chain describing that same sequence is

$$\begin{aligned} L(3) &= \Pr(\texttt{CTA})\Pr(\texttt{T}|\texttt{CTA})\Pr(\texttt{A}|\texttt{TAT})\Pr(\texttt{A}|\texttt{ATA}) \\ &\quad\times \Pr(\texttt{T}|\texttt{TAA})\Pr(\texttt{A}|\texttt{AAT})\Pr(\texttt{G}|\texttt{ATA}) \end{aligned}$$

The general expression, if $\Pr(a_1, a_2, \ldots, a_k)_F$ refers to the first k bases in the sequence, is

$$L(k) = \Pr(a_1, a_2, \ldots, a_k)_F \prod [\Pr(a_{k+1}|a_1, a_2, \ldots, a_k)]^{n_{a_1,a_2,\ldots,a_{k+1}}}$$

where the product is over all combinations of $k+1$ successive bases that have an observed count greater than zero. The logarithms of likelihoods for the first five orders, estimated for the sequence displayed in Figure 9.2, are shown in Table 9.7. As the order gets higher, more parameters are being fitted and better fits to the data are expected. For an order k chain, the number of independent parameters is 3×4^k. Differences in successive log-likelihoods correspond to likelihood ratios for testing the effectiveness of one order versus the previous one. Taking minus two times these differences provides test statistics that are approximately chi-square (under the hypotheses that successive orders fit equally well) with degrees of freedom equal to the difference in degrees of freedom for each of the two orders (Chapter 3). Such test statistics are shown in Table 9.7.

Approximate critical values for chi-squares with large degrees of freedom can be found from the normal distribution. If $\alpha\%$ of values of the standard normal variable (Appendix Table A.1) lie beyond $\pm z_{\alpha/2}$ then $\alpha\%$ of the values of a chi-square variable with ν degrees of freedom lie above $\frac{1}{2}(z_{\alpha/2} + \sqrt{2\nu+1})$ (Rohlf and Sokal 1995). Applying this approximate expression to the values in Table 9.7 shows that there is a significant difference (at the 5% level) in fit only between chains of order 0 and order 1. Higher-order chains seem to fit equally well.

In estimating the order k that best describes the sequence, account needs to be taken of the fact that higher-order chains use more parameters. Katz (1981) showed the validity of penalizing for an increased number of fitted parameters by using the Bayesian Information Criterion (BIC) defined as

$$\text{BIC}(k) = \text{Constant} - 2\ln L(k) + 3 \times 4^k \ln n_k$$

Table 9.7 Values of the Bayesian Information Criterion and the log-likelihoods for estimating the order of the Markov chain to represent the chicken β-globin sequence of Figure 9.2.

k	BIC(k)	Log-likelihood(k)	No. parameters	Chi-square (d.f.)
0	1213.04	−597.40	3	
1	1198.81	−562.93	12	68.94* (9)
2	1371.97	−540.12	48	45.62 (36)
3	2076.01	−454.77	192	170.70 (144)
4	5245.04	−290.52	768	328.50 (576)

* Significant at 5%.

where n_k is the number of subsequences of length $k+1$ that are found in the sequence. That k for which BIC(k) is minimized is taken as the estimate. Bearing in mind that the sequence in Figure 9.2 is really too short to allow the detection of chains of high order, the BIC values for that sequence are shown in Table 9.7. The lowest BIC value is for order $k = 1$. This reinforces the conclusion from the likelihood ratio tests that an order 1 chain is appropriate.

Once the order of the chain is established, such chains can be used to estimate the frequency of specific subsequences, as was done for yeast DNA by Arnold et al. (1988). They worked with chains of order three, and estimated the probability of the recognition sequence GCGGCCGC of the rare-cutting enzyme *Not*I as

$$\begin{aligned} \Pr(\texttt{GCGGCCGC}) &= \Pr(\texttt{GCG})\Pr(\texttt{G}|\texttt{GCG})\Pr(\texttt{C}|\texttt{CGG})\Pr(\texttt{C}|\texttt{GGC}) \\ &\quad \times \Pr(\texttt{G}|\texttt{GCC})\Pr(\texttt{C}|\texttt{CCG}) \\ &\hat{=} \frac{n_{\texttt{GCG}}}{n_{\bullet\bullet\bullet}} \frac{n_{\texttt{GCGG}}n_{\texttt{CGGC}}n_{\texttt{GGCC}}n_{\texttt{GCCG}}n_{\texttt{CCGC}}}{n_{\texttt{GCG}\bullet}n_{\texttt{CGG}\bullet}n_{\texttt{GGC}\bullet}n_{\texttt{GCC}\bullet}n_{\texttt{CCG}\bullet}} \end{aligned}$$

The term $n_{\bullet\bullet\bullet}$ in the estimator is the total number of trinucleotides in the sequence. Arnold et al. used 392 kb to provide estimates of tetranucleotide frequencies and estimated the probability of an *Not*I site as 2.0×10^{-6}. The expected length of the sequence between two points that has a probability of P of occurring is approximately $1/P$ for small P. In this case, the size of fragments produced by the *Not*I enzyme is expected to be $1/(2 \times 10^{-6}) =$

Table 9.8 Values and positions of 2-words for the sequence TGGAAATAAAACGTAAGTAG following the algorithm of Karlin et al. (1983).

Value	Position	Value	Position
1	4, 5, 8, 9, 10, 15	9	3
2	11	10	–
3	16, 19	11	2
4	6	12	13, 17
5	–	13	7, 14, 18
6	–	14	–
7	12	15	1
8	–	16	–

500 kb. Cuticchia et al. (1992) found a third-order chain also provided a satisfactory fit to 691 kb of *Drosophila melanogaster* sequence.

Patterns in a Single Sequence

Apart from characterizing the extent to which associations hold along sequences, there is often interest in looking for patterns such as direct repeats. An efficient algorithm for doing this was given by Karlin et al. (1983). The method requires the entire sequence to be searched just once, and uses words composed of particular sets of base letters. Each base is assigned a value, α, such as 0, 1, 2, 3 for A,C,G,T. Each different subsequence, or ***word***, of k letters $X_1, X_2, \ldots, X_k$ can be given a unique word value $1 + \sum_{i=1}^{k} \alpha_i 4^{k-i}$. Two subsequences with the same word value must be the same sequence of bases. These word values lie in the range from 1 to 4^k. The 5-word TGACC, for example, has value $1 + 3 \times 4^4 + 2 \times 4^3 + 0 \times 4^2 + 1 \times 4^1 + 1 \times 4^0 = 459$. Some low value of k is taken, and then the values of the k-words that start at each position in the sequence are recorded. Only those positions at which repeated k-words are found need be considered further in searches for repeated words of length greater than k.

For the sequence TGGAAATAAAACGTAAGTAG, the initial positions and values of all 2-words are shown in Table 9.8. Searches for ***direct repeats***, or subsequences that are repeated exactly, of length greater than 2 can be confined to the initial positions of 2-words that are repeated. Only four 2-words are repeated in this example. Words of value 1 are found at positions 4, 5, 8, 9, 10, and 15, for example. The 3-words that start with each of the

Table 9.9 Values and positions of 3-words that begin with repeated 2-words for the sequence TGGAAATAAAACGTAAGTAG.

Value	Position
1	4, 8, 9
2	10
3	15
4	5
45	13, 17
49	7, 14
51	18

repeated 2-words are shown in Table 9.9, where words of value 1, 45, and 49 are found to be repeated. An examination of the 4-words that start with each of the repeated 3-words fails to reveal any longer repeated values, so the longest direct repeats are AAA at positions 4, 8, and 9, GTA at positions 13 and 17, and TAA at positions 7 and 14. Applying the technique to the chicken β-globin sequence in Figure 9.2 locates the longest direct repeats shown in Table 9.10.

Karlin et al. (1983) discuss methods for assessing the statistical significance of the longest direct repeat found within a sequence. This addresses the question of whether a repeat of that length is expected to occur simply by chance. Under the assumption that nucleotide positions are independent (a Markov chain of order zero), they give asymptotic expressions for L_n, the longest direct repeat in a sequence of length n. The expected length is

$$\mu_L = \frac{0.6359 + 2\ln n + \ln(1-P)}{\ln(1/P)} - 1$$

which increases as the logarithm of sequence length, while the variance is independent of sequence length:

$$\sigma_L^2 = \frac{1.645}{(\ln P)^2}$$

In these expressions, P is the sum of squares of base frequencies in the sequence

$$P = \sum_{i=1}^{4} p_i^2$$

Table 9.10 Repeated words of lengths eight or more in the chicken β-globin sequence of Figure 9.2.

Length	Word	Starting Positions
8	GCCCTGGC	79, 418
	GCCAGGCT	85, 377
	CCAGGCTG	86, 378
	CAGGCTGC	87, 379
	TCCTTTGG	130, 208
	CCTTTGGG	131, 209
	TGGTCCGC	176, 398
	GGTCCGCG	177, 399
9	GCCAGGCTG	85, 377
	CCAGGCTGC	86, 378
	TCCTTTGGG	130, 208
	TGGTCCGCG	176, 398
10	GCCAGGCTGC	85, 377

Computing effort may be saved by starting the algorithm of Karlin et al. (1983) with a word length k as close as possible to the expected mean of the longest length.

An approximate test of the hypothesis that the sequence is a random collection of bases, with the longest direct repeat having mean and variance given by these last expressions, would be to compare the observed longest repeat with the sum of the mean and 1.645 times the standard deviation. By supposing the direct repeat lengths were normally distributed, this calculated value would be exceeded by chance only 5% of the time if the hypothesis was true. For the chicken β-globin sequence, the four bases have counts 87, 144, 118, 89 for `A`, `C`, `G`, `T`, so $P = 0.2614$ and the longest repeat has an expected length of 8.31 with an expected variance of 0.9138. With a probability of 95%, normal distribution theory says that the longest direct repeat is expected to be no longer than $9.88 \approx 10$.

Permutation Tests

A procedure that does not depend on asymptotic expressions, or assumptions of normality, is the ***permutation test***, performed by randomly permuting the

Table 9.11 Results of a permutation test of the significance of the longest direct repeat found in the chicken β-globin sequence of Figure 9.2.

Longest Length L	Frequency* f	fL	fL^2
7	22	154	1078
8	54	432	3456
9	16	144	1296
10	8	80	800
Total†	100	810	6630

* Among 100 permutations.
† Totals lead to sample mean of $\bar{L} = 8.10$ and sample variance of $s_L^2 = 0.6970$.

elements of the sequence a number of times, and searching for direct repeats in each permuted sequence. In this way a distribution of direct repeat lengths is built up, under the supposition that the sequence is a random collection of bases, and the hypothesis can be rejected at the 5% level if the observed longest repeat lies in the upper 5% of this distribution.

The permutations for the test preserve the base frequencies in the original sequence, and an algorithm for doing this is contained in Appendix B. More sophisticated permutations can be constructed to preserve the original dinucleotide or codon frequencies, or any other feature of the original sequence.

One hundred permutations were carried out on the chicken β-globin sequence of Figure 9.2, with the results displayed in Table 9.11. The observed longest direct repeat of 10 appears in the top 8% of those expected by chance alone. It would probably not be regarded as being "significant." These permutations ignored the codon structure in this particular sequence. The mean and variance calculated from the data in Table 9.11 are close to those predicted by Karlin et al. (1983).

COMPARING SEQUENCES

Dot Plots

Comparison of different sequences can be accomplished in many ways, the most direct being visually, with ***dot plots***. The sequences to be compared are arranged along the margins of a table, and a "dot" is placed in the table at the intersection of those rows and columns that are headed by the same letter. A diagonal stretch of dots will indicate equal subsequences in the two sequences. To reduce the number of dots, which may represent matches that have arisen by chance, dot plots can be enhanced by filtering (Maizel and Lenk 1981). Dots are placed only when a certain proportion of a small group of successive bases match. In Figure 9.3, the dot plot is shown for a comparison of the chicken β-globin sequence in Figure 9.2 with itself. Groups of 9 successive bases were compared, for all possible starting positions of groups of this length. If 7 or more bases matched in the groups starting at sites i and j, a dot was placed at position $i+4, j+4$ in the plot. When a sequence is compared with itself, there must be a solid line of dots on the diagonal, but several runs of "near" matches can be seen in Figure 9.3. The small circle indicates the regions of 10 matches, starting at positions 85 and 377, noted in Table 9.10. The larger circle indicates the regions between positions 72 and 89, and 411 and 428 that meet the 7 out of 9 criterion.

A region of high similarity between two sequences appears as a diagonal in the dot plot with a large number of runs of k matching bases. When many comparisons are being made, as is the case when a query sequence is compared to all entries in a databank, it is important to know how likely a set of runs on a diagonal is to have arisen by chance alone. A methodology for performing statistical tests on observed similarities in this context is described by Mott et al. (1989).

Exact Matches between Sequences

The simplest kind of similarity between two sequences is that of a region in one exactly matching a region in the other. As a basis for assessing the biological significance of detected similarities, a model is set up that supposes that a DNA sequence consists of a random array of nucleotides, each one with the same probability of having a particular base type. Under this random model, two sequences with base frequencies a_i and b_i give a constant probability $P = \sum_{i=1}^{4} a_i b_i$ that any two positions, one in each sequence, carry the same base. A run of ℓ matches, terminated by a mismatch, therefore has probability $P^{\ell}(1-P)$. A complete search along sequences of lengths n_1 and n_2 for matches of length ℓ will involve $(n_1 - \ell + 1)(n_2 - \ell + 1)$ starting positions, and there may be interest in the longest match, of length L, found in all of these comparisons. Approximate expressions can be found (Arratia and Waterman 1985) for the mean and variance of this longest match found

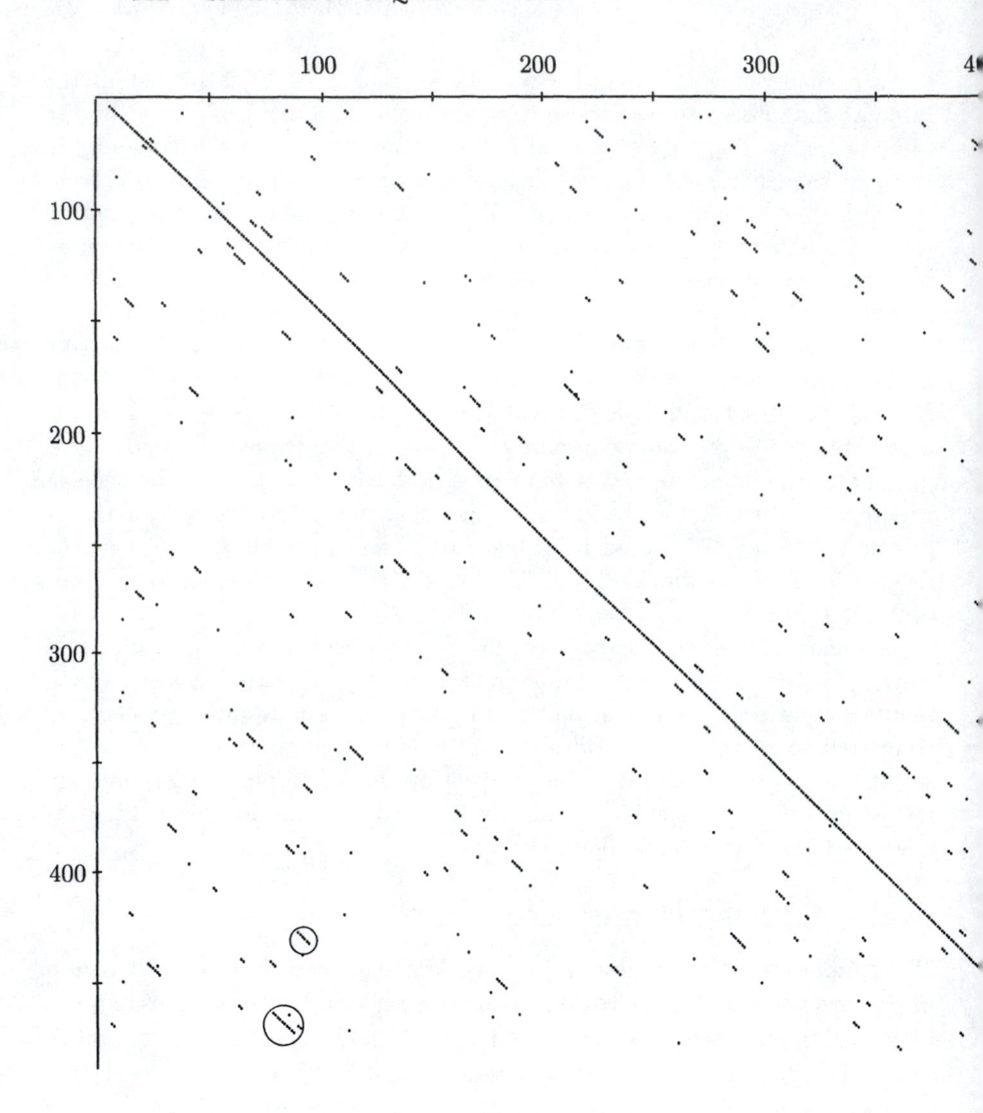

Figure 9.3 Dot-plot for comparison of sequence in Figure 9.2 with itself. Ci explained in text.

from all starting positions in two random sequences of length n

$$\mu_L = \frac{2\ln n}{\ln(1/P)}$$
$$\sigma_L^2 = \frac{1.645}{(\ln P)^2} + \frac{1}{12}$$

These values are substantially larger than those for the length ℓ expected to be found from any particular pair of starting positions

$$\mu_\ell = \frac{P}{1-P}$$
$$\sigma_\ell^2 = \frac{P}{(1-P)^2}$$

Two sequences of length 100 bases, with the four bases equally frequent in each, are expected to contain an exact match of 6.64 bases somewhere along their lengths. From any particular starting position, three fourths of the time the length is zero and the expected length of an exact match is only 0.33 bases.

To test whether an observed exact match could have arisen by chance, one procedure is to suppose that longest matches are normally distributed, and then reject the hypothesis when the observed value exceeds $\mu_L + 1.645\sigma_L$. Once again, however, normal theory approximations can be avoided by numerical studies. The two sequences could be shuffled and each shuffled pair searched for exact matches. Alternatively, benchmark values could be established by simulation. For any given base composition, pairs of sequences of specified lengths could be generated at random, and matches sought in each such simulated pair. With many simulations a distribution of longest match lengths could be generated.

Even if two sequences are known to be homologous, meaning that they have descended from a single ancestral sequence, it is unlikely that they will retain the same base composition or sequence length. Both features will change over time because of base substitution mutations or structural rearrangements, and searches for similarities between sequences should therefore accommodate some departures from exact matches. A variety of ad hoc procedures for searching for less-than-perfect matches have been proposed, and that due to Queen and Korn (1980) for ***local similarity*** was an early example. The Smith-Waterman algorithm for finding local similarities is described below.

Needleman-Wunsch Algorithm

Global searches for similarity, which take into account the total lengths of the sequences, are used to align sequences in such a way as to maximize the degree of ***global similarity***. Alignment of sequences of unequal length necessarily requires the introduction of loopouts on one sequence, or gaps on the other. An example of such procedures is provided by the algorithm of Needleman and Wunsch (1970). This algorithm was developed for amino acid sequences, but it can also be used for nucleotide sequences. As originally introduced, the algorithm sought to minimize a distance between two sequences. Although the elements of such distances are specified in an ad-hoc manner, the algorithm has optimal properties in that it determines the minimum distance. It is an example of a ***dynamic programming*** method, and the following description is based on that given by Kruskal (1983).

The two sequences to be aligned are placed along the axes of a two-way table. From any pair of elements (bases or amino acids), i.e., any cell in the table, an alignment can be extended in three possible ways: adding an element in each sequence, with a specified addition to the distance if the elements do not match, or adding an element in one sequence but a gap in the other, or vice versa. Introduction of a gap also contributes a specified amount to the distance. A cell in the table can therefore be reached from (at most) three neighboring cells, and the direction that results in the smallest distance to that cell from the upper left-hand corner is taken to be the direction of extending the similarity. Equal distances imply that two directions are possible. These directions are recorded and, when all cells have been investigated, a path is traced from the bottom right-hand corner (the end of the two sequences) to the top left-hand corner (the beginning of the two sequences) along the recorded directions. The resulting path(s) gives the alignments(s) of smallest distance.

The process is illustrated in Figure 9.4 for the two small sequences `CTATAATCCC` and `CTGTATC` and weights 1 for each mismatched element or 3 for each element opposite a gap. This figure shows that a boundary row and column are added to the table as initial conditions. In column 3 of row 5, corresponding to the second `T` in the shorter (the second) sequence and the first `T` in the longer (the first) sequence, the three possible increments to distance are shown. Adding the base `T` in each sequence, and moving from column 2 of row 4, contributes 0 to the distance. A horizontal movement, from column 2 of row 5, equivalent to adding the first (horizontal) sequence `T` but introducing a gap in (vertical) sequence 1, adds a penalty of 3 in this example. A vertical movement from column 3 of row 4 corresponds to adding the second sequence T but introducing a gap in sequence 1, and this also has a penalty of 3. The additions that result in the smallest distance for the cell are the diagonal and vertical ones, so arrows in those two directions are recorded on the table. Both result in a distance of 6 units to this cell from the top left-hand corner. Traversing the table along the arrows from bottom right to top left produces six possible alignments:

```
CTATAATCCC    CTATAATCCC    CTATAATCCC
CTGTA-TC--    CTGTA-T-C-    CTGTA-T--C

CTATAATCCC    CTATAATCCC    CTATAATCCC
CTGT-ATC--    CTGT-AT-C-    CTGT-AT--C
```

In each case the distance is 10, corresponding to six matches, one mismatch, and three gaps in the shorter sequence.

Algebraically, the algorithm can be described for sequences **a** and **b** with elements a_i and b_j. The distance between the sequences is $d(a, b)$ and is found recursively by evaluating the distances $d(a^i, b^j)$ that take into account the first i positions in **a** and the first j positions in **b**. If **a** and **b** are of lengths m and n, then the desired distance is $d(a^m, b^n)$. The introduced first row and column of the table are labeled with 0, corresponding to an empty sequence. Within cell (i, j), the additions to distance for the three possible events that lead to that cell are

1. Vertical movement from cell $(i-1, j)$ to cell (i, j) corresponds to extending the similarity by inserting a gap in sequence **b**. In other words, **b** has resulted from the deletion of a_i in **a**, and the weight for this event is written as $w_-(a_j)$.

2. Diagonal movement from cell $(i-1, j-1)$ to cell (i, j) corresponds to extending the similarity by adding elements a_i and b_j. In other words, **b** has resulted from the substitution of a_i in **a** by b_j, and the weight for this event is written as $w(a_i, b_j)$.

3. Horizontal movement from cell $(i, j-1)$ to cell (i, j) corresponds to extending the similarity by inserting a gap in sequence **a**. In other words, **b** has resulted from adding b_j, and the weight for this event is written as $w_+(b_j)$.

The distance $d(a^i, b^j)$ associated with cell (i, j) is taken to be the minimum of the distances in each of the three neighboring cells plus the associated

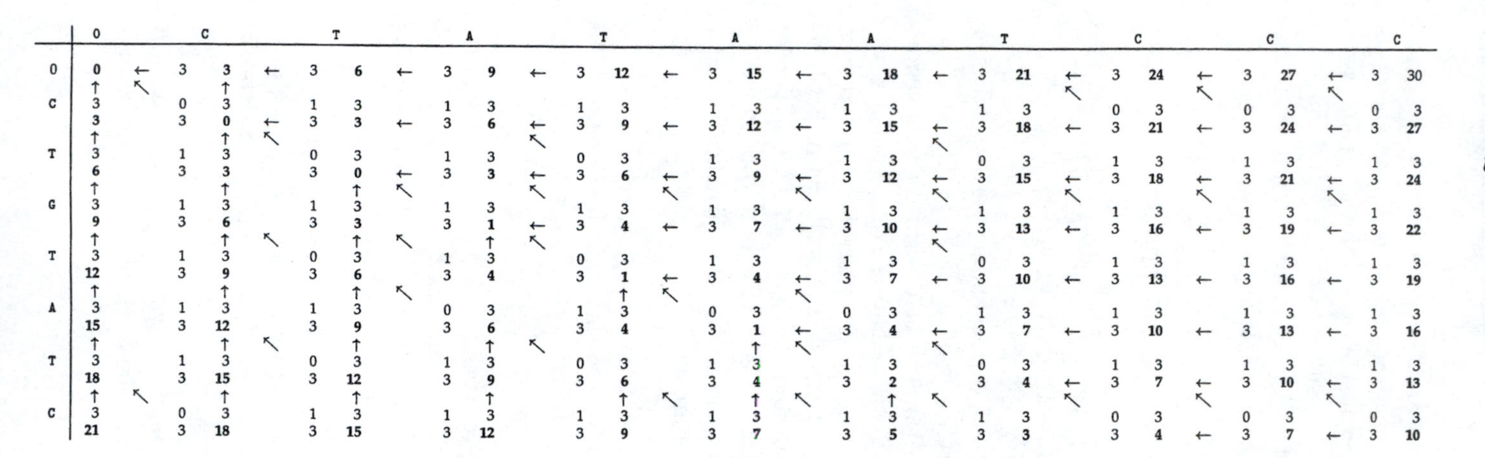

Figure 9.4 An implementation of the Needleman-Wunsch algorithm that weights mismatches by 1 and single deletions or insertions by 3.

weights:

$$d(a^i, b^j) = \min \begin{cases} d(a^{i-1}, b^j) + w_-(a_i) \\ d(a^{i-1}, b^{j-1}) + w(a_i, b_j) \\ d(a^i, b^{j-1}) + w_+(b_j) \end{cases}$$

and there is a need for initial conditions

$$d(a^0, b^0) = 0$$

$$d(a^0, b^j) = \sum_{k=1}^{j} w_+(b_k)$$

$$d(a^i, b^0) = \sum_{k=1}^{i} w_-(a_k)$$

In the example of Figure 9.5,

$$w_-(a_i) = 3 \text{ for every } i$$

$$w(a_i, b_j) = \begin{cases} 0 \text{ for every } i = j \\ 1 \text{ for every } i \neq j \end{cases}$$

$$w_+(b_j) = 3 \text{ for every } j$$

When two sequences have been aligned, an assessment of the minimum distance between them could be gained by applying the algorithm to shuffled versions of the sequences. If the actual alignment has a shorter distance than 95% of distances between the alignments of shuffled sequences, then it can be declared significant at the 5% level and is unlikely to have arisen by chance.

Smith-Waterman Algorithm

Concepts similar to those in the Needleman-Wunsch algorithm were included in the local-similarity algorithm of Smith and Waterman (1981). Because distantly-related proteins may share only isolated regions of similarity, searches for local similarity may sometimes be more appropriate than global searches. Smith and Waterman described an algorithm to find a pair of segments, one from each of two long sequences, such that there is no other pair of segments with greater similarity. They phrased the algorithm in terms of similarity, whereas Gotoh (1982) used distance as a criterion. For sequences $\mathbf{A} = (a_1, a_2, \ldots, a_m)$ and $\mathbf{B} = (b_1, b_2, \ldots, b_n)$, H_{ij} is the similarity of the two subsequences ending in a_i and b_j. As with the Needleman-Wunsch algorithm there is a recursive relationship for determining these H's. The initial values are

$$H_{i0} = 0, \; 0 \leq i \leq n \quad , \quad H_{0j} = 0, \; 0 \leq j \leq m$$

Similarities are built up with quantities $s(a_i, b_j)$, the similarity between sequence elements a_i and b_j, and gap weights $w_k = v + uk$ for gaps of length k. The algorithm operates to maximize similarity, and takes notice that the subsequences ending with a_i, b_j may result from adding elements to the subsequences ending with a_{i-1}, b_{j-1}, or a_i may follow a deletion of length k, or b_j may follow a deletion of length l. Gotoh introduced the quantities

$$\begin{aligned} P_{ij} &= \max(H_{i-1,j} - w_1, P_{i-1,j} - u) \\ Q_{ij} &= \max(H_{i,j-1} - w_1, Q_{i,j-1} - u) \end{aligned}$$

and showed that

$$H_{ij} = \max \begin{cases} H_{i-1,j-1} + s(a_i, b_j) \\ P_{ij} = \max_{1 \le k \le i}(H_{i-k,j} - w_k) \\ Q_{ij} = \max_{1 \le l \le j}(H_{i,j-l} - w_l) \\ 0 \end{cases}, \quad 1 \le i \le m, 1 \le j \le n$$

To use this method, initial values are set to $P_{0,0} = P_{0,j} = Q_{0,0} = Q_{i,0} = 0$. The algorithm guarantees that the maximum value of $H_{i,j}$ indicates the end points of the two subsequences with greatest similarity. Starting from this element, the path leading to it is traced back in the $H_{i,j}$ matrix until a negative value is reached. Dissimilarities between the two sequences outside the region of greatest similarity do not affect the score H of that region.

As an example, consider the same two short sequences, CTGTATC and CTATAATCCC used in Figure 9.5 to illustrate the Needleman-Wunsch algorithm. These sequences are arranged along the margins of Table 9.12, and the values of H_{ij}, P_{ij}, Q_{ij} are shown in the cells of the table. The weights are those used in an example by Smith and Waterman (1981):

$$\begin{aligned} s(a_i, b_j) &= \begin{cases} 1 & , a_i = b_j \\ -1/3 & , a_i \ne b_j \end{cases} \\ w_k &= 1 + k/3 \end{aligned}$$

The $s(a_i, b_j)$ values have an average of zero for long, random sequences in which the four bases are equally probable. The deletion value w_k must be at least as big as the difference between match and mismatch weights.

In Table 9.12, the maximum value of H_{ij} is 4.33 in row 8 and column 7. Asterisks indicate the traceback path for maximally similar segments:

```
C T G T A - T C
C T A T A A T C
```

Table 9.12 Example for Smith-Waterman Algorithm.

			$j=0$	$j=1$	$j=2$	$j=3$	$j=4$	$j=5$	$j=6$	$j=7$
			0	C	T	G	T	A	T	C
$i=0$	0	H_{ij}	0	0	0	0	0	0	0	0
		P_{ij}	0	0	0	0	0	0	0	0
		Q_{ij}	0	0	0	0	0	0	0	0
$i=1$	C	H_{ij}	0	1.00*	.00	.00	.00	.00	.00	1.00
		P_{ij}	0	−.33	−.33	−.33	−.33	−.33	−.33	−.33
		Q_{ij}	0	−.33	−.33	−.67	−1.00	−1.33	−1.33	−1.33
$i=2$	T	H_{ij}	0	.00	2.00*	.67	1.00	.00	1.00	.00
		P_{ij}	0	−.33	−.67	−.67	−.67	−.67	−.67	−.33
		Q_{ij}	0	−.33	−.67	.67	.33	.00	−.33	−.33
$i=3$	A	H_{ij}	0	.00	.67	1.67*	.33	2.00	.67	.67
		P_{ij}	0	−.67	.67	−.67	−.33	−1.00	−.33	−.67
		Q_{ij}	0	−.33	−.67	−.67	.33	.00	.67	.33
$i=4$	T	H_{ij}	0	.00	1.00	.33	2.67*	1.33	3.00	1.67
		P_{ij}	0	−1.00	.33	.33	.67	.67	−.67	−.67
		Q_{ij}	0	−.33	−.67	−.33	−.67	1.33	1.00	1.67
$i=5$	A	H_{ij}	0	.00	.00	.67	1.33	3.67*	2.33	2.67
		P_{ij}	0	−1.33	.00	.00	1.33	.33	1.67	.33
		Q_{ij}	0	−.33	−.67	−1.00	−.67	.00	2.33	2.00
$i=6$	A	H_{ij}	0	.00	.00	.00	1.00	2.33*	3.33	2.00
		P_{ij}	0	−1.33	−.33	−.33	1.00	2.33	1.33	1.33
		Q_{ij}	0	−.33	−.67	−1.00	−1.33	−.33	1.00	2.00
$i=7$	T	H_{ij}	0	.00	1.00	.00	1.00	2.00	3.33*	3.00
		P_{ij}	0	−1.33	−.67	−.67	.67	2.00	2.00	1.00
		Q_{ij}	0	−.33	−.67	−.33	−.67	−.33	.67	2.00
$i=8$	C	H_{ij}	0	1.00	.00	.67	.33	1.67	2.00	4.33*
		P_{ij}	0	1.33	−.33	−1.00	.33	1.67	2.00	1.67
		Q_{ij}	0	−.33	−.33	−.67	−.67	−1.00	.33	.67
$i=9$	C	H_{ij}	0	1.00	.67	.00	.33	1.33	1.67	3.00
		P_{ij}	0	−.33	−.67	−.67	.00	1.33	1.67	3.00
		Q_{ij}	0	−.33	−.33	−.67	−1.00	−1.00	.00	.33
$i=10$	C	H_{ij}	0	1.00	.67	.33	.00	1.00	1.33	2.67
		P_{ij}	0	−.33	−.67	−1.00	−.33	1.00	1.33	2.67
		Q_{ij}	0	−.33	−.33	−.67	−1.00	−1.33	−.33	.00

PROTEIN SEQUENCE DATA

Many of the algorithms applied to DNA sequences were originally derived for sequences of amino acids. Because proteins are likely to be the target of natural selection, it might be argued that analyses of these sequences have more relevance than do analyses of DNA sequences. Protein analyses certainly avoid problems of degeneracy in the genetic code, whereby several triplets may encode the same amino acid. As protein sequences can be written in a 20-letter alphabet (Table 1.4), there is a smaller probability that two proteins will have the same letter (amino acid) by chance alone at any position than is the case for DNA sequences.

There is a need for more refined analyses that reflect varying degrees of similarity between amino acids, and consequent varying likelihoods of one being substituted for another during the course of evolution. It may be convenient, for example, to classify amino acids into such groups as neutral and hydrophobic (G, A, V, L, I, F, P, M), neutral and polar (S, T, Y, W, N, E, C), basic (K, R, H), or acidic (D, E) (Lewin 1994) in the one-letter notation of Table 1.4. Other classifications have been given (e.g., Dayhoff et al. 1979 use hydrophobic, aromatic, basic, acid or acid-amide, cysteine, other hydrophylic).

Quantification of the similarity between amino acids is by means of ***scoring matrices***. The 20×20 matrices, relating each amino acid to every amino acid, in most common use fall into the PAM and BLOSUM classes, and are now described. A general discussion of scoring matrices was given by Jones et al. (1992) and by Altschul et al. (1994).

PAM Matrices

Alignment of protein sequences can take account of the differential rates at which amino acids substitute for each other, and the following discussion is based on Dayhoff (1979), Dayhoff et al. (1979), and Schwartz and Dayhoff (1979). On the basis of comparisons among many pairs of very similar protein sequences, Dayhoff and her colleagues constructed a ***mutation probability matrix*** $\mathbf{M}$. They began by comparing many pairs of protein sequences to determine the empirical frequencies with which one amino acid is replaced by others during evolution. Specifically, they inferred common ancestral sequences for pairs of present-day sequences and compared each of the pair to this ancestor. On the basis of 1572 observed substitutions, they found the "accepted point mutation matrix" $\mathbf{A}$ shown in Table 9.13 (note that rounding errors have prevented the entries in that table from adding exactly to 1572). The empirical frequency with which amino acid type i is replaced by type j (or vice versa) is found from that table, and is written as A_{ij}. Matrix $\mathbf{A}$ may be called the raw PAM matrix. Jones et al. (1992) used a more extensive data set with 59,190 accepted point mutations, and their $\mathbf{A}$ matrix did not have any zero elements.

Table 9.13 Numbers of substitutions between amino acids observed in 1572 exchanges. Each entry has been multiplied by 10. Fractional exchanges occur when ancestral sequences are ambiguous.

	A	R	N	D	C	Q	E	G	H	I	L	K	M	F	P	S	T	W	Y
R	30																		
N	109	17																	
D	154	0	532																
C	33	10	0	0															
Q	93	120	50	76	0														
E	266	0	94	831	0	422													
G	579	10	156	162	10	30	112												
H	21	103	226	43	10	243	23	10											
I	66	30	36	13	17	8	35	0	3										
L	95	17	37	0	0	75	15	17	40	253									
K	57	477	322	85	0	147	104	60	23	43	39								
M	29	17	0	0	0	20	7	7	0	57	207	90							
F	20	7	7	0	0	0	0	17	20	90	167	0	17						
P	345	67	27	10	10	93	40	49	50	7	43	43	4	7					
S	772	137	432	98	117	47	86	450	26	20	32	168	20	40	269				
T	590	20	169	57	10	37	31	50	14	129	52	200	28	10	73	696			
W	0	27	3	0	0	0	0	0	3	0	13	0	0	10	0	17	0		
Y	20	3	36	0	30	0	10	0	40	13	23	10	0	260	0	22	23	6	
V	365	20	13	17	33	27	37	97	30	661	303	17	77	10	50	43	186	0	17

Source: Dayhoff et al. (1979.)

For the mutation probability matrix, element M_{ij} in row i and column j is the (empirical) probability that the amino acid in column j will be replaced by that in row i after a specified time interval. Dayhoff et al. measured time in units of ***percent accepted mutations*** or ***point accepted mutations per 100 residues*** (PAM's). Assuming that mutation does not affect the same site twice, 1 PAM is the time over which one amino acid replacement is expected to occur in a 100 amino acid polypeptide chain of average composition. Translation from the A_{ij} terms to the M_{ij}'s depends on the number of PAM's for which the matrix $\mathbf{M}$ is appropriate.

Dayhoff et al. (1979) defined the relative "mutability" m_j of amino acid j as being proportional to the number of observed changes of the amino acid divided by its frequency of occurrence f_j in the aligned sequences

$$m_j \propto \sum_{i \neq j} A_{ij}/f_j$$

This takes into account all of the ways in which amino acid a_j may change. These quantities are normalized, and numerical values shown in Table 9.14.

The probability that the jth amino acid changes is $1 - M_{jj}$ and this must be proportional to the mutability

$$1 - M_{jj} \propto m_j$$

or

$$M_{jj} = 1 - \lambda m_j$$

to define the proportionality constant λ. Similarly, the element M_{ij} for a specific change from j to i is

$$M_{ij} \propto m_j A_{ij}$$

Because M_{jj} and $\sum_{k \neq j} M_{kj}$ must sum to one

$$M_{ij} = \lambda m_j A_{ij} / \sum_{k \neq j} A_{kj}$$

Further, because 1 PAM refers to an expected change of 1 amino acid per 100 amino acids, there are expected to be 99 unchanged

$$99 = 100 \sum_i f_i M_{ii}$$

$$\lambda = \frac{1}{100 \sum_i m_i f_i}$$

Table 9.14 Relative mutabilities m_i and frequencies f_i of amino acids in the accepted point mutation data.

	m_i	f_i		m_i	f_i
A	100	0.087	L	40	0.085
R	65	0.041	K	56	0.081
N	134	0.040	M	94	0.015
D	106	0.047	F	41	0.040
C	20	0.033	P	56	0.051
Q	93	0.038	S	120	0.070
E	102	0.050	T	97	0.058
G	49	0.089	W	18	0.010
H	66	0.034	Y	41	0.030
I	96	0.037	V	20	0.065

Source: Dayhoff et al. (1979).

Unlike matrix **A**, matrix **M** is not symmetric. Premultiplying the vector **f** of 20 frequencies f_i by the matrix **M** leaves it unchanged. Matrices representing larger evolutionary distances may be derived from the 1 PAM matrix **M** by matrix multiplication.

Schwartz and Dayhoff (1979) found that 250 PAM's was an appropriate time for studying evolutionary relationships among distantly related proteins. The resulting matrix is the 250th power of **M**, and is shown in Table 9.15. Rounding errors prevent Table 9.15 following exactly from Tables 9.13 and 9.14.

Finally, Dayhoff et al. (1979) defined a relatedness odds matrix **R** with elements $R_{ij} = M_{ij}/f_i$ (where the M's now refer to the 250 PAM matrix). This odds matrix is symmetric and its elements give the probabilities of replacement of amino acid j by amino acid i per occurrence of i. This matrix, after logarithms have been taken of each element, is shown in Table 9.16 in a way that groups amino acids that are more likely to be interchangeable. Once again, rounding errors prevent an exact matching of Tables 9.13 – 9.16. Elements of the matrix are used in alignment procedures, for example, in a Needleman-Wunsch-like algorithm as the weights w_{ij} for substituting amino acid i by amino acid j, although Dayhoff et al. recommended adding a bias term B to each element of the matrix. A penalty for an insertion or deletion also needs to be defined. Dayhoff et al. gave a constant gap penalty for gaps of any length. A typical value for the bias correction and the gap penalty is

Table 9.15 Mutation probability matrix for the evolutionary time of 250 PAM's. Each element has been multiplied by 100.

	A	R	N	D	C	Q	E	G	H	I	L	K	M	F	P	S	T	W	Y	V
A	13	6	9	9	5	8	9	12	6	8	6	7	7	4	11	11	11	2	4	9
R	3	17	4	3	2	5	3	2	6	3	2	9	4	1	4	4	3	7	2	2
N	4	4	6	7	2	5	6	4	6	3	2	5	3	2	4	5	4	2	3	3
D	5	4	8	11	1	7	10	5	6	3	2	5	3	1	4	5	5	1	2	3
C	2	1	1	1	52	1	1	2	2	2	1	1	1	1	2	3	2	1	4	2
Q	3	5	5	6	1	10	7	3	7	2	3	5	3	1	4	3	3	1	2	3
E	5	4	7	11	1	9	12	5	6	3	2	5	3	1	4	5	5	1	2	3
G	12	5	10	10	4	7	9	27	5	5	4	6	5	3	8	11	9	2	3	7
H	2	5	5	4	2	7	4	2	15	2	2	3	2	2	3	3	2	2	3	2
I	3	2	2	2	2	2	2	2	2	10	6	2	6	5	2	3	4	1	3	9
L	6	4	4	3	2	6	4	3	5	15	34	4	20	13	5	4	6	6	7	13
K	6	18	10	8	2	10	8	5	8	5	4	24	9	2	6	8	8	4	3	5
M	1	1	1	1	0	1	1	1	1	2	3	2	6	2	1	1	1	1	1	2
F	2	1	2	1	1	1	1	1	3	5	6	1	4	32	1	2	2	4	20	3
P	7	5	5	4	3	5	4	5	5	3	3	4	3	2	20	6	5	1	2	4
S	9	6	8	7	7	6	7	9	6	5	4	7	5	3	9	10	9	4	4	6
T	8	5	6	6	4	5	5	6	4	6	4	6	5	3	6	8	11	2	3	6
W	0	2	0	0	0	0	0	0	1	0	1	0	0	1	0	1	0	55	1	0
Y	1	1	2	1	3	1	1	1	3	2	2	1	2	15	1	2	2	3	31	2
V	7	4	4	4	4	4	4	5	4	15	10	4	10	5	5	5	7	2	4	17

Source: Dayhoff et al. (1979).

Table 9.16 Log-odds matrix for 250 PAM's. Elements are shown multiplied by 10.

	C	S	T	P	A	G	N	D	E	Q	H	R	K	M	I	L	V	F	Y	W
C	12																			
S	0	2																		
T	-2	1	3																	
P	-3	1	0	6																
A	-2	1	1	1	2															
G	-3	1	0	-1	1	5														
N	-4	1	0	-1	0	0	2													
D	-5	0	0	-1	0	1	2	4												
E	-5	0	0	-1	0	0	1	3	4											
Q	-5	-1	-1	0	0	-1	1	2	2	4										
H	-3	-1	-1	0	-1	-2	2	1	1	3	6									
R	-4	0	-1	0	-2	-3	0	-1	-1	1	2	6								
K	-5	0	0	-1	-1	-2	1	0	0	1	0	3	5							
M	-5	-2	-1	-2	-1	-3	-2	-3	-2	-1	-2	0	0	6						
I	-2	-1	0	-2	-1	-3	-2	-2	-2	-2	-2	-2	-2	2	5					
L	-6	-3	-2	-3	-2	-4	-3	-4	-3	-2	-2	-3	-3	4	2	6				
V	-2	-1	0	-1	0	-1	-2	-2	-2	-2	-2	-2	-2	2	4	2	4			
F	-4	-3	-3	-5	-4	-5	-4	-6	-5	-5	-2	-4	-5	0	1	2	-1	9		
Y	0	-3	-3	-5	-3	-5	-2	-4	-4	-4	0	-4	-4	-2	-1	-1	-2	7	10	
W	-8	-2	-5	-6	-6	-7	-4	-7	-7	-5	-3	2	-3	-4	-5	-2	-6	0	0	17

Source: Dayhoff et al. 1979.

6. The alignment of fish protamine sequences shown in Figure 1.2 followed from such an algorithm. The relatedness odds for each alignment position are multiplied together to provide a total alignment odds value. A more general discussion on odds scores is given below in the section on statistical significance of similarity scores, and the use of PAM matrices in the BLAST alignment program is also described below.

BLOSUM Matrices

The PAM matrices introduced by Dayhoff are constructed from the amino acid replacements inferred from alignments of protein sequences that are at least 85% identical. These replacement rates are extrapolated to more distantly related sequences by taking powers of the PAM matrix. An alternative approach, leading to ***blocks substitution matrices*** (BLOSUM), has been described by Henikoff and Henikoff (1992). These authors considered blocks, or highly conserved regions, in aligned protein sequences.

Local alignments can be represented as ungapped blocks, where rows are the protein segments within protein sequences and columns are the aligned positions. For a block of width w amino acids, and depth s sequences, there are $s(s-1)/2$ pairs of amino acids for each column and $ws(s-1)/2$ pairs overall. Of this number, suppose f_{ij} are of amino acid types i,j (ignoring order). The proportion of i,j pairs is

$$q_{ij} = f_{ij}/\sum_{i,j} f_{ij}$$

whereas the expected proportion under independence is

$$e_{ij} = \begin{cases} p_i^2 & i = j \\ 2p_ip_j & i \neq j \end{cases}$$

and

$$p_i = q_{ii} + \frac{1}{2}\sum_{j \neq i} q_{ij}$$

The i,jth element of the BLOSUM matrix is defined as

$$s_{ij} = 2\log_2(q_{ij}/e_{ij})$$

All pairs from all blocks in a set of reference protein sequences contribute to this matrix. The BLOSUM62 matrix is derived from a database of blocks in which sequence segments that are identical for at least 62% of aligned positions are clustered.

Henikoff and Henikoff (1994) describe a means for searching databases of blocks of locally aligned subsequences to reveal global similarity and hence classify protein sequences. This work recognizes that protein families may be characterized by multiple local motifs. If a query sequence belongs to a family with multiple blocks, then at least some of these blocks should score highly when that sequence is compared to the block database.

FASTA Algorithm

Dayhoff's PAM matrices can be used in the FASTA protein sequence comparison program (Pearson and Lipman 1988). When a sequence is compared to all sequences in a database, there is an implicit search for relatedness or descent from a common ancestral sequence. Alignment of two such presumed homologous sequences amounts to identifying the most likely set of mutations since the ancestral sequence, and this should be based on a model of molecular evolution. FASTA can compare either DNA or protein sequences, but distant sequence relationships can best be identified with protein sequences. The FASTA program originally used the PAM250 matrix as an empirical evolutionary model to weight amino acid replacements and construct a similarity score for alternative alignments, although Altschul (1991) argues that the PAM120 matrix is more appropriate for general protein similarity searches, with at least three searches using PAM40, PAM120 and PAM250 being preferred. Altschul (1993) later presented an "All-PAM" matrix that is sensitive at all detectable evolutionary distances. The following description of the FASTA program is taken from Lipman and Pearson (1985) and Pearson (1990).

FASTA uses four steps to calculate three scores that characterize sequence similarity:

- Step 1. Identify regions in the two sequences with the highest density of matches.

  A lookup table is constructed for each sequence that lists the positions of all sets of *ktup* amino acids (generally *ktup* is 1 or 2), in a table analogous to Table 9.8. The sequence `FLWRTWS` would list `L` as appearing at position 2, `W` at positions 3 and 6, `R` at position 4, and so on. Comparison proceeds by looking up each of the amino acids of sequence 2 in the table of positions for sequence 1. In particular, matches with the same degree of offset (the difference in position numbers of matching amino acids in the two sequences) are identified. If the second sequence is `SWKTWT`, the offset of the matching `W`'s at positions 3 and 2 is 1, as is that for the matching `T`'s at positions 5 and 4. Elements `W, T` would lie on the same diagonal in a dot matrix comparison, and the comparison score for the subsequences `WRT` and `WKT` would need to include a penalty for the mismatching of the `R` and `K` pair. This procedure identifies regions of a diagonal with the highest density of *ktup* matches, and the program saves the (say) ten best local regions, whether they are on the same or different diagonals.

- Step 2. Rescore the ten regions with the highest density of identities using a scoring matrix. Trim the ends of the region to include only those positions contributing to the highest score. Each region is a partial alignment without gaps.

  The PAM250 scoring matrix shown in Table 9.16 gives positive scores for replacements that commonly occur among related proteins and negative scores for unlikely replacements. For each of the best diagonal regions rescored with the scoring matrix, a subregion with the maximal score is identified.

- Step 3. If there are several initial regions with scores greater than some cutoff value, check to see whether the trimmed initial regions can be joined to form an approximate alignment with gaps. Calculate a similarity score that is the sum of scores for the joined initial regions minus a penalty for each gap. This initial similarity score (*initn*) is used to rank the library of sequences being compared to the query sequence of interest. The score of the best initial region found in step 2 is reported (*init1*).

This joining step increases the ability to identify distantly related sequences, or ***sensitivity*** of the search method, because it allows for insertions and deletions as well as conservative replacements. The modification decreases ***selectivity***, the avoidance of false positives, however. The loss of selectivity is limited by considering only the initial regions with scores higher than some threshold value.

- Step 4. Construct a Needleman-Wunsch optimal alignment of the query sequence and the library sequence, considering only those positions that lie in a band 32 positions wide centered on the best initial region found in step 2. The score for this alignment is the optimized (*opt*) score.

  The Needleman-Wunsch algorithm is used in this step with the same weights as in the earlier steps. It is usual to report the mean of the similarity scores for the query sequence against all the library sequences, along with the standard deviation of these scores. The significance of the matches found may be found by permutation procedures, or may be indicated by reporting the number of standard deviations above the mean is the score of a particular match.

The FASTA program has been modified to apply to DNA sequences. Instead of a PAM or BLOSUM matrix, replacement matrices that allow separate penalties for transitions and transversions, for example, can be used.

BLAST Program

Another means for rapid detection of local similarities was given by Altschul et al. (1990). They described a ***Basic Alignment Search Tool*** (BLAST), a heuristic method designed to detect alignments that optimize a measure of local similarity known as the maximal segment pair (MSP) score. Gaps are not allowed.

A sequence segment is a contiguous stretch of residues, and the segment score is the sum of scores for each pair of aligned residues. These scores may come from the PAM250 matrix, although Altschul et al. used the PAM120 matrix for protein sequences. The BLAST programs provided by the National Center for Biotechnology Information (World Wide Web URL http://www.ncbi.nlm.nih.gov) use BLOSUM62 as the default scoring matrix. For DNA sequences, the NCBI BLAST programs score matches as +5 and mismatches as −4. The ***maximal segment pair*** MSP is the highest scoring pair of segments of identical length chosen from the two sequences and a segment pair is locally maximal if its score cannot be improved by either extending or shortening both segments.

Few, if any, of the sequences in a database will be homologous to any particular query sequence, so it makes sense to examine only those sequences with some minimal MSP score, S. Even these high scoring sequences will include cases where the score arises at random rather than through homology. The theory of Karlin and Altschul (1990) discussed in the next section allows prediction of the highest MSP score for pairs of random sequences, and BLAST operates to minimize time spent on sequence regions whose similarity with the query sequence has little chance of exceeding this score.

The method has three steps: compiling a list of high-scoring words, scanning the database for hits, and extending hits. Details are:

- Step 1. Compile a list of high-scoring words (subsequences), of length w. This was illustrated in the first step of Karlin's algorithm, and typically $w = 4$. Altschul et al. (1990) found that the list typically contains 50 words for each residue in the query sequence.

- Step 2. Scan each sequence in the database to see if it contains a word of length w that can pair with the query sequence to produce a word pair with a score greater than some threshold T. Each database word is used to look up the list generated in step 1, and matches are seen immediately.

- Step 3. Extend such "hits" (matching word pairs) to determine if they are contained within a segment pair whose score is greater or equal to S. The process of extending in one direction stops when a segment pair is reached whose score falls a certain amount below the best score found for shorter extensions.

Statistical Significance of Local Similarities

When a local similarity is found, it is important to be able to determine if such a similarity could have arisen by chance. This may be achieved with permutation procedures, or it may make use of a quantitative description of the similarity. Quantitative measures also allow a comparison of the quality of different similarities. The theory of Karlin and Altschul (1990), reviewed in Altschul (1991), gives a method for constructing such measures. Before describing that, the general discussion of Altschul et al. (1994) will be reviewed.

Because the score of an optimal local alignment is the maximum of the sums of many essentially independent scores, the theory of ***extreme values*** is appropriate. An extreme value distribution is described by a characteristic value u and a decay constant λ, and this distribution provides the number of sequence pairs having a score of at least x by chance as $e^{-\lambda(x-u)}$. Furthermore, the probability of a score X being greater than or equal to x, by chance, is

$$\Pr(X \geq x) \quad = \quad 1 - \exp[-e^{-\lambda(x-u)}]$$

The parameters u, λ depend on the sequences being compared and the scoring system being used. For ungapped alignments, and a comparison of sequences of lengths m, n

$$u \quad = \quad \frac{\ln(Kmn)}{\lambda}$$

where K depends on the sequence composition and scoring method.

The probability p of a given local-alignment score is for a pair of sequences. If one sequence is being compared to all D sequences in a database, there needs to be an adjustment for multiple comparisons (see Chapter 3). From a Poisson distribution assumption, the probability of no pair of sequences having an alignment score of p is e^{-p}. The probability of all D pairs not having such a score after the database search is e^{-pD}, so the probability P of at least one value of p is

$$P \quad = \quad 1 - e^{-pD}$$

Instead of declaring an alignment significant if $p \leq 0.05$, it may be better to require that $p \leq 0.05/D$. Alternatively, Altschul et al. (1994) suggest that D be replaced by N/n where n is the length of the aligned database sequence and N is the total number of residues in the database.

Aligned amino acids a_i and a_j are given a replacement score of s_{ij}. Relatively similar amino acids have positive scores and dissimilar amino acids have negative scores. Two amino acid segments of the same length that, when aligned, have the greatest aggregate score provide the maximal segment pair as discussed above. If amino acid a_i occurs by chance, in either sequence, with probability p_i then two assumptions are made about replacement scores. There must be at least one positive score to allow alignments of more than one amino acid in length. The expected score $\sum_{i,j} p_i p_j s_{ij}$ at any position must be negative, otherwise longer similarities would tend to have greater scores and *local* alignments would not result.

Karlin and Altschul (1990) introduced the parameter $\lambda > 0$ which satisfies

$$\sum_{i,j} p_i p_j e^{\lambda s_{ij}} = 1$$

and showed that, for random sequences, amino acids a_i and a_j are aligned with "target" frequencies approaching

$$q_{ij} = p_i p_j e^{\lambda s_{ij}}$$

Karlin and Altschul argue that a scoring scheme is optimal for identifying a region of biological interest only if there is no significant difference between the composition of high-scoring chance segments and the composition of similarly scoring true segments (i.e. those of interest). Only then are the scoring parameters optimal for distinguishing distant local homologies from similarities due to chance. This leads to the optimal scores

$$s_{ij} = \frac{1}{\lambda} \ln\left(\frac{q_{ij}}{p_i p_j}\right)$$

In other words, the score s_{ij} for a pair of amino acids depends on the target frequency of that pair and the background frequency with which that pair occurs. Since the ratio of these frequencies compares probabilities of an event under alternative hypotheses, it is called a likelihood ratio or odds ratio. Taking logarithms gives log-odds scores. Any substitution matrix used for assessing local alignments is implicitly a log-odds matrix (Altschul 1993). The target frequencies q_{ij} for closely related sequences will differ from those for sequences that are greatly diverged, and a single scoring matrix of odds ratio type will not be optimal for all evolutionary distances. The PAM120 and BLOSUM62 matrices are designed for moderately diverged sequences.

The matrix in Table 9.16 consists of log-odds scores. Although PAM matrices were developed in the context of global sequence comparisons, they are actually of optimal form for local similarity searches. Scoring matrices can be obtained from appropriate random and target distributions. Related sets of segments from protein superfamilies provide amino acid replacement frequencies q_{ij} for evolution over substantial periods of time. Individual amino acid frequencies can also be found from the same sequences, and then log-odds scores calculated from $\ln(q_{ij}/p_i p_j)$.

DNA SEQUENCE DISTANCES

The distance employed in the Needleman-Wunsch algorithm was meant to accommodate the mutational events of base substitution or insertion and deletion, but the relative weights of these events were arbitrary. Such distances would not be used in the construction of evolutionary trees. The earliest distance for sequences, introduced for amino acid sequences, was given by Jukes and Cantor (1969). It is based on the quantity q discussed in Chapter 5.

Jukes-Cantor Distance

Although the construction of evolutionary trees is taken up in Chapter 10, this is a good place to repeat the observation of Felsenstein (1988) that alignment of sequences and construction of trees are not really separate issues. They are being treated separately here by way of introduction only. Sankoff et al. (1973) presented alignments of some 5S rRNA sequences by taking explicit account of their evolutionary history. Thompson et al. (1994) gave a method for multiple sequence alignment following the branching order in a phylogenetic tree, and made use of the neighbor- joining tree construction method treated in Chapter 10.

When a sequence is available from each of two populations, q is the probability that homologous nucleotides have the same base. Note that there is an explicit assumption that the sequences are homologous, having a single ancestral sequence. The comparison of sequences to allow q to be estimated as the proportion of equal homologous bases requires a preliminary alignment of the sequences. As shown in Chapter 5, q decreases over time because of base substitution, and has a value in generation t since divergence from the ancestral sequence of

$$q_t = \frac{1}{4} + \frac{3}{4}\left(1 - \frac{8\mu}{3}\right)^t$$

where μ is the rate of base substitution. The quantity

$$K = \frac{3}{4}\ln\left(\frac{3}{4q-1}\right) \approx 2\mu t$$

can serve as a distance measure.

Distance K is appropriate for indicating the time since two sequences diverged from an ancestral sequence, and can lead to evolutionary trees between sequences (gene trees) as described in Chapter 10. Care does need to be taken when the time of divergence is very large. Even though K increases

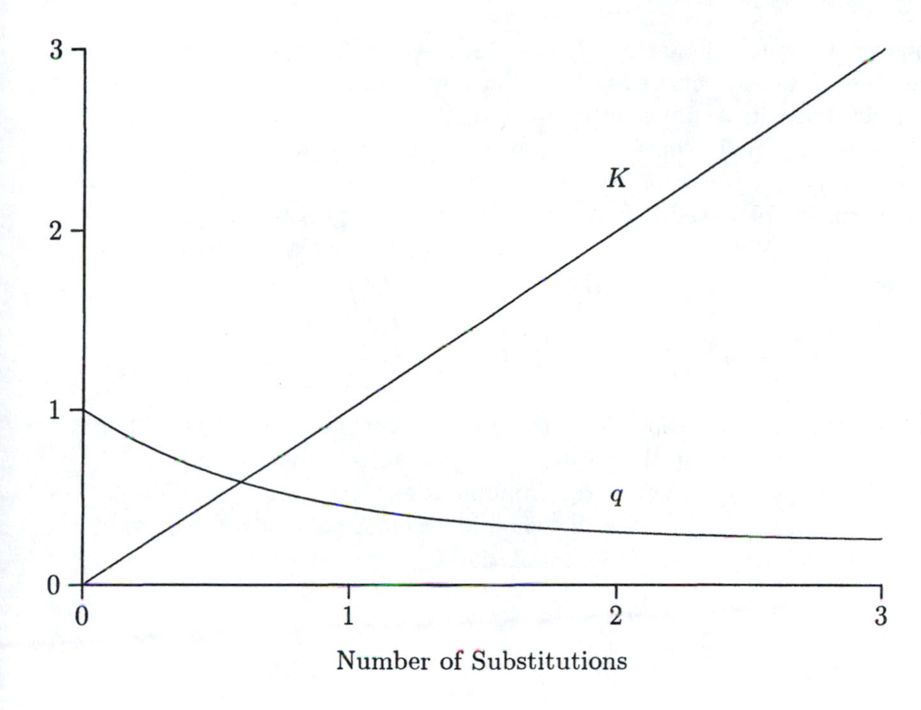

Figure 9.5 Probability q that homologous bases are the same, and Jukes-Cantor distance K, as a function of the number of base substitutions $2\mu t$.

linearly over all time, there is a limit to how different sequences may become. Eventually, with equal base frequencies, there is just a probability of 0.25 that homologous bases are the same. The quantity q can then no longer discriminate between sequences, and any discrimination based on K may be misleading. The behavior of both q and K, as a function of $2\mu t$, is shown in Figure 9.5. Pearson (1990) considers that DNA sequences rarely allow demonstration of common ancestry for sequences that diverged more than 100 to 200 million years ago, whereas common ancestry can be demonstrated for protein sequences with a divergence time of 1 to 2 billion years. Changes in DNA sequence may not change encoded proteins, and there is biochemical information in the amino acid itself, such as arginines being similar to lysines but glycines and isoleucines being very different.

When several sequences are available from each of two species, it is possible to construct species trees allowing for the ancestral species being polymorphic. Two sequences, within or between species, may have descended from different sequences in that ancestral population. A measure of within-species sequence similarity, Q, was discussed in Chapter 5 and it depends on the population size as well as the mutation rate. When the ancestral population is in equilibrium between drift and mutation, the quantity Q is expected to remain constant at $\hat{Q}$ over time and to furnish the initial value of q. Although this constant value is not known, it can be estimated from the variation observed in each of two extant species (Cockerham 1984, Weir and Basten 1990). A modified version of the Jukes-Cantor distance, taking account of within-species variation, is then

$$K_W = \frac{3}{4}\ln\left(\frac{4\hat{Q}-1}{4q-1}\right)$$

This is still directly proportional to time, but time now refers to that between the extant and ancestral *populations* rather than sequences.

If n_i sequences are observed in population i, and $r_{ijj'}$ is the proportion of homologous bases that is the same between sequences j and j', then the sample similarity within that population is

$$\tilde{Q}_i = \frac{1}{n_i(n_i-1)}\sum_{j=1}^{n_i}\sum_{j'\neq j} r_{ijj'}$$

and $\hat{Q}$ could be estimated as

$$\tilde{Q} = \frac{1}{2}(\tilde{Q}_1 + \tilde{Q}_2)$$

If $s_{jj'}$ is the number of like bases between sequence j in population 1 and sequence j' in population 2, then the between-population similarity is estimated as

$$\tilde{q} = \frac{1}{n_1 n_2}\sum_{j=1}^{n_1}\sum_{j'=1}^{n_2} s_{jj'}$$

Putting these together

$$\tilde{K}_W = \frac{3}{4}\ln\left(\frac{4\tilde{Q}-1}{4\tilde{q}-1}\right)$$

Kimura Two-parameter Distance

There have been several attempts to generalize such distances by allowing for less restrictive mutation models. Use of the Jukes-Cantor measure assumes that each base is equally likely to mutate to each of the three other bases with a constant rate $\mu/3$. Kimura (1980) allowed ***transitions*** (mutation between two pyrimidines or two purines: $T \leftrightarrow C$, or $A \leftrightarrow G$) and ***transversions*** (mutation between a pyrimidine and a purine: $T, C \leftrightarrow A, G$) to have different rates, α and β, respectively.

After the sequences have been aligned, the proportions of positions having bases that differ by a transition (a type I change), or differ by a transversion (a type II change), or are the same are denoted by $\tilde{p}_{\rm I}$, $\tilde{p}_{\rm II}$, or $1-\tilde{p}_{\rm I}-\tilde{p}_{\rm II}$, respectively. Under his mutation model, Kimura (1980) showed that the corresponding probabilities changed over the time t since divergence according to

$$p_{{\rm I}_t} = \frac{1}{4}\left(1 - 2e^{-4(\alpha+\beta)t} + e^{-8\beta t}\right)$$

$$p_{{\rm II}_t} = \frac{1}{4}\left(1 - e^{-8\beta t}\right)$$

If $k = \alpha + 2\beta$ is the total rate of base substitution per unit time, an appropriate measure of distance for a sequence tree is

$$K = -\frac{1}{2}\ln\left[(1 - 2p_{\rm I} - p_{\rm II})\sqrt{1 - 2p_{\rm II}}\right] \approx 2kt \tag{9.1}$$

This quantity is the Kimura two-parameter distance, and may be estimated with sample proportions substituted into Equation 9.1. J. Felsenstein (personal communication) points out that such estimates are not maximum likelihood, and that numerical methods will be needed to find the ML estimate of K.

Kimura used β-globin sequences from rabbit and chicken as an example, and these sequences are displayed in Figure 9.6. The sequence lengths are 438 bp, and there are 58 type I changes and 63 type II changes. Therefore $\tilde{p}_{\rm I} = 0.1324$ and $\tilde{p}_{\rm II} = 0.1438$. Kimura's distance is 0.3513, which is not substantially different from the Jukes-Cantor distance of 0.3446 based only on the proportion $\tilde{q} = 0.7237$ of identical bases.

For the five mitochondrial sequences in Figure 1.3, the distance matrix $D_K = \{d_{xy}\}$, where d_{xy} is the Kimura distance between sequences x, y is

$$D_K = \begin{bmatrix} .000 & .015 & .045 & .149 & .210 \\ .015 & .000 & .030 & .130 & .189 \\ .045 & .030 & .000 & .094 & .189 \\ .149 & .130 & .094 & .000 & .186 \\ .210 & .189 & .189 & .186 & .000 \end{bmatrix}$$

By way of contrast, the distance matrix D_{JK} based on the Jukes-Cantor

```
type I        **     **            *   *  *    *
Chicken  GTGCACTGGA CTGCTGAGGA GAAGCAGCTC ATCACCGGCC TCTGGGGCAA GGTCAATGTG 60
Rabbit   GTGCATCTGT CCAGTGAGGA GAAGTCTGCG GTCACTGCCC TGTGGGGCAA GGTGAATGTG
type II         * *    *            *** *        *    *            *

type I                  *  *                    * *
Chicken  GCCGAATGTG GGGCCGAAGC CCTGGCCAGG CTGCTGATCG TCTACCCCTG GACCCAGAGG 120
Rabbit   GAAGAAGTTG GTGGTGAGGC CCTGGGCAGG CTGCTGGTTG TCTACCCATG GACCCAGAGG
type II   **   **    * *            *                       *

type I        *             *           *         **  *   **    *  *
Chicken  TTCTTTGCGT CCTTTGGGAA CCTCTCCAGC CCCACTGCCA TCCTTGGCAA CCCCATGGTC 180
Rabbit   TTCTTCGAGT CCTTTGGGGA CCTGTCCTCT GCAAATGCTG TTATGAACAA TCCTAAGGTG
type II         *                 *   ** * * *        * *           *   *

type I    *   *  *         *       * *      **             *
Chicken  CGCGCCCACG GCAAGAAAGT GCTCACCTCC TTTGGGGATG CTGTGAAGAA CCTGGACAAC 240
Rabbit   AAGGCTCATG GCAAGAAGGT GCTGGCTGCC TTCAGTGAGG GTCTGAGTCA CCTGGACAAC
type II  * *                      *   *        *  *  *  *   **

type I        ***       *  *   *     *         *                *     *  *
Chicken  ATCAAGAACA CCTTCTCCCA ACTGTCCGAA CTGCATTGTG ACAAGCTGCA TGTGGACCCC 300
Rabbit   CTCAAAGGCA CCTTTGCTAA GCTGAGTGAA CTGCACTGTG ACAAGCTGCA CGTGGATCCT
type II  *               *  *      **

type I                         **  *      * *            *  *   *  **    *
Chicken  GAGAACTTCA GGCTCCTGGG TGACATCCTC ATCATTGTCC TGGCCGCCCA CTTCAGCAAG 360
Rabbit   GAGAACTTCA GGCTCCTGGG CAACGTGCTG GTTATTGTGC TGTCTCATCA TTTTGGCAAA
type II                              *  *         *    *  **

type I                  *              *                 *
Chicken  GACTTCACTC CTGAATGCCA GGCTGCCTGG CAGAAGCTGG TCCGCGTGGT GGCCCATGCC 420
Rabbit   GAATTCACTC CTCAGGTGCA GGCTGCCTAT CAGAAGGTGG TGGCTGGTGT GGCCAATGCC
type II    *          *  ***            *       *     ***  **       *

type I          *    *
Chicken  CTGGCTCGCA AGTACCAC
Rabbit   CTGGCTCACA AATACCAC
type II
```

Figure 9.6 Chicken and rabbit β-globin sequences. Asterisks denote differences. Source: GenBank database loci *CHKHBBM* and *RABHBB1A1*, accession numbers J00860 and J00659.

distance measure is

$$D_{JK} = \begin{bmatrix} .000 & .015 & .045 & .143 & .198 \\ .015 & .000 & .030 & .126 & .179 \\ .045 & .030 & .000 & .092 & .179 \\ .143 & .126 & .092 & .000 & .179 \\ .198 & .179 & .179 & .179 & .000 \end{bmatrix}$$

Adding an extra parameter has allowed the Kimura distance to distinguish between the distances of Chimpanzee and Gorilla to Gibbon, whereas these distances are the same with Jukes-Cantor distance.

Other Distances

Reducing all the information in two DNA sequences to a single measure of distance between them can obviously be done in many ways, and the Jukes-Cantor and Kimura two-parameter distances are merely the simplest two. The search for the distance best suited for phylogeny reconstruction (Chapter 10) continues. Lockhart et al. (1994) presented a distance designed to allow for sequences with different nucleotide compositions, resulting from asymmetric mutation processes that vary between lineages.

For sequences x and y, Lockhart et al. (1994) first form a divergence matrix F_{xy} in which the i, jth element is the proportion of sites in which sequence x has character state (e.g. nucleotide or amino acid) i and sequence y has character state j. The sum of all the elements in this matrix is 1. The ***LogDet distance*** between the sequences is then defined as

$$d_{xy} = \ln(\det F_{xy})$$

where det denotes the determinant of a matrix. To give a distance of zero between a sequence and itself, this quantity is modified, for DNA sequences, to

$$d'_{xy} = -\frac{1}{4}\ln\left(\frac{\det F_{xy}}{\sqrt{(\det F_{xx})(\det F_{yy})}}\right)$$

If the four base frequencies are all equal, $\det F_{xx} = \det F_{yy} = (1/4)^4$, and the expected value of the LogDet distance is the mean number of substitutions per site.

For the five mitochondrial sequences in Figure 1.3, the distance matrix

$D_L = \{d'_{xy}\}$ is

$$D_L = \begin{bmatrix} .000 & .015 & .050 & .177 & .239 \\ .015 & .000 & .033 & .154 & .209 \\ .050 & .033 & .000 & .117 & .218 \\ .177 & .154 & .117 & .000 & .222 \\ .239 & .209 & .218 & .222 & .000 \end{bmatrix}$$

There is some similarity of this matrix to those shown above for Jukes-Cantor and Kimura distances, except there are now different distances from Gibbon to Chimpanzee, Gorilla and Orangutan.

It is possible to construct distances between sequences that incorporate some of the features within each sequence (Karlin and Ladunga 1994), rather than treating nucleotides as being independent. Karlin et al. (1994) define some dinucleotide relative abundance distance measures. If f_{XY} is the frequency of dinucleotide XY in a sequence, and f_X, f_Y are the frequencies of single nucleotides X, Y, then the odds ratio $\rho_{XY} = f_{XY}/f_X f_Y$ will deviate from 1 when there are dependencies between adjacent nucleotides. This quantity is modified to accommodate double-stranded DNA by combining the sequence with its inverted complement sequence. This changes $f_{\mathtt{A}}$ to $f^*_{\mathtt{A}} = f^*_{\mathtt{T}} = (f_{\mathtt{A}} + f_{\mathtt{T}})/2$ and $f_{\mathtt{C}}$ to $f^*_{\mathtt{C}} = f^*_{\mathtt{G}} = (f_{\mathtt{C}} + f_{\mathtt{G}})/2$. Also, $f^*_{\mathtt{GT}} = (f_{\mathtt{GT}} + f_{\mathtt{AC}})/2$ and so on. The symmetrized dinucleotide odds ratio for double-stranded DNA is therefore

$$\rho^*_{\mathtt{AC}} = \rho^*_{\mathtt{GT}} = \frac{f^*_{\mathtt{GT}}}{f^*_{\mathtt{G}} f^*_{\mathtt{T}}} = \frac{f_{\mathtt{GT}} + f_{\mathtt{AC}}}{(f_{\mathtt{G}} + f_{\mathtt{C}})(f_{\mathtt{A}} + f_{\mathtt{T}})}$$

with similar expressions for other dinucleotides. The departure of this quantity from 1 measures the dinucleotide bias of GT/AC. The fact that these ratios may differ between sequences suggests a sequence distance measure. For nucleotides X, Y in sequences i, j, the absolute difference of the two odds ratios is calculated. This quantity $|\rho^*_{XYi} - \rho^*_{XYj}|$ is then averaged over all 16 X, Y pairs to give the δ-distance of Karlin et al.

$$\delta_{ij} = \frac{1}{16} \sum_{X,Y} |\rho^*_{XYi} - \rho^*_{XYj}|$$

This distance should not be too much affected by differences in base compositions in the two sequences.

SUMMARY

Restriction site data may generally be analyzed with the techniques introduced in earlier chapters for such Mendelizing units as allozymes. When conclusions depend critically on knowing the lengths of restriction fragments, it must be remembered that experimental error affects the relationship between lengths and gel migration distances.

Sequence data analyses currently take one of two forms. In the first place, analyses proceed with little in the way of formal genetic models. The assessment of the significance of the length of longest direct DNA repeats, for example, may rest on the assumption that nucleotides are independent and that bases have equal frequencies at all sites. Biological experiments are the ultimate way of assessing significance, but statistical analyses can indicate whether observed similarities are unlikely to have arisen by chance and therefore of possible biological interest. Methods for finding local similarities or aligning whole sequences are often ad hoc. Although there may be a biological basis for considering base substitutions and deletions as different events, there is little basis for the particular weights assigned to these events in the various algorithms. Some progress is being made though by using Hidden Markov models in papers such as those by Baldi et al. (1994) and Krogh et al. (1994), and by the use of Gibbs sampling (Lawrence et al. 1993). When coding regions are being studied, a greater degree of biological relevance is probably attained when amino acid sequences, rather than DNA sequences, are used. These protein sequences are more likely to reflect biological constraints.

For evolutionary studies, it is essential that a genetic model be employed. Genetic distances between sequences should attempt to reflect an appropriate mutation model, for example. Sequence alignments that take evolutionary relatedness into account are necessarily more complicated than those that do not. Maximum likelihood alignments based on models of base substitution, insertion and deletion, have been discussed by Bishop and Thompson (1986) and Thorne et al. (1991, 1992). These last authors point out that the weights to be assigned to base substitutions, insertions or deletions, should be based on the probabilities of these evolutionary events for the pair of sequences being aligned, and hence can differ for every pair of sequences.

EXERCISES

Exercise 9.1

Find the order of the Markov chain that best fits the rabbit β-globin sequence in Figure 9.6.

Exercise 9.2

Calculate the distance matrix for the five mitochondrial sequences of Figure 1.3, using Karlin's δ-distance.

Chapter 10

Phylogeny Reconstruction

INTRODUCTION

Taxonomy is concerned with grouping organisms into a manageable number of groups whose members have the same or very similar characters. This enables the classification of species. For evolutionary studies, the classification also involves the reconstruction of phylogenies. These activities may shed light on the question of whether inferred relationships of species suggest anything about the nature of evolutionary forces. Fine summaries of the process of constructing phylogenies are given by Nei (1987) and Li and Graur (1991), and useful collections of papers were edited by Hillis et al. (1996) and Miyamoto and Cracraft (1991). The Hillis et al. volume contains the excellent treatment by Swofford et al. (1996). Only a hint of the many methods and complexities is given in this chapter.

A distinction can be made between ***phenetic*** and ***cladistic*** data. Phenetic relationships are defined by Sneath and Sokal (1973) as similarities based on a set of phenotypic characters for the objects under study, whereas cladistic relationships contain information about ancestry and so can be used to study evolutionary pathways. Both relationships are best represented as phylogenetic trees or ***dendrograms***. The terms ***phenogram*** and ***cladogram*** have been used for trees based on phenetic and cladistic relationships, respectively. A cladogram may display evolutionary times between events or groups (although most do not), but no concept of time is needed for a phenogram. In this chapter, no great attention will be paid to the distinction. As Nei (1987) points out, if measures of phenotypic similarity imply degrees of evolutionary similarity, then phenetic methodology can provide cladistic trees.

Phylogenetic trees represent the evolutionary pathways, and a distinction between ***species trees*** and ***gene trees*** was made by Tateno et al. (1982). Branches in a species tree join extant species to an ancestral species. The

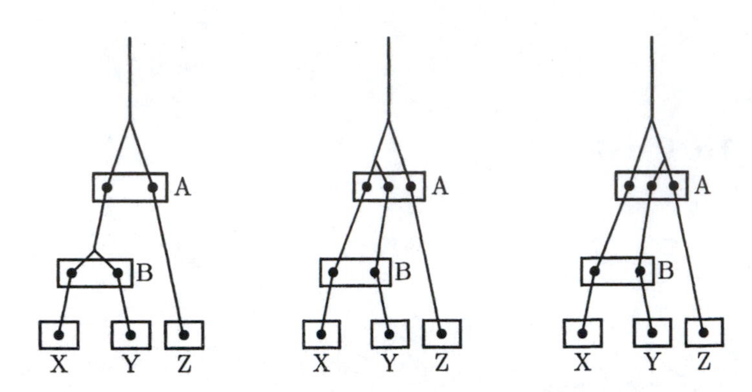

Figure 10.1 A species tree links species A, B, X, Y, and Z. Three possible gene trees, indicated by solid lines, link the genes indicated by solid circles. Source: Nei (1987).

species tree in Figure 10.1 has species B ancestral to species X, Y, and species A ancestral to species B, Z. However, the data used to construct a tree are often from a single region of the genome of those species. A tree constructed from the DNA sequence of a short region, for example, is called a gene tree and may not be the same as the species tree. Two species may carry genes that diverged prior to the species split, or introgression may have resulted in genes having diverged after the species split. Possible differences between species and gene trees are shown in Figure 10.1, taken from Nei (1987) (see also Pamilo and Nei 1988). The left and center of the three gene trees have genes in species X, Y more closely related to each other than to the gene in species Z whereas the gene tree on the right has the Y, Z genes being the most closely related. For times of divergence between species that are long compared to the ***coalescent*** times between genes (i.e., the times at which two genes split from a single ancestral gene), the two trees will be similar.

Another distinction among trees is between ***rooted*** and ***unrooted*** trees. In Figure 10.2 two of the 15 possible rooted trees and one of the three possible unrooted trees for four species are shown. A rooted tree conveys the notion of temporal ordering of the species or genes on a tree, while an unrooted tree merely reflects distances between units with no notion of which was ancestral to which.

The data used for ordering species in a phylogenetic history are of two types. ***Character data*** provide information about attributes of genes, individ-

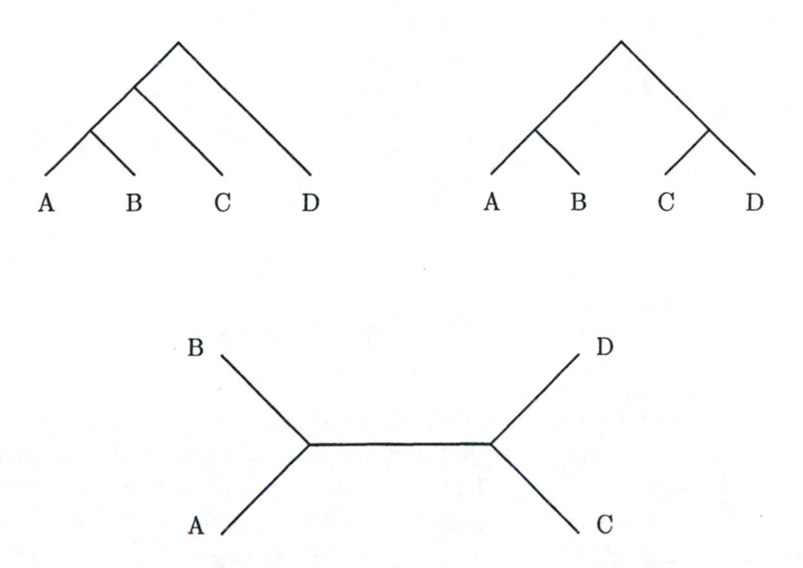

Figure 10.2 Two of the 15 rooted trees and one of the three unrooted trees for four species.

uals, populations, or species, whereas ***distance data*** or ***similarity data*** refer to pairs of genes, individuals, populations, or species. Distances or similarities can be constructed from character values, but the reverse is not true. In each case the data can be arranged in matrices, but the appropriate methods for phylogenetic tree construction are different.

Distance matrix methods are based on the set of distances calculated between each pair of taxonomic units. Distances are generally based on genetic models, and usually refer to the number of changes between the units. The quality of the resulting trees depends on the quality of the distance estimate. A distance matrix method in great use is that of ***neighbor-joining*** (Saitou and Nei 1987).

For character data, a matrix can also be used to display the data for each character in each taxonomic unit. Several methods are applicable to such data. ***Maximum parsimony*** methods make little explicit mention of evolutionary assumptions, although they may require approximately constant rates of change (Nei 1987). Heavy dependence on models is a feature of ***maximum likelihood*** methods. Although this class of methods is computationally demanding, it can provide a basis for statistical inference.

Swofford et al. (1996) make a useful distinction between two approaches

Table 10.1 Matrix of distances between all pairs of n operational taxonomic units.

		OTU's				
		1	2	3	...	n
OTU's	1	–	d_{12}	d_{13}	...	d_{1t}
	2	d_{21}	–	d_{23}	...	d_{2t}
	3	d_{31}	d_{32}	–	...	d_{3t}
	...	...	...	...	...	...
	n	d_{t1}	d_{t2}	d_{t3}	...	–

to selecting a phylogenetic tree on the basis of data. There may be an algorithm, such as the neighbor-joining method, that leads to determination of a tree, or there may be a criterion by which alternative trees can be compared. For example, the maximum likelihood approach proceeds by finding the tree with greatest likelihood. A feature of the algorithmic approach is that only one tree is determined, and there is no immediate information about other trees that may provide (nearly) as good a representation of the data. Criterion-based approaches generally examine many different trees as part of the search for the one with the optimal value of the criterion. In this way information is gained about alternative trees.

DISTANCE MATRIX METHODS

Phylogenetic trees may be based on distance matrices, here of genetic distances between all pairs of ***operational taxonomic units (OTU's)***. For n OTU's the matrix of distances between each pair is shown in Table 10.1.

Using these distances to group OTU's in a phenetic context may employ ***clustering***, and possible approaches to clustering were given by Sneath and Sokal (1973). Clustering may be interpreted loosely as the process of identifying groups of "close" OTU's. When distances are used to construct additive trees, as with the Fitch-Margoliash algorithm described below, the process is said to use a pairwise method rather than clustering.

Average Linkage Clustering (UPGMA)

There is a class of strategies used for finding clusters, called sequential, agglomerative, hierarchical, and nonoverlapping by Sneath and Sokal (1973). The most widely used is UPGMA (unweighted pair-group method using an

Table 10.2 Numbers of differences (below diagonal) and Jukes-Cantor distances (above diagonal) for the five mitochondrial sequences of Figure 1.3.

	Human	Chimpanzee	Gorilla	Orangutan	Gibbon
Human	–	0.015	0.045	0.143	0.198
Chimpanzee	1	–	0.030	0.126	0.179
Gorilla	3	2	–	0.092	0.179
Orangutan	9	8	6	–	0.179
Gibbon	12	11	11	11	–

arithmetic average). It defines the intercluster distance as the average of all the pairwise distances for members of two clusters.

As an example, consider the mitochondrial DNA sequence data displayed in Figure 1.3. Jukes-Cantor distances (Chapter 9) between each pair of these sequences depend on the observed numbers of nucleotide differences between each pair. If a proportion $\tilde{q}$ of the bases in two sequences is the same, then the distance K is estimated as

$$\tilde{K} = \frac{3}{4}\ln\left(\frac{3}{4\tilde{q}-1}\right)$$

Differences and Jukes-Cantor distances for the sequences are shown in Table 10.2.

The closest pair is human and chimpanzee, and these are joined to form a cluster. The distance of each other sequence from this cluster is found as the average distance from the sequence to members of the cluster:

$$\begin{aligned} d_{(\mathrm{hu-ch}),\mathrm{go}} &= \frac{1}{2}(d_{\mathrm{hu,go}} + d_{\mathrm{ch,go}}) = 0.037 \\ d_{(\mathrm{hu-ch}),\mathrm{or}} &= \frac{1}{2}(d_{\mathrm{hu,or}} + d_{\mathrm{ch,or}}) = 0.135 \\ d_{(\mathrm{hu-ch}),\mathrm{gi}} &= \frac{1}{2}(d_{\mathrm{hu,gi}} + d_{\mathrm{ch,gi}}) = 0.189 \end{aligned}$$

The distance matrix can be reduced to

	(hu-ch)	go	or	gi
(hu-ch)		0.037	0.135	0.189
go			0.092	0.179
or				0.179
gi				

and this has smallest distance between the human-chimpanzee cluster and gorilla. These are clustered, and the new distances required are

$$d_{(\text{hu}-\text{ch}-\text{go}),\text{or}} = \frac{1}{3}(d_{\text{hu,or}} + d_{\text{ch,or}} + d_{\text{go,or}}) = 0.121$$
$$d_{(\text{hu}-\text{ch}-\text{go}),\text{gi}} = \frac{1}{3}(d_{\text{hu,gi}} + d_{\text{ch,gi}} + d_{\text{go,gi}}) = 0.185$$

The next reduction in the distance matrix is

	(hu-ch-go)	or	gi
(hu-ch-go)		0.121	0.185
or			0.179
gi			

As the smallest distance is now that between the human-chimpanzee-gorilla cluster and orangutan, these sequences are clustered and the distance from this cluster to the gibbon sequence is

$$d_{(\text{hu}-\text{ch}-\text{go}-\text{or}),\text{gi}} = \frac{1}{4}(d_{\text{hu,gi}} + d_{\text{ch,gi}} + d_{\text{go,gi}} + d_{\text{or,gi}}) = 0.183$$

The results of the clustering can be represented as the dendrogram shown in Figure 10.3. In constructing this dendrogram, branchpoints are placed midway between two sequences, or clusters. The distance between a pair of sequences in the figure is the sum of the branch lengths.

The UPGMA method is widely used for matrices of distances. Degens (1983) has discussed conditions under which the estimated branch lengths are maximum likelihood. Nei et al. (1983) conducted simulation studies to compare different methods of constructing trees, and found that UPGMA generally performed well when the substitution rates were the same along all branches of the trees. It must be emphasized that the assumption of (nearly) equal rates is crucial for the UPGMA method to be appropriate. Other simulation studies (e.g., Kim and Burgman 1988) have demonstrated that the method is not satisfactory when the rates are unequal in different lineages. When rates are equal, a ***molecular clock*** is said to be operating.

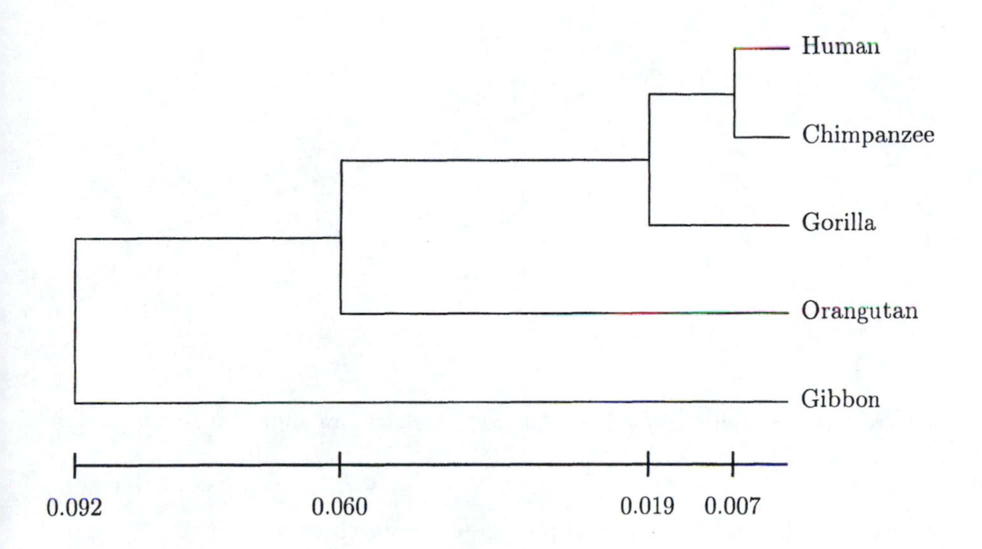

Figure 10.3 UPGMA dendrogram for the mitochondrial sequence data of Figure 1.3.

Fitch-Margoliash Algorithm

The UPGMA method implicitly assumes a constant rate of change along all branches in a tree. This assumption is removed by an algorithm developed by Fitch and Margoliash (1967). Their method proceeds by inserting "missing" OTU's as common ancestors of later OTU's and fits branch lengths to groups of three OTU's at a time. A similar additive approach was first discussed by Cavalli-Sforza and Edwards (1967). The Fitch-Margoliash algorithm is now illustrated for the mitochondrial data of Figure 1.3.

The OTU's are divided into three groups: the closest pair, A = human and B = chimpanzee, and the remaining set, X = (gorilla, orangutan, gibbon). Node C, which is the immediate ancestor of A and B, is introduced. The branch lengths from C to A, B are a, b, while the branch from C to X is x (Figure 10.4). As a further notational device, d_{UV} means the distance from node U to node V, $d_{U\bar{V}}$ means the average distance from node U to all nodes other than node V, and d_{U^*V} means the average distance from all tip nodes below U to V. An asterisked letter, U^*, denotes the set of tips descended from the node, U, of the same letter. The three pairwise distances

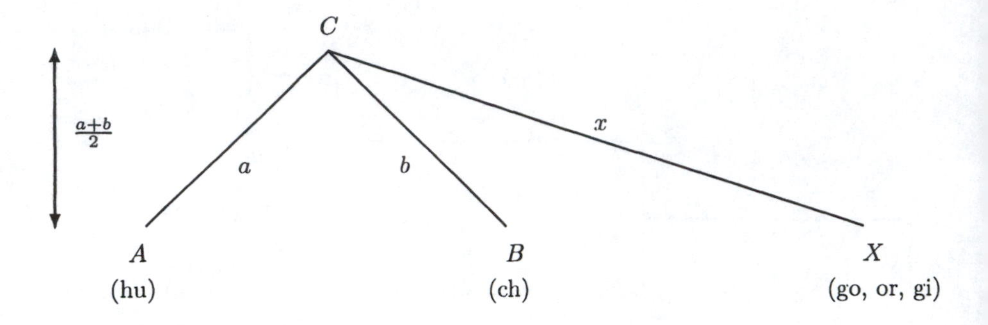

Figure 10.4 Initial step for Fitch-Margoliash algorithm applied to the mitochondrial data of Figure 1.3.

between A, B, and C provide three equations in three unknowns:

$$\begin{aligned} a + x &= d_{AX} = d_{A\bar{B}} \\ &= \frac{1}{3}(0.045 + 0.143 + 0.198) = 0.129 \\ b + x &= d_{BX} = d_{B\bar{A}} \\ &= \frac{1}{3}(0.030 + 0.126 + 0.179) = 0.112 \\ a + b &= d_{AB} \\ &= 0.015 \end{aligned}$$

The first equation uses the average distance from A to each member of set X. Solving these three equations gives

$$a = 0.016 \quad , \quad b = -0.001$$

By convention, negative estimates are set equal to zero so that $b = 0$ here. The average of a, b is called the height of node C. This value is 0.008. Replacing the pair A, B by C, and recalculating the distances as was done for the UPGMA case, gives the next closest pair of C and $D =$ gorilla. Node E is introduced as the immediate ancestor of C and D. As shown in Figure 10.5, the branch lengths for C^* and E, D and E, and E and X are c, d, and x, where the set X is now orangutan and gibbon. The three equations to be solved are

$$\begin{aligned} c + d &= d_{C^*D} \\ &= \frac{1}{2}(0.045 + 0.030) \;=\; 0.037 \end{aligned}$$

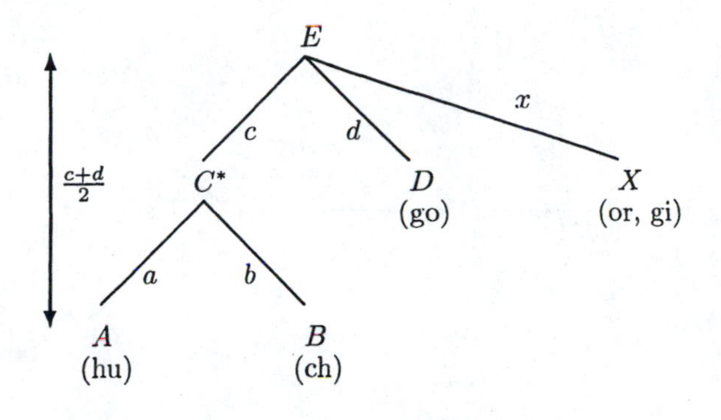

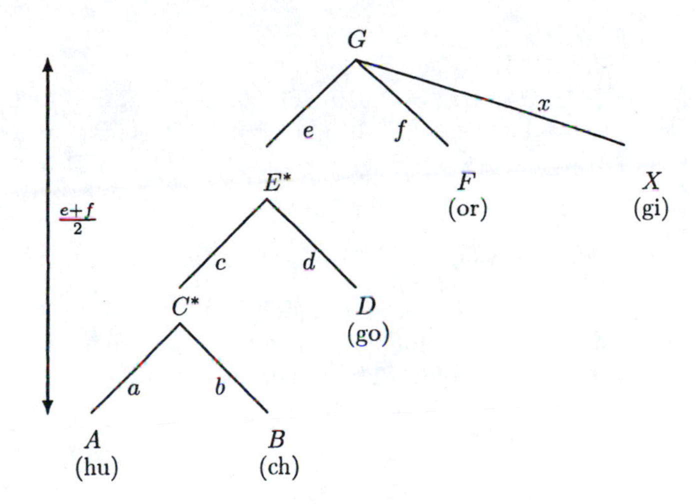

Figure 10.5 Intermediate steps for Fitch-Margoliash algorithm applied to mitochondrial sequence data of Figure 1.3.

$$\begin{aligned} c+x &= d_{C^*X} = d_{(AB)X} \\ &= \frac{1}{4}(0.143+0.198+0.126+0.179) \ = \ 0.162 \\ d+x &= d_{DX} \\ &= \frac{1}{2}(0.092+0.179) \ = \ 0.136 \end{aligned}$$

so that

$$c = 0.032 \quad , \quad d = 0.006$$

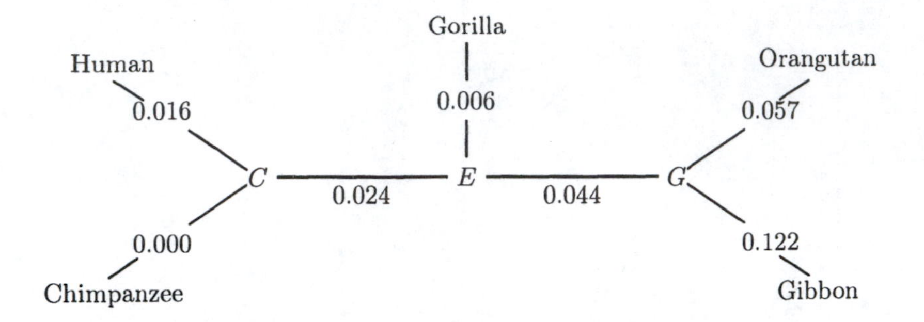

Figure 10.6 Fitch-Margoliash unrooted tree for mitochondrial sequence data in Figure 1.3.

The height of node E is $(c+d)/2 = 0.019$. The branch length c' between nodes C and E is given by c minus the height of node C, since c measures the distance from C to E *and* the average distance from A and B to C. In other words

$$c' = 0.032 - 0.008 = 0.024$$

With the OTU's reduced to E, orangutan and gorilla, the closest pair is E and F = orangutan. Node G is introduced as their immediate ancestor, and the remainder is now X = gibbon. The equations to be solved for branch lengths are

$$\begin{aligned} e+f &= d_{E^*F} \\ &= \frac{1}{3}(0.143+0.126+0.092) = 0.121 \\ e+x &= d_{E^*X} \\ &= \frac{1}{3}(0.198+0.179+0.179) = 0.185 \\ f+x &= d_{FX} = 0.179 \end{aligned}$$

with solutions

$$e = 0.063 \quad , \quad f = 0.057$$

The height of node G is $(e+f)/2 = 0.060$, and the branch length e' from E to G is the difference of e and the height of E, or $0.063 - 0.019 = 0.044$. The process could stop here, and the unrooted tree is shown in Figure 10.6.

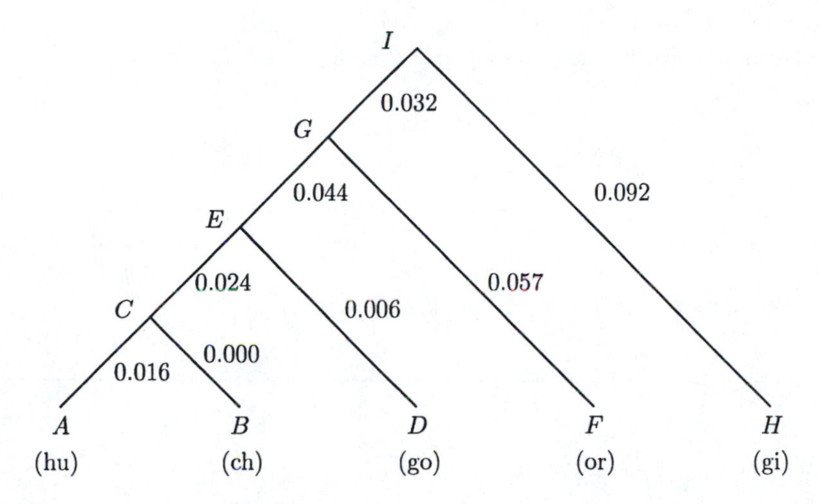

Figure 10.7 Fitch-Margoliash rooted tree for the mitochondrial sequence data in Figure 1.3.

Without extra information, or without assuming the same rates of change along all branches, the Fitch-Margoliash algorithm can lead only to an unrooted tree. Additional information, from a distant sequence (an ***outgroup***), would allow the tree to be rooted. The same topology as provided by the UPGMA method will result if the root I is placed so that the distances g, from G^* to I and h from H to I are equal (i.e., assuming equal rates of change in the two branches from I to the extant sequences). Since

$$\begin{aligned} g + h &= d_{G^*H} \\ &= \frac{1}{4}(0.198 + 0.179 + 0.179 + 0.179) = 0.184 \end{aligned}$$

$g = h = 0.092$, and the distance g' from G to I is g minus the height of G, or 0.032. Putting all these branch lengths together produces the tree shown in Figure 10.7.

Fitch and Margoliash allow for the possibility that the topology found by their algorithm is incorrect, and they suggest that other topologies be examined. Different trees are compared on the basis of a measure of goodness of fit, called the "percent standard deviation" by Fitch and Margoliash (1967). The best tree will have the smallest percent standard deviation. If d_{ij} is the observed distance pair, i and j, of n OTU's and e_{ij} is the sum of

Table 10.3 Jukes-Cantor distances (above diagonal) and calculated distances (below diagonal) from Fitch-Margoliash algorithm for the five mitochondrial sequences of Figure 1.3.

	Human	Chimpanzee	Gorilla	Orangutan	Gibbon
Human	-	0.015	0.045	0.143	0.198
Chimpanzee	0.016	-	0.030	0.126	0.179
Gorilla	0.046	0.030	-	0.092	0.179
Orangutan	0.141	0.125	0.107	-	0.179
Gibbon	0.208	0.192	0.174	0.181	-

the branch lengths between them on the tree,

$$s = \left\{ \frac{\sum_{i,j}[(d_{ij} - e_{ij})/d_{ij}]^2}{n(n-1)} \right\}^{1/2} \times 100$$

is the percent standard deviation. Note the additivity assumption, whereby the distance between any two nodes is the sum of the branch lengths between them. For the tree in Figure 10.6, the observed distances and branch lengths are shown in Table 10.3 and the percent standard deviation is 1.94. It may also be possible to adjust branch lengths in the fitted tree to reduce the quantity s.

Choosing trees on the basis of the percent standard deviation should be referred to as using the Fitch-Margoliash criterion, and the best such tree may not be the same as found by the Fitch-Margoliash algorithm. Use of the criterion is expected to give similar results to the UPGMA method when there is a molecular clock. If there is not a clock, so that rates of change are different in different lineages, the Fitch-Margoliash criterion will do much better than UPGMA. (J. Felsenstein, personal communication).

Other trees can be chosen for examination by selecting a different initial pair of OTU's. That tree with the lowest percent standard deviation is considered to be the best tree, and this criterion is the basis on which the Fitch-Margoliash algorithm operates. For example, human and gorilla could be grouped first, and then chimpanzee, orangutan, and gibbon added in that order. However, in that case the height of the second interior node E is less than the height of the first interior node C and the fit between observed and calculated distances is not nearly as good as for the first order.

For the data of Figure 1.3, the trees found by UPGMA and Fitch-

Margoliash were very similar. If the correct topology is used, UPGMA gives least-squares estimates of branch lengths (Chakraborty 1977), whereas Fitch-Margoliash uses a weighted least-squares approach. The two methods are not expected to differ very much when there is a clock.

Neighbor-joining Method

Saitou and Nei (1987) described a method for identifying closest pairs, or neighbors, of taxonomic units in a way to minimize the total length of a tree. A pair of neighbors is defined to be two units connected through a single node in an unrooted, bifurcating tree (two branches joining at each interior node). In Figure 10.6, human and chimpanzee are neighbors, but human and gorilla are not. If human and chimpanzee are combined into one unit, then this combined unit and gorilla become neighbors. In general, it is possible to define the topology of a tree by successively joining pairs of neighbors to form new neighbor pairs.

The method begins with a star-like phylogeny, as shown in Figure 10.8(a) for eight species. Neighbors are the pair of species that, when joined, result in a tree of shortest total length. These are joined to form a combined unit. The procedure of identifying the neighbors among the reduced set of units is repeated until there are just three (combined) units left. At that point there is only one unrooted tree.

A convenient summary of the neighbor-joining method was given by Swofford et al. (1996):

1. For the ith terminal node (i.e., taxonomic unit i) calculate its net divergence r_i from all other taxa

   $$r_i = \sum_{k=1}^{N} d_{ik}$$

   Here N is the number of terminal nodes in the current distance matrix, and d_{ik} is the distance between nodes i and k. Note that $d_{ik} = d_{ki}$.

2. Find the pair i, j that minimizes the rate-corrected distance M_{ij}, defined as

   $$M_{ij} = d_{ij} - \frac{r_i + r_j}{N - 2}$$

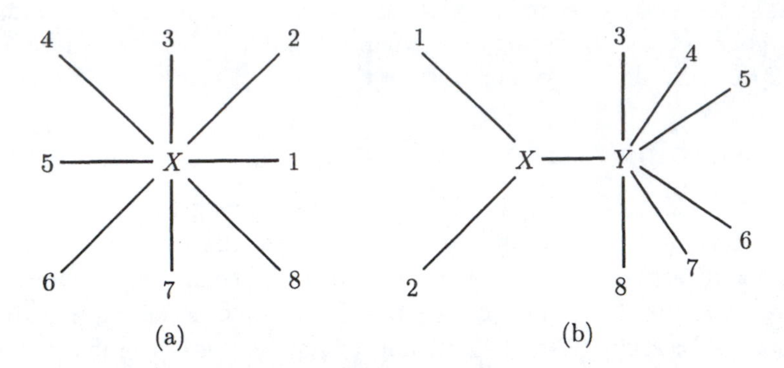

Figure 10.8 (a) A starlike tree with no hierarchical structure; (b) a tree in which units 1 and 2 have been clustered. Source: Saitou and Nei (1987).

3. Define a new node u that joins nodes i, j and the rest of the tree. The branch lengths from nodes i, j to u are

$$\begin{aligned} s_{iu} &= \frac{d_{ij}}{2} + \frac{r_i - r_j}{2(N-2)} \\ s_{ju} &= d_{ij} - s_{iu} \end{aligned}$$

   and from u to each other terminal node k

$$d_{ku} = \frac{d_{ik} + d_{jk} - d_{ij}}{2}$$

4. Remove distances to nodes i, j from the data matrix, and decrease N by 1.

5. If more than two nodes remain, go to step 1. Otherwise the tree is fully defined, except for the length of the branch joining the remaining two nodes i, j. This length is $s_{ij} = d_{ij}$.

   Each step has generated one internal node and has estimated the lengths of two of the branches connected to that node. This allows the tree to be drawn. The angles between the branches in these drawings is arbitrary.

As an example, consider again the Jukes-Cantor distances displayed in Table 10.2. The calculations are set out in Table 10.4, where the minimum value of M_{ij} is shown in bold at each step. For the first step, orangutan and gibbon have the smallest value of M_{ij}, and they are replaced by node 1 for the second step. The distances from each to this new node are

$$\begin{aligned} d_{\text{Or,Node 1}} &= \frac{1}{2}d_{\text{Or,Gi}} + \frac{r_{\text{Or}} - r_{\text{Gi}}}{6} = 0.057 \\ d_{\text{Gi,Node 1}} &= d_{\text{Or,Gi}} - d_{\text{Or,Node 1}} = 0.122 \end{aligned}$$

and the distances of the remaining three taxa to node 1 are shown in the second step matrix of Table 10.4. At that step, human and chimpanzee have the smallest value of M_{ij} and they are replaced by node 2 for the third step. The required distances are

$$\begin{aligned} d_{\text{Hu,Node 2}} &= \frac{1}{2}d_{\text{Hu,Ch}} + \frac{r_{\text{Hu}} - r_{\text{Ch}}}{4} = 0.016 \\ d_{\text{Ch,Node 2}} &= d_{\text{Hu,Ch}} - d_{\text{Hu,Node 2}} = -0.001 \end{aligned}$$

Kuhner and Felsenstein (1994) set negative lengths to zero, and transferred the negative value to the adjacent branch length to preserve the total distance between the terminal nodes. In the present example, this means that chimpanzee and node 2 coincide.

The distance from gorilla to node 2 is 0.030, since

$$d_{\text{Go,Node 2}} = \frac{1}{2}(d_{\text{Hu,Go}} + d_{\text{Ch,Go}} - d_{\text{Hu,Ch}}) = 0.030$$

With only three nodes remaining; gorilla, node 1 and node 2, there is only one unrooted tree possible and all the rate-corrected distances have the same value of $-.141$. A third interior node can be introduced to replace nodes 1 and 2, and the distance from gorilla to this node 3 is

$$d_{\text{Go,Node 3}} = \frac{1}{2}(d_{\text{Go,Node 1}} + d_{\text{Go,Node 2}} - d_{\text{Node 1,Node 2}}) = 0.005$$

Also, the distances from nodes 1 and 2 to node 3 are

$$\begin{aligned} d_{\text{Node 1,Node 3}} &= \frac{1}{2}d_{\text{Node 1,Node 2}} + \frac{r_{\text{Node 1}} - r_{\text{Node 2}}}{2} = 0.040 \\ d_{\text{Node 2,Node 3}} &= d_{\text{Node 1,Node 2}} - d_{\text{Node 1,Node 3}} = 0.025 \end{aligned}$$

The neighbor-joining tree is displayed in Figure 10.9. There is a strong resemblance to the Fitch-Margoliash tree in Figure 10.6.

Saitou and Nei (1987) establish that the neighbor-joining method produces the correct tree for purely additive data, where the distance between

Table 10.4 Neighbor-joining calculations for mitochondrial sequences of Figure 1.3. d_{ij} above diagonal, M_{ij} below diagonal.

		Hu $j=1$	Ch $j=2$	Go $j=3$	Or $j=4$	Gi $j=5$	r_i
Hu	$i=1$	.000	.015	.045	.143	.198	.401
Ch	$i=2$	−.235	.000	.030	.126	.179	.350
Go	$i=3$	−.204	−.202	.000	.092	.179	.346
Or	$i=4$	−.171	−.171	−.203	.000	.179	.540
Gi	$i=5$	−.181	−.183	−.181	**−.246**	.000	.735

		Hu $j=1$	Ch $j=2$	Go $j=3$	Node 1 $j=4$	r_i
Hu	$i=1$	.000	.015	.045	.081	.141
Ch	$i=2$	**−.110**	.000	.030	.063	.108
Go	$i=3$	−.086	−.084	.000	.046	.121
Node 1	$i=4$	−.085	−.086	−.110	.000	.190

		Go $j=1$	Node 1 $j=2$	Node 2 $j=3$	r_i
Go	$i=1$	.000	.046	.030	.076
Node 1	$i=2$	**−.141**	.000	.065	.111
Node 2	$i=3$	**−.141**	**−.141**	.000	.095

		Go $j=1$	Node 3 $j=2$
Go	$i=1$	.000	.005
Node 3	$i=2$		.000

each pair of taxa is the sum of the lengths of the branches joining those taxa in the tree. Actual data, such as those in Table 10.3, however, are not additive and the correct topology cannot be guaranteed. Simulations suggest the method to be among the best of the distance matrix methods. In the additive case the neighbor- joining tree is the ***minimum evolution*** tree. In the language of Rzhetsky and Nei (1992) this means that the sum of deviations between taxon-pair distances and tree path-length distances is minimized.

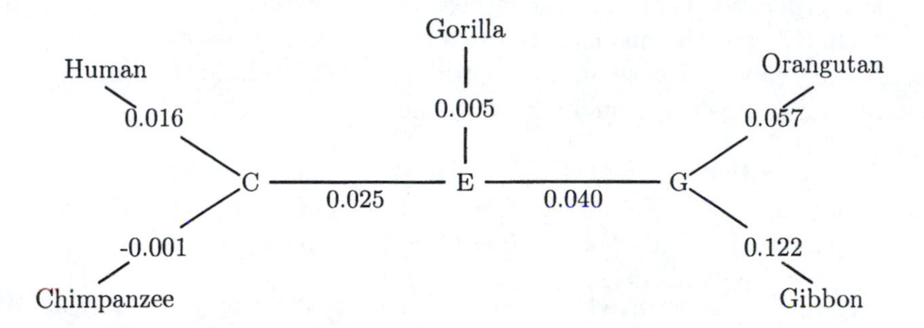

Figure 10.9 Neighbor-joining unrooted tree for mitochondrial sequence data in Figure 1.3.

PARSIMONY METHODS

Parsimony methods take explicit notice of the character values observed for each species, rather than working with the distances between sequences that summarize differences between character values. They do not require explicit models of evolutionary change. The approach was introduced for gene frequency data by Edwards and Cavalli-Sforza (1963) under the name "principle of minimum evolution," although this term has a different meaning from that of Rzhetsky and Nei (1992) given in the previous section. If sequences are available for a set of species then the most parsimonious topology linking them is sought. Branch lengths are generally not obtained. Swofford et al. (1996) point out the need to distinguish between the parsimony optimality criteria, such as the minimal tree length under specified restrictions on character state changes, and the algorithms used to search for optimal trees.

For each possible topology, the sequences at each node are inferred to be those that require the least number of changes to give each of the two immediately descendant sequences. The total number of changes required to traverse the whole tree is then found, and that tree with the minimum total is the most parsimonious. To illustrate the procedure, consider the example given by Fitch (1971). Sequences from six species, A–F, are available, and at one particular position they have bases `C`, `T`, `G`, `T`, `A`, `A`, respectively. There are many possible topologies, and one of them is shown in Figure 10.10. Each of the nodes 1–5, starting with those nearest the extant sequences, is considered in turn. At each node the "parsimony" (W. M. Fitch, unpublished) of the two descendant sequences is written down. This operation, denoted here by $\diamond$, is a set operation defined as the intersection of two sets if this intersection is not empty, and the union of the two sets if their intersection is empty. For distinct sets (sequences) X, Y, Z, the usual set operations of union and intersection can be contrasted with parsimony:

intersection	$[X,Y] \cap [X,Z] = [X]$	$[X] \cap [Y] = \phi$
union	$[X,Y] \cup [X,Z] = [X,Y,Z]$	$[X] \cup [Y] = [X,Y]$
parsimony	$[X,Y] \diamond [X,Z] = [X]$	$[X] \diamond [Y] = [X,Y]$

The minimum number of changes in the tree is the number of unions at internal nodes in the tree.

If two sequences have the same base at a position, no change is required if their immediate ancestor also has this base. If they have different bases, minimal changes require that their ancestor has either of those two. In Figure 10.10, nodes 1 and 2 are `(CT)` and `(GT)`, meaning that either of the two bases shown will give the smallest number of changes. There are three possibilities for node 3, but only one for node 4, and then two for node 5. The smallest number of changes, four, for this topology results if nodes 1–5 all have base `T`. As Nei (1987) points out, however, the same minimum number results if each node has base `A`. There are a further nine possibilities, each requiring four changes: those in which the five nodes have bases `TTTTA`, `TAAAA`, `CAAAA`, `AGAAA`, `ATAAA`, `CGAAA`, `CTAAA`, `TTAAA`, `TGAAA`.

The process is repeated for other topologies, and for other sites, and that topology requiring the smallest total number of changes is taken as the final tree. For maximum parsimony, it is necessary only to consider the ***informative sites***: those which allow some trees to be declared better than others. For DNA sequences, these are sites at which there are at least two different bases in the sequences being considered, with each different base occurring at least twice among the extant sequences. A site that has a different base in only one sequence is not informative since that unique base can be assumed to have arisen by a single change on the branch leading immediately to the sequence in which it is carried. Such a change is compatible with any topology.

For the mitochondrial sequences in Figure 1.3, there are five informative sites: 25, 39, 41, 44, and 55. A parsimonious tree based on these five positions is shown in Figure 10.11. It has six base changes, as do other possible trees. Although this has the same topology as the trees found using the matrix of distances, the very limited amount of data has resulted in some surprising effects. The branch between nodes E and G in Figure 10.7 was found to be shorter than that between nodes G and F, yet among the informative sites there have been three base changes on the former and none on the latter.

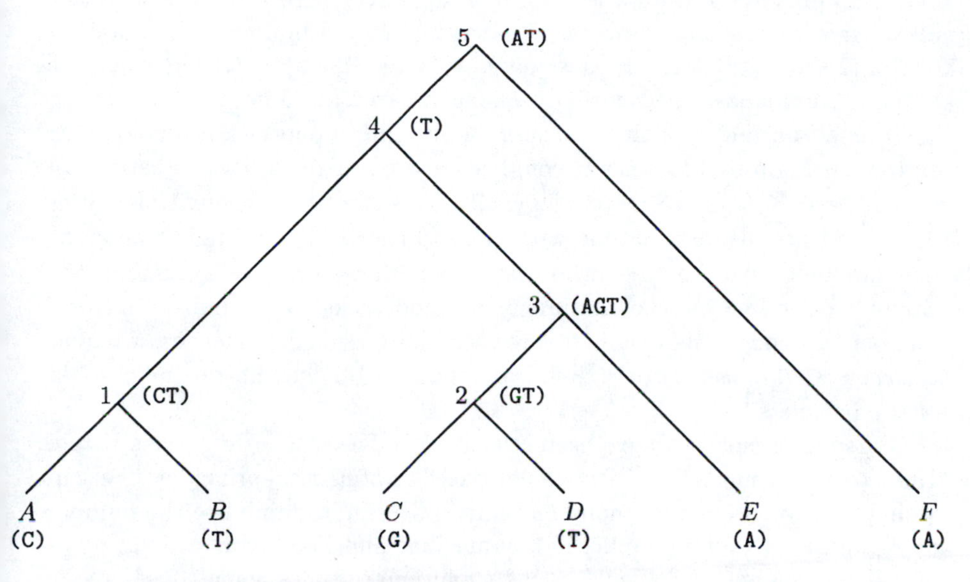

Figure 10.10 Illustration of procedure for finding maximum parsimony tree for one site in six sequences. Source: Fitch (1971).

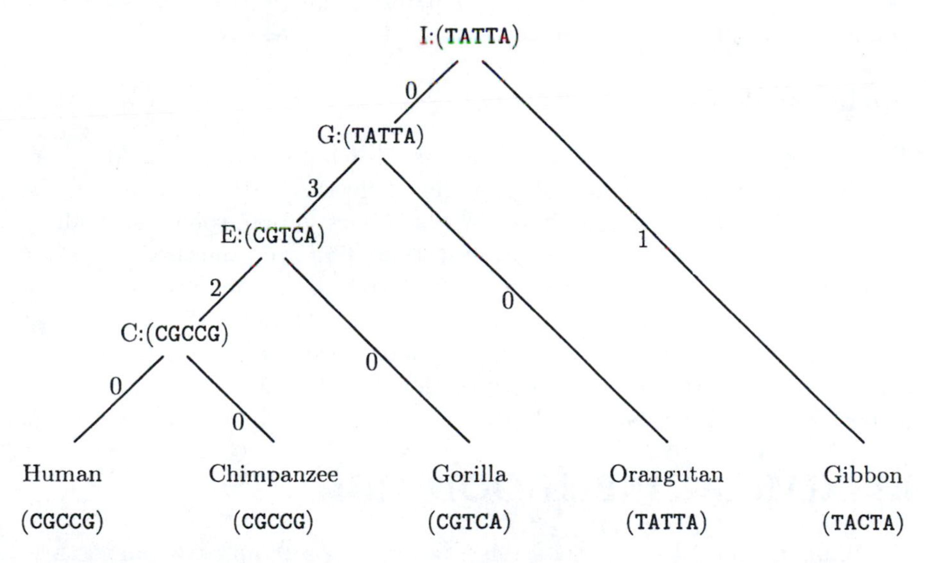

Figure 10.11 Maximum parsimony tree for mitochondrial sequence data of Figure 1.3. Figures are the numbers of changes between nodes.

The mitochondrial data leading to Figure 10.11 are typical in that not all the informative sites support the same topology. Employing the notation of Nei (1987), site 25 can be designated as (Hu-Ch-Go) – (Or-Gi) since the human, chimpanzee, and gorilla sequences have the same base and so do the orangutan and gibbon sequences. This split supports a topology such as that in Figure 10.11 since it could result from a single base substitution on the path E–G. For this topology, site 25 is said to be compatible. Sites 39 and 44 are also compatible with that topology. For a single change to be consistent with the observations at site 41, however, a different topology would be necessary. The parsimony method seeks to minimize the total number of changes necessary to produce the observed sequences, whereas the ***compatibility method*** (e.g., Estabrook et al. 1975) maximizes the number of compatible sites.

Parsimony methods have been criticized by Felsenstein (e.g., Felsenstein 1983) on the grounds that they are not based on statistical principles. Felsenstein points out that parsimony methods, in trying to minimize the number of evolutionary events, implicitly assume that multiple events are improbable. If the amount of change over the evolutionary times being considered is small, the parsimony methods will be well- justified, but for cases in which there are large amounts of change, parsimony methods can give estimated trees that actually are more likely to be wrong as more data are used (Felsenstein 1978b). A demonstration that parsimony methods can be cast in a maximum likelihood light is given at the end of the next section.

Fitch and Wagner Parsimony

The descriptions given so far have been for "Fitch parsimony" in that there are no constraints on permissible changes in character states. This is to be contrasted with "Wagner parsimony," which does impose some constraints. Wagner parsimony assumes that any transformation from one character state to another also implies a transformation through any intervening states, whereas Fitch parsimony allows any state to transform directly to any other state (Swofford et al. 1996). Both methods allow free reversibility, with character state changes in either direction being equally probable. This means that a tree can be rooted at any node with no change in tree length.

MAXIMUM LIKELIHOOD METHODS

Maximum likelihood methods of constructing trees attempt to avoid the limitations of other methods, although they may require a prohibitive amount of computing. Unlike distance matrix methods, maximum likelihood methods try to make explicit and efficient use of all the data, instead of reducing the data to a set of distances. They differ from parsimony methods by employing standard statistical methods for a probabilistic model of evolution (Felsenstein 1981). Maximum likelihood methods consider that changes are more likely along long branches than short ones, and estimation of branch lengths is an important component of the methods, whereas parsimony ignores information on branch lengths when evaluating a tree (Swofford et al. 1996).

As they are presently implemented, maximum likelihood methods of tree construction assume the form of the tree, and then choose the branch lengths to maximize the probability of the data given that tree. These probabilities are then compared over different trees, and the tree with the greatest probability is taken as the best estimate. An immediate problem is that the number of possible trees increases very quickly as the number of OTU's increases. The number of unrooted bifurcating trees (two branches joining at each interior node) with n OTU's at the tips is $(2n-5)!/[(n-3)!2^{n-3}] = \prod_{i=3}^{n}(2i-5)$ (Edwards and Cavalli-Sforza 1964, Felsenstein 1978a). The first few values of this number are 1, 3, 105, 10,395, and 2,027,025 for n=3, 4, 6, 8, and 10. The number of rooted trees with n tips is the same as the number of unrooted trees with n+1 tips (Felsenstein 1978a). In practice, only a subset of all trees is examined.

Likelihood Model for DNA Sequences

For DNA sequence data, the model on which the likelihood is based specifies the probability of one sequence changing by mutation to another sequence in a specified time. Although neighboring nucleotides in a DNA sequence are not independent, the models do assume independence of evolution at different sites so that the probability of a set of sequences for some tree is the product of the probabilities for each of the sites in the sequences. At any single site, the models work with probabilities $P_{ij}(T)$ that base i will have changed to base j after a time T. The subscripts i, j take the values 1,2,3,4 for bases `A,C,G,T`.

The simplest base-substitution mutation model assumes a constant rate u of mutation. If a base mutates, it changes to type i with a constant probability π_i. This includes the case in which a base mutates to the same type, although this type of substitution is not observable. With u being the rate of base substitution per unit time (generation), the probability of no mutations at a site after T generations is $(1-u)^T$, so that the probability p

of there being mutation is

$$p = 1 - (1-u)^T \approx 1 - e^{-uT}$$

The probabilities of changes from base i to base j after time T can therefore be written as (Felsenstein 1981)

$$\begin{aligned} P_{ii}(T) &= (1-p) + p\pi_i \\ P_{ij}(T) &= p\pi_j, \ j \neq i \end{aligned}$$

When the π_i's are all set equal to 1/4, this is just the Jukes-Cantor mutation model, but with a slightly different interpretation for the mutation rate. The rate u of this model is for all base substitutions and is equal to 4/3 times the rate μ for detectable substitutions in the Jukes-Cantor model (Chapter 9).

Note that the probabilities involve mutation rate and time only through their product. Neither can be estimated separately by the methods discussed here, so interest is confined to the products $v = uT$, which are the expected number of substitutions along the branches. If base substitutions occur at the same rate in all branches of the tree, then the branch lengths will indicate the relative times between each pair of nodes in the tree.

The likelihood method assumes the structure of the tree. The extant sequences form the tips of the tree, but none of the sequences at other nodes is known. The probability of the data for the tree must therefore consider all possibilities for the unknown sequences.

Under the one-parameter mutation model described here, each of the four base types is expected to become equally frequent, suggesting that π_i be set to 0.25 for i=1,2,3,4. An alternative would be to use the average base frequencies found in the set of sequences for which a tree is being constructed.

Instantaneous Rate Matrices

The derivation of transition probabilities in the previous section was for time measured in discrete generations. A more general treatment works with continuous time, and uses instantaneous probabilities of transition from one base to another. For the Felsenstein model discussed in the last section, during the very small time interval dT, the probability $P_{ij}(dT)$ of changing from base i to base j is

$$P_{ij}(dT) = u\pi_j dT$$

and the probability of base i not changing in time dT is

$$P_{ii}(dT) = 1 - u\sum_{j\neq i} \pi_j dT$$

$$= \quad 1 - u(1 - \pi_i)dT$$

The probability that base i has changed to base j at the end of time $T + dT$ can be decomposed into the changes by the end of time T plus the additional changes during time dT:

$$P_{ij}(T + dT) \quad = \quad \sum_k P_{ik}(T)P_{kj}(dT)$$

The sixteen transition probabilities $\{P_{ij}(T)\}$ can be written in a 4×4 matrix $\boldsymbol{P}$

$$\boldsymbol{P}(T) \quad = \quad \begin{bmatrix} P_{AA}(T) & P_{AC}(T) & P_{AG}(T) & P_{AT}(T) \\ P_{CA}(T) & P_{CC}(T) & P_{CG}(T) & P_{CT}(T) \\ P_{GA}(T) & P_{GC}(T) & P_{GG}(T) & P_{GT}(T) \\ P_{TA}(T) & P_{TC}(T) & P_{TG}(T) & P_{TT}(T) \end{bmatrix}$$

The Felsenstein probabilities change according to the equation

$$\boldsymbol{P}(T + dT) \quad = \quad \boldsymbol{P}(T) + \boldsymbol{P}(T)\boldsymbol{A}dT$$

where the ***instantaneous rate matrix*** **A** of transition probabilities is

$$\boldsymbol{A} \quad = \quad \begin{bmatrix} -u(1 - \pi_A) & u\pi_C & u\pi_G & u\pi_T \\ u\pi_A & -u(1 - \pi_C) & u\pi_G & u\pi_T \\ u\pi_A & u\pi_C & -u(1 - \pi_G) & u\pi_T \\ u\pi_A & u\pi_C & u\pi_G & -u(1 - \pi_T) \end{bmatrix}$$

This leads to an easily-solved differential equation

$$\frac{d\boldsymbol{P}(T)}{dT} \quad = \quad \boldsymbol{P}(T)\boldsymbol{A}$$

$$\boldsymbol{P}(T) \quad = \quad e^{AT} = \boldsymbol{I} + \boldsymbol{A}T + (\boldsymbol{A}T)^2/2! + \cdots$$

Here $\boldsymbol{I}$ denotes the identity matrix. Setting each of the base frequencies to 1/4, leads back to Felsenstein's solutions.

There are many other patterns of probabilities available for matrix **A**. Muse and Weir (1992) discussed the two-parameter model of Hasegawa et al. (1985), in which the instantaneous rate matrix is

$$\begin{bmatrix} -\beta\pi_C - \alpha\pi_G - \beta\pi_T & \beta\pi_C & \alpha\pi_G & \beta\pi_T \\ \beta\pi_A & -\beta\pi_A - \beta\pi_G - \alpha\pi_T & \beta\pi_G & \alpha\pi_T \\ \alpha\pi_A & \beta\pi_C & -\alpha\pi_A - \beta\pi_C - \beta\pi_T & \beta\pi_T \\ \beta\pi_A & \alpha\pi_C & \beta\pi_G & -\beta\pi_A - \alpha\pi_C - \beta\pi_G \end{bmatrix}$$

This model reduces to that of Kimura discussed in Chapter 9 when the base frequencies π are equal, and to that of Felsenstein when the difference between transitions and transversions is ignored ($\alpha = \beta = u$). When both these simplifications are made, the Jukes-Cantor model results.

Two Sequences

There is one rooted tree for two sequences, shown in Figure 10.12. For nucleotide position j in these sequences, the observed bases are s_1, s_2 and the base in the unobserved root sequence is set to k. Adding over all possible values of k, the likelihood $L(j)$ is

$$L(j) = \sum_{k=1}^{4} \pi_k P_{ks_1}(v_1) P_{ks_2}(v_2)$$

and over all m sites the likelihood is

$$L = \prod_{j=1}^{m} L(j)$$

This likelihood is a function of two unknown branch lengths v_1, v_2.

As there is only one set of observed transitions, from sequence 1 to sequence 2, the interior node 0 cannot be located uniquely. A demonstration of this follows from Felsenstein's (1981) "pulley principle." Consider the likelihood for site j written out to show the sum over the four bases at the interior node in the case where sequence 1 has base A and sequence 2 has

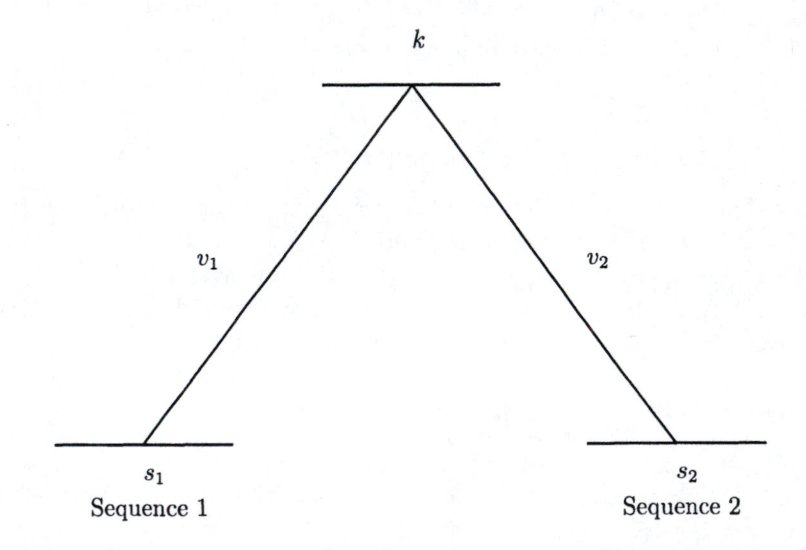

Figure 10.12 Rooted tree for two sequences. At site j, the two sequences have bases s_1, s_2 and the node has base k.

base C at that site

$$\begin{aligned} L(j) &= \pi_A P_{AA}(v_1)P_{AC}(v_2) + \pi_C P_{CA}(v_1)P_{CC}(v_2) + \pi_G P_{GA}(v_1)P_{GC}(v_2) \\ &\quad + \pi_T P_{TA}(v_1)P_{TC}(v_2) \\ &= \pi_A[(1-p_1) + p_1\pi_A]p_2\pi_C + \pi_C p_1 \pi_A[(1-p_2) + p_2\pi_C] \\ &\quad + \pi_G p_1 \pi_A p_2 \pi_C + \pi_T p_1 \pi_A p_2 \pi_C \\ &= \pi_A(p_1 + p_2 - p_1 p_2)\pi_C \\ &= \pi_A p_{12} \pi_C \end{aligned}$$

In other words, the likelihood involving two paths (k to A and k to C) with mutation probabilities p_1 and p_2 is the same as that involving one path (A to C) with probability p_{12}. Note that

$$p_{12} = p_1 + p_2 - p_1 p_2 = 1 - e^{-(v_1+v_2)}$$

so that the likelihood for the tree in Figure 10.12 depends only on the total branch length $v_1 + v_2$ between the two species, 1 and 2, and not on the location of the node 0. It is not possible to estimate v_1 and v_2 separately, and the tree reduces to a single branch between the two sequences. In other words the estimable tree is unrooted.

When the four base probabilities are equal, π_i=1/4, i=1, 2, 3, 4, and the likelihood for this one-branch tree reduces to

$$L = \left(\frac{4-3p}{64}\right)^s \left(\frac{p}{64}\right)^{m-s}$$

where p is the mutation probability for the branch and s of the m sites in the two sequences have the same base. This likelihood is maximized for

$$\hat{p} = \frac{4(m-s)}{3m}$$

The MLE for the branch length is therefore

$$\hat{v} = \ln\left(\frac{3}{4\tilde{q}-1}\right)$$

where

$$\tilde{q} = \frac{s}{m}$$

Recall that u corresponds to $4\mu/3$ in the Jukes-Cantor model, and that the time T between the two sequences was written as $2t$ in that previous model (twice the time from each sequence to the ancestral sequence). These relations show that the branch length could also have been recovered from the Jukes-Cantor distance K between the two sequences:

$$v = uT = \ln\left(\frac{3}{4q-1}\right)$$

$$K = 2\mu t = \frac{3}{4}\ln\left(\frac{3}{4q-1}\right)$$

Length v is the expected number of all substitutions, while length K is for detectable substitutions, and $v = 4K/3$. The estimate could have been found more directly from Bailey's method (Chapter 2) since there is one unknown (K) to estimate from one observed distance.

Three Sequences

There are three rooted trees for three sequences, one of which is shown in Figure 10.13. In addition to the three observed sequences, there are unobserved sequences at the nodes 0 and 4, and there are four branch lengths to be determined. Each of the three trees could be considered in turn, and the one giving the greatest maximum likelihood taken to be the estimated tree. In fact, however, it is not necessary to do this as all three trees have the same likelihood.

For the arrangement shown in Figure 10.13, the likelihood for site j is expressed with base ℓ at node 4 and base k at node 0:

$$L(j) = \sum_k \sum_\ell \pi_k P_{k\ell}(v_4) P_{ks_3}(v_3) P_{\ell s_1}(v_1) P_{\ell s_2}(v_2)$$

Applying Felsenstein's pulley principle leaves the likelihood unchanged if node 0 is moved to any position between nodes 3 and 4. The likelihood depends only on the total distance $v_3 + v_4$. If nodes 0 and 4 are made coincident, the likelihood can be written as

$$L(j) = \sum_k \pi_k P_{ks_1}(v_1) P_{ks_2}(v_2) P_{ks_3}(v_3)$$

There is no way of locating the root 0 uniquely, and only the star phylogeny of Figure 10.14 needs to be considered for three sequences.

Under the assumption of equal base frequencies, since there are three unknown branch lengths, and three pairwise Jukes-Cantor distances available, Bailey's method allows the maximum likelihood estimates to be found from solving

$$\hat{v}_1 + \hat{v}_2 = K_{12}$$

$$\hat{v}_1 + \hat{v}_3 = K_{13}$$

$$\hat{v}_2 + \hat{v}_3 = K_{23}$$

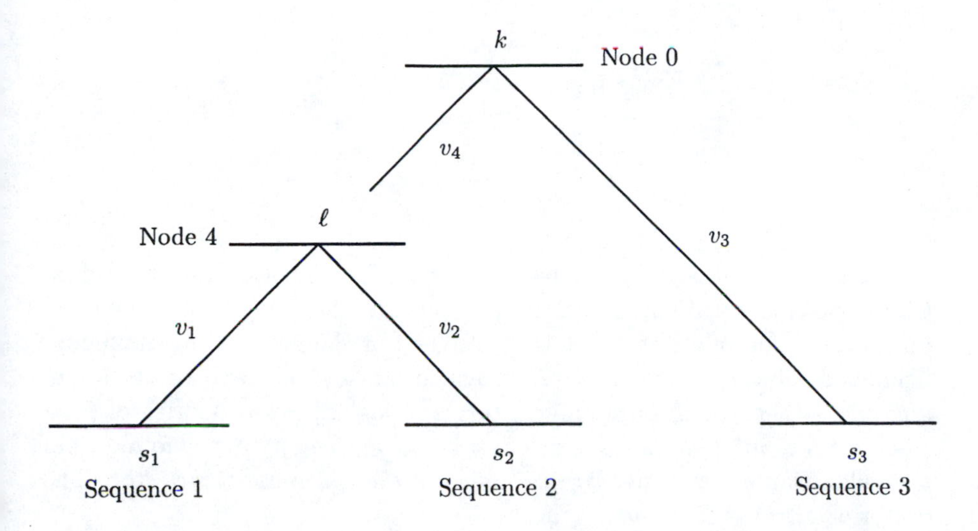

Figure 10.13 One of the three rooted trees for three sequences. At site j, the three sequences have bases s_1, s_2, s_3 and the nodes 0 and 4 have bases k and ℓ.

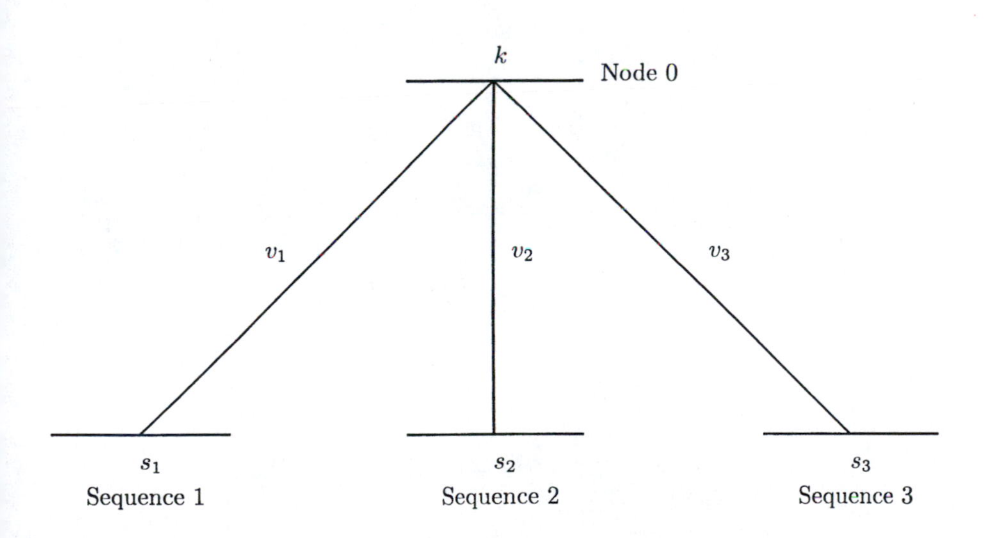

Figure 10.14 Star phylogeny for three sequences. Each of the three sequences 1, 2, 3 descends from the same ancestral sequence 0.

The estimates are

$$\begin{aligned}
\hat{v}_1 &= \frac{1}{2}(K_{12} + K_{13} - K_{23}) \\
\hat{v}_2 &= \frac{1}{2}(K_{12} + K_{23} - K_{13}) \\
\hat{v}_3 &= \frac{1}{2}(K_{13} + K_{23} - K_{12})
\end{aligned}$$

Actual sequences do not have equal base frequencies, and the Jukes-Cantor distances will not maximize the likelihood, but they do provide good initial values for an iterative solution. A direct method for finding maximum likelihood solutions numerically is provided by Newton-Raphson iteration, and this is phrased most simply in terms of looking for estimates of $p_i = 1 - e^{-v_i}$. An initial column vector of estimates $p = (p_1, p_2, p_3)'$ is found from the Jukes-Cantor estimates, where $'$ denotes the transpose of a vector. The vector of scores is

$$S = \begin{bmatrix} \dfrac{\partial \ln L}{\partial p_1} \\ \\ \dfrac{\partial \ln L}{\partial p_2} \\ \\ \dfrac{\partial \ln L}{\partial p_3} \end{bmatrix}$$

The necessary derivatives are

$$\begin{aligned}
\frac{\partial L(j)}{\partial p_1} &= \sum_k \pi_k \frac{\partial P_{ks_1}(p_1)}{\partial p_1} P_{ks_2}(p_2) P_{ks_3}(p_3) \\
\frac{\partial L(j)}{\partial p_2} &= \sum_k \pi_k P_{ks_1}(p_1) \frac{\partial P_{ks_2}(p_2)}{\partial p_2} P_{ks_3}(p_3) \\
\frac{\partial L(j)}{\partial p_3} &= \sum_k \pi_k P_{ks_1}(p_1) P_{ks_2}(p_2) \frac{\partial P_{ks_3}(p_3)}{\partial p_3} \\
\frac{\partial^2 L(j)}{\partial p_i^2} &= 0, \ i = 1, 2, 3 \\
\frac{\partial^2 L(j)}{\partial p_1 \partial p_2} &= \sum_k \pi_k \frac{\partial P_{ks_1}(p_1)}{\partial p_1} \frac{\partial P_{ks_2}(p_2)}{\partial p_2} P_{ks_3} \\
\frac{\partial^2 L(j)}{\partial p_1 \partial p_3} &= \sum_k \pi_k \frac{\partial P_{ks_1}(p_1)}{\partial p_1} P_{ks_2} \frac{\partial P_{ks_3}(p_3)}{\partial p_3}
\end{aligned}$$

$$\frac{\partial^2 L(j)}{\partial p_2 \partial p_3} = \sum_k \pi_k P_{ks_1}(p_1) \frac{\partial P_{ks_2}(p_2)}{\partial p_2} \frac{\partial P_{ks_3}(p_3)}{\partial p_3}$$

The elements of the score vector are

$$S_1 = \frac{\partial \ln L}{\partial p_1} = \sum_j \frac{1}{L(j)} \frac{\partial L(j)}{\partial p_1}$$

$$S_2 = \frac{\partial \ln L}{\partial p_2} = \sum_j \frac{1}{L(j)} \frac{\partial L(j)}{\partial p_2}$$

$$S_3 = \frac{\partial \ln L}{\partial p_3} = \sum_j \frac{1}{L(j)} \frac{\partial L(j)}{\partial p_3}$$

The probabilities $P_{ij}(v)$ are linear functions of the p_i's and so have very simple derivatives.

The information matrix consists of the derivatives of the scores with respect to each p_i

$$I = \begin{bmatrix} -\frac{\partial S_1}{\partial p_1} & -\frac{\partial S_1}{\partial p_2} & -\frac{\partial S_1}{\partial p_3} \\ -\frac{\partial S_2}{\partial p_1} & -\frac{\partial S_2}{\partial p_2} & -\frac{\partial S_2}{\partial p_3} \\ -\frac{\partial S_3}{\partial p_1} & -\frac{\partial S_3}{\partial p_2} & -\frac{\partial S_3}{\partial p_3} \end{bmatrix}$$

and, for example,

$$\frac{\partial S_1}{\partial p_1} = -\sum_j \frac{1}{L(j)^2} \left[\frac{\partial L(j)}{\partial p_1}\right]^2$$

$$\frac{\partial S_1}{\partial p_2} = -\sum_j \frac{1}{L(j)^2} \left[\frac{\partial L(j)}{\partial p_1}\right] \left[\frac{\partial L(j)}{\partial p_2}\right] + \sum_j \frac{1}{L(j)} \frac{\partial^2 L(j)}{\partial p_1 \partial p_2}$$

The initial set of estimates is then modified to a new set p' by

$$p' = p + I^{-1} S$$

where all the terms on the right-hand side are evaluated with the initial estimates. The process is continued until successive values of the estimates are sufficiently close that convergence can be considered to have been achieved. As discussed in Chapter 2, the final value of the information matrix provides estimates of the variances and the covariance of the two estimates. An example of the convergence process is shown in Table 10.5 for the mitochondrial sequences for human (1), gorilla (2) and gibbon (3) in Figure 1.3. The overall average base frequencies among these three sequences were used as the π_i's in the probability terms of the model.

Table 10.5 Successive iterates for branch lengths in the unconstrained maximum likelihood tree for human, gorilla, and gibbon mitochondrial sequences from Figure 1.3.

Iterate	v_1	v_2	v_3	$\ln L$
Initial	0.0423	0.0174	0.2215	
1	0.0420	0.0196	0.2230	−148.44
2	0.0420	0.0199	0.2299	−148.42
3	0.0420	0.0199	0.2299	−148.42
Std. Dev.	0.0297	0.0218	0.0600	

Parsimony as Maximum Likelihood

The method of maximum parsimony seeks to minimize the number of changes over an evolutionary tree. Goldman (1990) clarifies the underlying implicit model of change for Wagner parsimony when evolution proceeds by DNA substitutions. For simplicity he considered only whether character states were the same or different at two ends of a branch. If the probability of a discernible change is π, a data set with S branches having the same character states at the ends and D branches having different states at the ends leads to the likelihood

$$\begin{aligned} L(\pi) &\propto \pi^D(1-\pi)^S \\ &= (1-\pi)^{S+D}[\pi/(1-\pi)]^D \end{aligned}$$

The first term in this last expression is a constant for all trees with $S+D$ branches. The second term is less than one, provided no change is more likely than change ($\pi < 0.5$). The likelihood is therefore maximized when D is minimized. In other words, the total number of observed changes is to be minimized, meaning that the maximum likelihood and maximum parsimony criteria are the same under this model. Goldman's argument makes it explicit that the maximum parsimony model imposes the same probability of evident change for every branch, regardless of branch length.

Relative Rate Test

Within the likelihood framework there is the opportunity for hypothesis testing by comparing likelihoods under different models. One such situation concerns the ***relative rate test*** (Wu and Li 1985). The concept of testing for differences in evolutionary rates without referring to geological times was first discussed by Sarich and Wilson (1973) for amino acid sequences. The work of Wu and Li was for DNA sequences. The hypothesis to be tested is that the rates of base substitution are the same in the branches leading to sequences 1 and 2 in Figure 10.12. As shown above, this is equivalent to testing that lengths v_1, v_2 are the same in Figure 10.13. Using the methodology described in the previous section, the unconstrained likelihood L_1 can be found. Under the constraint of the hypothesis the likelihood is L_0, and the quantity $-2\ln(L_0/L_1)$ may be used as a single degree-of-freedom chi-square test statistic, as discussed in Chapter 3.

Under the hypothesis, the probabilities $P_{ks_1}(v_1)$ and $P_{ks_2}(v_2)$ are both functions of the same parameter $v_1 = v_2$. Numerical methods are again needed to find the MLE's of branch lengths. The methodology of the previous section can be used with some minor changes.

The vector of scores now has two elements:

$$S = \begin{bmatrix} \dfrac{\partial \ln L}{\partial p_1} \\ \\ \dfrac{\partial \ln L}{\partial p_3} \end{bmatrix}$$

Introducing the quantity X as

$$X = P_{ks_1}(p_1)\frac{\partial P_{ks_2}(p_1)}{\partial p_1} + P_{ks_2}(p_1)\frac{\partial P_{ks_1}(p_1)}{\partial p_1}$$

allows the necessary derivatives to be written as

$$\begin{aligned}
\frac{\partial L(j)}{\partial p_1} &= \sum_k \pi_k X P_{ks_3}(p_3) \\
\frac{\partial L(j)}{\partial p_3} &= \sum_k \pi_k P_{ks_1}(p_1) P_{ks_2}(p_1)\frac{\partial P_{ks_3}(p_3)}{\partial p_3} \\
\frac{\partial^2 L(j)}{\partial p_1^2} &= 2\sum_k \pi_k \left[\frac{\partial P_{ks_1}(p_1)}{\partial p_1}\right]\left[\frac{\partial P_{ks_2}(p_1)}{\partial p_1}\right] P_{ks_3}(p_3) \\
\frac{\partial^2 L(j)}{\partial p_1 \partial p_3} &= \sum_k \pi_k X \frac{\partial P_{ks_3}(p_3)}{\partial p_3} \\
\frac{\partial^2 L(j)}{\partial p_3^2} &= 0
\end{aligned}$$

The terms in the vector of scores are

$$S_1 = \frac{\partial \ln L}{\partial p_1}$$

$$S_2 = \frac{\partial \ln L}{\partial p_3}$$

The information matrix is now 2×2

$$I = \begin{bmatrix} -\frac{\partial S_1}{\partial p_1} & -\frac{\partial S_1}{\partial p_3} \\ -\frac{\partial S_2}{\partial p_1} & -\frac{\partial S_2}{\partial p_3} \end{bmatrix}$$

with, for example,

$$\frac{\partial S_1}{\partial p_1} = -\sum_j \frac{1}{L(j)^2}\left[\frac{\partial L(j)}{\partial p_1}\right]^2 + \sum_j \frac{1}{L(j)}\frac{\partial^2 L(j)}{\partial p_1^2}$$

$$\frac{\partial S_1}{\partial p_3} = -\sum_j \frac{1}{L(j)^2}\left[\frac{\partial L(j)}{\partial p_1}\right]\left[\frac{\partial L(j)}{\partial p_3}\right] + \sum_j \frac{1}{L(j)}\frac{\partial^2 L(j)}{\partial p_1 \partial p_3}$$

The same Newton-Raphson iterative equation as for the unconstrained model is used, and the results for the human (1), gorilla (2), and gibbon (3) mitochondrial sequences, with equal rates assigned to the branches from the common ancestor to human and gorilla, are shown in Table 10.6. The common initial value of v_1 and v_2 was set to the average of the initial values in the unconstrained case.

Twice the difference of the log-likelihoods in Tables 10.5 and 10.6 is only 0.31, indicating no significant differences in the rates of evolution leading to human and gorilla from their common ancestor on the basis of this very small set of data. Of course this result was expected, as the estimated rates in Table 10.5 for those two lineages were within one standard deviation of each other.

Properties of the relative rate test using likelihood ratios were described by Muse and Weir (1992). For sequences of 250 bases or more, the test statistic does appear to have a chi-square distribution with one d.f. under the null hypothesis of equal rates. Muse and Weir also used the two-parameter base-substitution model of Hasegawa et al. (1985) and were able to test hypotheses that made separate statements about transitions and transversions. Gaut and Weir (1994) extended the analysis to test for heterogeneity in nucleotide substitution rate among regions of a DNA sequence. Each of the three sequences in Figure 10.13 is partitioned into a number of regions, with

Table 10.6 Successive iterates for two branch lengths in the constrained maximum likelihood tree for human, gorilla, and gibbon mitochondrial sequences from Figure 1.3.

Iterate	v_1	v_3	$\ln L$
Initial	0.0299	0.2215	
1	0.0309	0.2300	−148.59
2	0.0309	0.2303	−148.58
3	0.0309	0.2303	−148.58
Std. Dev.	0.0175	0.0601	

substitution rates depending on both the sequence and the region. Tests for equality of rates within and between sequences can be established as appropriate likelihood ratios.

Muse and Gaut (1994) applied likelihood relative-rate tests to coding sequences, with substitutions being among the 61 nontermination codons. The instantaneous probability matrix is 61 × 61, and two different substitution rates were allowed, one for synonymous and one for non-synonymous changes. Muse (1995) discussed maximum likelihood estimation of these two rates, and Muse and Gaut tested hypotheses that made separate statements about these two rates.

Many Sequences

Construction of trees with several sequences as tips follows the same general procedure as outlined above. A tree is specified, and then the branch lengths estimated by Newton-Raphson iteration of equations resulting from the likelihood of the tree. In theory, all possible trees should be studied to find the one with the highest likelihood.

In his computer programs, Felsenstein (1981) adopts a strategy that examines at least $2n^2 - 9n + 8$ trees for n sequences, but this can be a very much smaller number than the total number of trees possible. He proceeds by successively adding sequences to a tree already constructed, and starts with a two-sequence tree. The kth sequence can be added to any of $2k - 5$ segments on the tree found for the previous $k - 1$ sequences. This is why there are $\prod_{i=3}^{n}(2i - 5)$ unrooted trees for n taxa. For $k > 4$, local rearrangements of the tree are carried out to see if this improves the likelihood before

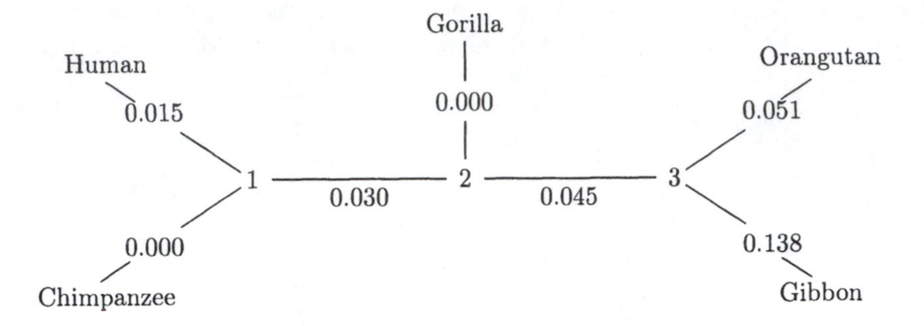

Figure 10.15 Maximum likelihood tree for mitochondrial sequence data of Figure 1.3. Constructed with PHYLIP package of J. Felsenstein.

the next sequence is added. Of course this strategy depends on the order in which sequences are added to the tree, and Felsenstein suggests that several different orderings be tried. He also uses an EM-algorithm approach by which one branch length is found at a time, instead of the Newton-Raphson method for all branches simultaneously. Application of his algorithm for the five mitochondrial sequences of Figure 1.3 results in the unrooted tree shown in Figure 10.15.

Other Substitution Models

It is unlikely that all types of base substitution occur at the same rate, and it would be desirable to include different rates in the likelihood model. Felsenstein allows for different rates of transitions (`A` $\leftrightarrow$ `G`, `C` $\leftrightarrow$ `T`) and transversions (`A,G` $\leftrightarrow$ `C,T`) in current versions of his computer programs (Felsenstein 1993, WWW URL http://evolution.genetics.washington.edu/phylip.html).

Bootstrapping Phylogenies

Within any specified tree topology, maximum likelihood is known to provide consistent estimates of branch lengths, meaning that the estimates approach the true values as the amount of data increases. Moreover, the information matrix provides an indication of the sampling variances of these estimates. What about properties of the tree found to have the largest likelihood over all topologies? In what sense can it be regarded as estimating the true tree? Although this is a difficult theoretical question, empirical evidence can be provided in practice by numerical resampling.

Felsenstein (1985) suggested that bootstrapping be performed over the sites in the sequences being studied. If the bases in a set of s sequences of length m are arranged in an $s \times m$ matrix, a bootstrap sample consists of a new matrix with m columns sampled with replacement from the original set of m columns. Each bootstrap sample is then subjected to the same likelihood estimation algorithm as was the original data set. Attention is paid to sets of monophyletic species over all the bootstrap samples. A group of species can be declared monophyletic at the 5% significance level if it is found together in 95% of the bootstrap trees. There is also the useful concept of a "majority rule" consensus tree (Margush and McMorris 1981) that consists of all groups of species that appear in a majority of the trees found from the bootstrap samples.

Related work has used permutation of the bases at one site among the sequences. Such permutations, performed independently for each site, destroy any phylogenetic structure in the data. Test statistics evaluated from permuted data sets have been proposed for testing hypotheses about monophyly. In criticizing some applications of this approach, Swofford et al. (1996) caution that permutation tests need to yield a test statistic distribution consistent with the hypothesis being tested. The absence of phylogenetic structure is not the same as absence of monophyly.

SUMMARY

The preceding discussion is a very brief look at the rich field of phylogeny reconstruction. As with many areas of statistical analysis of genetic data, there are conflicts between computational expediency and biological realism. The magnitude of DNA sequence data makes it presently difficult to adopt approaches with the otherwise desirable properties of maximum likelihood. Fast algorithms have been developed to build trees from pairwise distances, and such methods have many advantages, even though they use only a summary of the data available. Parsimony methods can also be fast, but may be appropriate only for very slow rates of evolutionary change.

EXERCISES

Exercise 10.1

The following data are for the shrub *Lisianthius skinneri* (Sytsma and Schaal 1985). Rogers' distances were calculated among 7 populations on the basis of frequencies at 12 isozyme loci.

Population	LS4	LS6	LS7	LJ1	LH1	LP1	LA1
LS4	–	0.083	0.250	0.458	0.509	0.563	0.399
LS6		–	0.167	0.392	0.425	0.563	0.337
LS7			–	0.392	0.342	0.479	0.307
LJ1				–	0.400	0.424	0.473
LH1					–	0.384	0.489
LP1						–	0.321

Apply the UPGMA and neighbor-joining methods to construct a tree from this distance matrix. Slight differences may be found from the branch lengths reported by Sytsma and Schaal.

Exercise 10.2

The mitochondrial data described by Brown et al. (1982) have two other *tRNA* genes, *ser* and *leu*. These data are now shown:

```
Leu-tRNA
Hu ACTTTTAAAG GATAACAGCT ATCCATTGGT CTTAGGCCCC AAAAATTTTG GTGCAACTCC AAATAAAAGT
Ch ACTTTTAAAG GATAACAGCC ATCCGTTGGT CTTAGGCCCC AAAAATTTTG GTGCAACTCC AAATAAAAGT
Go ACTTTTAAAG GATAACAGCT ATCCATTGGT CTTAGGACCC AAAAATTTTG GTGCAACTCC AAATAAAAGT
Or GCTTTTAAAG GATAACAGCT ATCCCTTGGT CTTAGGATCC AAAAATTTTG GTGCAACTCC AAATAAAAGT
Gi ACTTTTAAAG GATAACAGCT ATCCATTGGT CTTAGGACCC AAAAATTTTG GTGCAACTCC AAATAAAAGT

Ser-tRNA
Hu  GAGAAAGCTC ACAAGAACTG CTAACTCATG CCCCCATGTC TGACAACATG GCTTTCTCA
Ch  GAGAAAGCTT ATAAGAACTG CTAATTCATA TCCCCATGCC TGACAACATG GCTTTCTCA
Go  GAGAAAGCTC GTAAGAGCTG CTAACTCATA CCCCCGTGCT TGACAACATG GCTTTCTCA
Or  GAGAAAGCTC ACAAGAACTG CTAACTCTCA CT-CCATGTG TGACAACATG GCTTTCTCA
Gi  GAGAAAGCCC ACAAGAACTG CTAACTCACT ATCCCATGTA TGACAACATG GCTTTCTCA
```

The GenBank references for these sequences are: HUMMTTRPR (L00016 V00658), CHPMTTRCH (V00672), GORMTTGO (L00015 V00658), ORAMTTROR (V00675), GIBMTTGI (V00659). Note the gap in the orangutan sequence.

Use these data to construct distance matrix (UPGMA and neighbor-joining) and maximum parsimony trees. Either use these data alone, or add them to the *his* – *tRNA* sequences of Figure 1.3.

Appendix A

Statistical Tables

NORMAL DISTRIBUTION

The probability that a normal random variable lies between x and ∞ is given by the area under the normal density curve $f(x)$, to the right of x. A standard normal variable has mean of 0 and a variance of 1, and a density function of

$$f(x) \quad = \quad \frac{1}{\sqrt{2\pi}} e^{-\frac{x^2}{2}}$$

This curve is plotted in Figure A.1, and the areas for positive x values shown in Table A.1 were calculated by an approximate polynomial expression given by Abramowitz and Stegun (1970 – formula 26.2.17).

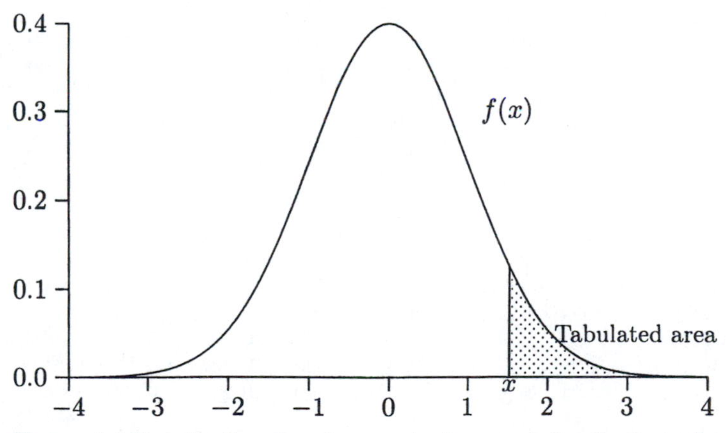

Figure A.1 Density function for standard normal distribution, showing areas tabulated in Table A.1.

Table A.1 Areas under the standard normal density curve, beyond value x.

x	0.00	0.01	0.02	0.03	0.04	0.05	0.06	0.07	0.08	0.09
0.0	0.5000	0.4960	0.4920	0.4880	0.4840	0.4801	0.4761	0.4721	0.4681	0.4641
0.1	0.4602	0.4562	0.4522	0.4483	0.4443	0.4404	0.4364	0.4325	0.4286	0.4247
0.2	0.4207	0.4168	0.4129	0.4090	0.4052	0.4013	0.3974	0.3936	0.3897	0.3859
0.3	0.3821	0.3783	0.3745	0.3707	0.3669	0.3632	0.3594	0.3557	0.3520	0.3483
0.4	0.3446	0.3409	0.3372	0.3336	0.3300	0.3264	0.3228	0.3192	0.3156	0.3121
0.5	0.3085	0.3050	0.3015	0.2981	0.2946	0.2912	0.2877	0.2843	0.2810	0.2776
0.6	0.2743	0.2709	0.2676	0.2643	0.2611	0.2578	0.2546	0.2514	0.2483	0.2451
0.7	0.2420	0.2389	0.2358	0.2327	0.2296	0.2266	0.2236	0.2206	0.2177	0.2148
0.8	0.2119	0.2090	0.2061	0.2033	0.2005	0.1977	0.1949	0.1922	0.1894	0.1867
0.9	0.1841	0.1814	0.1788	0.1762	0.1736	0.1711	0.1685	0.1660	0.1635	0.1611
1.0	0.1587	0.1562	0.1539	0.1515	0.1492	0.1469	0.1446	0.1423	0.1401	0.1379
1.1	0.1357	0.1335	0.1314	0.1292	0.1271	0.1251	0.1230	0.1210	0.1190	0.1170
1.2	0.1151	0.1131	0.1112	0.1093	0.1075	0.1056	0.1038	0.1020	0.1003	0.0985
1.3	0.0968	0.0951	0.0934	0.0918	0.0901	0.0885	0.0869	0.0853	0.0838	0.0823
1.4	0.0808	0.0793	0.0778	0.0764	0.0749	0.0735	0.0721	0.0708	0.0694	0.0681
1.5	0.0668	0.0655	0.0643	0.0630	0.0618	0.0606	0.0594	0.0582	0.0571	0.0559
1.6	0.0548	0.0537	0.0526	0.0516	0.0505	0.0495	0.0485	0.0475	0.0465	0.0455
1.7	0.0446	0.0436	0.0427	0.0418	0.0409	0.0401	0.0392	0.0384	0.0375	0.0367
1.8	0.0359	0.0351	0.0344	0.0336	0.0329	0.0322	0.0314	0.0307	0.0301	0.0294
1.9	0.0287	0.0281	0.0274	0.0268	0.0262	0.0256	0.0250	0.0244	0.0239	0.0233
2.0	0.0228	0.0222	0.0217	0.0212	0.0207	0.0202	0.0197	0.0192	0.0188	0.0183
2.1	0.0179	0.0174	0.0170	0.0166	0.0162	0.0158	0.0154	0.0150	0.0146	0.0143
2.2	0.0139	0.0136	0.0132	0.0129	0.0125	0.0122	0.0119	0.0116	0.0113	0.0110
2.3	0.0107	0.0104	0.0102	0.0099	0.0096	0.0094	0.0091	0.0089	0.0087	0.0084
2.4	0.0082	0.0080	0.0078	0.0075	0.0073	0.0071	0.0069	0.0068	0.0066	0.0064
2.5	0.0062	0.0060	0.0059	0.0057	0.0055	0.0054	0.0052	0.0051	0.0049	0.0048
2.6	0.0047	0.0045	0.0044	0.0043	0.0041	0.0040	0.0039	0.0038	0.0037	0.0036
2.7	0.0035	0.0034	0.0033	0.0032	0.0031	0.0030	0.0029	0.0028	0.0027	0.0026
2.8	0.0026	0.0025	0.0024	0.0023	0.0023	0.0022	0.0021	0.0021	0.0020	0.0019
2.9	0.0019	0.0018	0.0018	0.0017	0.0016	0.0016	0.0015	0.0015	0.0014	0.0014
3.0	0.0013	0.0013	0.0013	0.0012	0.0012	0.0011	0.0011	0.0011	0.0010	0.0010
3.1	0.0010	0.0009	0.0009	0.0009	0.0008	0.0008	0.0008	0.0008	0.0007	0.0007
3.2	0.0007	0.0007	0.0006	0.0006	0.0006	0.0006	0.0006	0.0005	0.0005	0.0005
3.3	0.0005	0.0005	0.0005	0.0004	0.0004	0.0004	0.0004	0.0004	0.0004	0.0003
3.4	0.0003	0.0003	0.0003	0.0003	0.0003	0.0003	0.0003	0.0003	0.0003	0.0002
3.5	0.0002	0.0002	0.0002	0.0002	0.0002	0.0002	0.0002	0.0002	0.0002	0.0002
4.0	0.0000	0.0000	0.0000	0.0000	0.0000	0.0000	0.0000	0.0000	0.0000	0.0000

CHI-SQUARE DISTRIBUTION

The probability that a chi-square random variable lies between x and ∞ is given by the area under the chi-square density curve $f(x)$, to the right of x. For n degrees of freedom, this density function is

$$f(x) = \frac{1}{2^{n/2}\Gamma(n/2)} x^{n/2-1} e^{-x/2}$$

where the gamma function $\Gamma(n/2)$ is given by

$$\Gamma(n/2) = \begin{cases} \sqrt{\pi} & n = 1 \\ 1 & n = 2 \\ (m-1)! & n = 2m \\ (2m)!\sqrt{\pi}/2^{2m}m! & n = 2m+1 \end{cases}$$

and m is a positive integer. The density curve for $n = 10$ d.f. is shown in Figure A.2. The probabilities in Table A.2 were calculated with an algorithm given by Posten (1989).

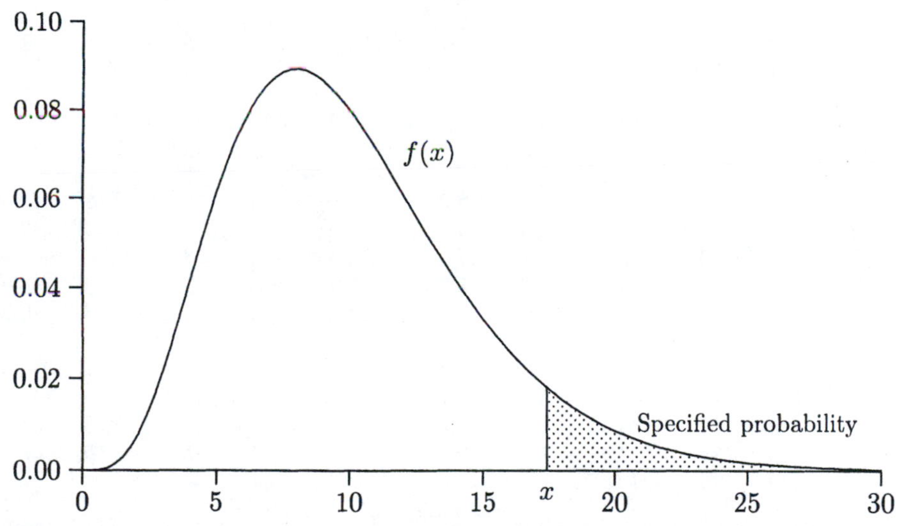

Figure A.2 Density function for chi-square distribution with 10 degrees of freedom, showing values exceeded with specified probabilities.

Table A.2 Chi-square values that are exceeded with specified probabilities.

	Probability									
d.f.	0.995	0.990	0.975	0.950	0.900	0.100	0.050	0.025	0.010	0.005
1	0.00	0.00	0.00	0.00	0.02	2.71	3.84	5.02	6.63	7.88
2	0.01	0.02	0.05	0.10	0.21	4.61	5.99	7.38	9.21	10.6
3	0.07	0.12	0.22	0.35	0.58	6.25	7.81	9.35	11.3	12.8
4	0.21	0.30	0.48	0.71	1.06	7.78	9.49	11.1	13.3	14.9
5	0.41	0.55	0.83	1.15	1.61	9.24	11.1	12.8	15.1	16.7
6	0.68	0.87	1.24	1.64	2.20	10.6	12.6	14.4	16.8	18.5
7	0.99	1.24	1.69	2.17	2.83	12.0	14.1	16.0	18.5	20.3
8	1.34	1.65	2.18	2.73	3.49	13.4	15.5	17.5	20.1	22.0
9	1.73	2.09	2.70	3.33	4.17	14.7	16.9	19.0	21.7	23.6
10	2.16	2.56	3.25	3.94	4.87	16.0	18.3	20.5	23.2	25.2
11	2.60	3.05	3.82	4.57	5.58	17.3	19.7	21.9	24.7	26.8
12	3.07	3.57	4.40	5.23	6.30	18.5	21.0	23.3	26.2	28.3
13	3.57	4.11	5.01	5.89	7.04	19.8	22.4	24.7	27.7	29.8
14	4.07	4.66	5.63	6.57	7.79	21.1	23.7	26.1	29.1	31.3
15	4.60	5.23	6.26	7.26	8.55	22.3	25.0	27.5	30.6	32.8
16	5.14	5.81	6.91	7.96	9.31	23.5	26.3	28.8	32.0	34.3
17	5.70	6.41	7.56	8.67	10.1	24.8	27.6	30.2	33.4	35.7
18	6.26	7.01	8.23	9.39	10.9	26.0	28.9	31.5	34.8	37.2
19	6.84	7.63	8.91	10.1	11.7	27.2	30.1	32.9	36.2	38.6
20	7.43	8.26	9.59	10.9	12.4	28.4	31.4	34.2	37.6	40.0
21	8.03	8.90	10.3	11.6	13.2	29.6	32.7	35.5	38.9	41.4
22	8.64	9.54	11.0	12.3	14.0	30.8	33.9	36.8	40.3	42.8
23	9.26	10.2	11.7	13.1	14.8	32.0	35.2	38.1	41.6	44.2
24	9.89	10.9	12.4	13.8	15.7	33.2	36.4	39.4	43.0	45.6
25	10.5	11.5	13.1	14.6	16.5	34.4	37.7	40.6	44.3	46.9
26	11.2	12.2	13.8	15.4	17.3	35.6	38.9	41.9	45.6	48.3
27	11.8	12.9	14.6	16.2	18.1	36.7	40.1	43.2	47.0	49.6
28	12.5	13.6	15.3	16.9	18.9	37.9	41.3	44.5	48.3	51.0
29	13.1	14.3	16.0	17.7	19.8	39.1	42.6	45.7	49.6	52.3
30	13.8	15.0	16.8	18.5	20.6	40.3	43.8	47.0	50.9	53.7
40	20.7	22.2	24.4	26.5	29.1	51.8	55.8	59.3	63.7	66.8
50	28.0	29.7	32.4	34.8	37.7	63.2	67.5	71.4	76.2	79.5
60	35.5	37.5	40.5	43.2	46.5	74.4	79.1	83.3	88.4	92.0
70	43.3	45.4	48.8	51.7	55.3	85.5	90.5	95.0	100.4	104.2
80	51.2	53.5	57.2	60.4	64.3	96.6	101.9	106.6	112.3	116.3
100	67.3	70.1	74.2	77.9	82.4	118.5	124.3	129.6	135.8	140.2

NONCENTRAL CHI-SQUARE DISTRIBUTION

The probability that a random variable with the noncentral chi-square distribution exceeds a value x is given by the sum of terms that are the probabilities that central (i.e., usual) chi-square variables exceed the same value. Writing these latter probabilities as $\Pr(\chi^2|n)$ for a central chi-square with n d.f., the required probability for the noncentral chi-square χ'^2 with n d.f. and noncentrality parameter ν is

$$\Pr(\chi'^2|n,\nu) = \sum_{j=0}^{\infty} e^{-\nu/2}\frac{(\nu/2)^j}{j!}\Pr(\chi^2|n+2j)$$

The noncentrality parameter values ν shown in Table A.3 were calculated with the algorithm of Posten (1989). These were chosen to give a probability β of the noncentral chi-square exceeding the $(1-\alpha)$th percentile of the central chi-square (i.e., that value of the central chi-square exceeded with probability α). When $\alpha = 0.05, \beta = 0.90$, for example, the probability is 0.90 that a noncentral chi-square exceeds the 95% percentile value of 3.84 if $\nu = 10.5$.

If a chi-square statistic is being calculated to perform a test, and large values of the statistic will cause rejection of a hypothesis, then α is the significance level of the test. It is the probability of rejection when the hypothesis is true. The probability β of rejecting the hypothesis when it is false is the power of the test. Table A.3 therefore relates significance level, power and noncentrality parameter.

Table A.3 Noncentrality parameters for power β and significance level α.

	$\alpha = 0.10$			$\alpha = 0.05$			$\alpha = 0.01$		
d.f.	β=0.50	β=0.90	β=0.99	β=0.50	β=0.90	β=0.99	β=0.50	β=0.90	β=0.99
1	2.70	8.56	15.8	3.84	10.5	18.4	6.64	14.9	24.0
2	3.56	10.5	18.6	4.96	12.7	21.4	8.19	17.4	27.4
3	4.18	11.8	20.5	5.76	14.2	23.5	9.31	19.2	29.8
4	4.69	12.9	22.1	6.42	15.4	25.2	10.2	20.7	31.8
5	5.14	13.8	23.4	6.99	16.5	26.7	11.0	22.0	33.5
6	5.53	14.7	24.6	7.50	17.4	28.0	11.8	23.2	35.0
7	5.90	15.4	25.7	7.97	18.3	29.2	12.4	24.2	36.4
8	6.24	16.1	26.8	8.41	19.1	30.4	13.0	25.2	37.7
9	6.55	16.8	27.7	8.81	19.8	31.4	13.6	26.1	38.9
10	6.85	17.4	28.6	9.19	20.5	32.4	14.1	27.0	40.0
11	7.13	18.0	29.4	9.56	21.2	33.3	14.6	27.8	41.1
12	7.40	18.5	30.2	9.90	21.8	34.2	15.1	28.6	42.1
13	7.66	19.1	31.0	10.2	22.4	35.0	15.6	29.3	43.1
14	7.91	19.6	31.7	10.6	23.0	35.8	16.0	30.0	44.0
15	8.15	20.1	32.4	10.9	23.6	36.6	16.5	30.7	44.9
16	8.38	20.5	33.1	11.2	24.1	37.3	16.9	31.4	45.8
17	8.60	21.0	33.8	11.4	24.7	38.1	17.3	32.0	46.6
18	8.82	21.4	34.4	11.7	25.2	38.8	17.7	32.7	47.5
19	9.03	21.9	35.0	12.0	25.7	39.4	18.1	33.3	48.3
20	9.24	22.3	35.6	12.3	26.2	40.1	18.5	33.9	49.0
21	9.44	22.7	36.2	12.5	26.6	40.7	18.8	34.4	49.8
22	9.64	23.1	36.8	12.8	27.1	41.4	19.2	35.0	50.5
23	9.83	23.5	37.3	13.0	27.5	42.0	19.5	35.5	51.2
24	10.0	23.9	37.9	13.3	27.9	42.6	19.9	36.1	51.9
25	10.2	24.3	38.4	13.5	28.4	43.2	20.2	36.6	52.6
26	10.4	24.6	38.9	13.7	28.8	43.7	20.5	37.1	53.3
27	10.6	25.0	39.5	14.0	29.2	44.3	20.8	37.6	53.9
28	10.7	25.3	40.0	14.2	29.6	44.9	21.1	38.1	54.6
29	10.9	25.7	40.4	14.4	30.0	45.4	21.5	38.6	55.2
30	11.1	26.0	40.9	14.6	30.4	45.9	21.8	39.1	55.9
40	12.6	29.1	45.4	16.6	33.9	50.8	24.5	43.5	61.6
50	14.0	31.9	49.3	18.3	37.1	55.1	27.0	47.3	66.6
60	15.2	34.3	52.8	19.9	39.9	59.0	29.2	50.8	71.1
70	16.3	36.6	56.0	21.3	42.5	62.5	31.2	54.0	75.3
80	17.4	38.7	59.0	22.7	44.9	65.8	33.1	57.0	79.1
100	19.3	42.6	64.5	25.1	49.3	71.8	36.6	62.4	86.2

Appendix B

Random Numbers

GENERATING RANDOM NUMBERS

Several procedures described in this book require the drawing of random numbers. It is not possible to specify a procedure for generating truly random numbers, but there are algorithms that give very long sequences of "pseudo-random" numbers that have the appearance of randomness. A complete discussion of the generation and testing of random numbers is given by Knuth (1981).

Random numbers uniformly distributed on the interval [0, 1] are often sought. As a first step, integers X are generated between 0 and m, where m is the largest integer that can be handled by the computer being used. For computers with a word length of 32, this number is

$$m = 2^{31} - 1 = 2,147,483,647$$

The required numbers R are then

$$R = \frac{X}{m}$$

The most common algorithm uses the ***linear congruential method*** to generate a sequence $\{R_n\}$ of random numbers. The random integers X_n are found from

$$X_{n+1} = (aX_n + c) \bmod m$$

with a specified starting value X_0 and constants a, c. A particular case that has been found to have a long ***period***, before the numbers start repeating themselves, was reported by Lewis et al. (1969) and has been used widely since then. This algorithm uses a "seed" a and is

$$X_0 = a$$

$$\begin{aligned} X_{n+1} &= 7^5 X_n \bmod (2^{31} - 1) \\ R_n &= 2^{-31} X_n \end{aligned}$$

A convenient seed would be the date. With $a = 951231$, for example

$$\begin{aligned} X_0 &= 951231 \\ X_1 &= 16,807 \times 951,231 \,(\mathrm{mod}\; 2,147,483,647) \\ &= 954,953,888 \\ R_1 &= 0.444685 \end{aligned}$$

To three decimal places the first random number is 0.445. The seed is usually chosen to be odd.

The set of random digits in Table B.1 was generated by this method. A set of 2,000 numbers was drawn, each multiplied by 10 and then truncated as an integer.

This procedure was modified by Schrage (1979) to allow FORTRAN programs to be more readily portable to a range of computers. It is necessary that the computer be able to represent all integers in the range $[0, 2^{31} - 1]$, and then the generator is full cycle – i.e., every integer from 1 to $2^{31} - 1$ is generated exactly once in a cycle.

Random Normal Deviates

Tables A.1 and B.1 can be used to draw a random normal deviate "by hand." First, a random number from the uniform distribution on the interval [0, 1] can be found from Table B.1. Successive sets of four digits can be regarded as decimal fraction by writing a decimal point in front of them. The first two such numbers in Table B.1 are 0.3024 and 0.6861. This number P is taken to represent the (cumulative) probability of a normal variable being less than the desired value X. For a P value less than 0.5, the standard normal variable Z is found as the marginal value in Table A.1 corresponding to the tabulated number P, with a change of sign. For 0.3024, the closest entry in Table A.1 is 0.3015, with a marginal entry of 0.52, so that $Z = -0.52$. For a random uniform number P greater than 0.5, the entry $1 - P$ in Table A.1 is sought and then the corresponding margin entry Z used with sign preserved. For $P = 0.6861$, the nearest entry to $1 - P = 0.3139$ is 0.3156, with a marginal entry of $Z = 0.48$.

Table B.1 Two thousand random digits.

	5	10	15	20	25	30	35	40	45	50
1	30246	86149	45548	80480	85924	02411	46456	23952	55145	18300
2	02806	20733	30853	08034	21238	39933	90958	87912	82486	96960
3	84868	17425	91536	08208	44761	40101	74109	08696	73249	10885
4	65043	86343	36953	04658	42008	84984	49584	53872	52737	24217
5	59792	12608	73246	57277	29384	02608	78779	59311	08421	72618
6	29008	02705	38780	09675	32573	74039	85654	12731	36846	21341
7	74800	20695	99211	38699	28454	21400	11524	81212	55327	93367
8	45715	29459	60745	64762	81553	00401	21852	65586	51269	73813
9	70056	78054	16563	32244	81117	26808	94318	00873	00154	81690
10	30072	38515	52181	21872	17193	57361	16000	51633	70345	48725
11	19490	00789	48629	84877	18858	73868	05461	57469	58009	23998
12	79558	05067	71799	72777	45475	39847	14211	09764	38988	94242
13	18072	34286	46778	95843	31600	57151	89995	58712	46820	81464
14	09933	43223	27657	00697	84736	96171	18120	74205	86558	72670
15	68396	26040	44227	73036	11903	59352	73105	88131	25523	48473
16	76023	01624	74545	18347	66573	79479	24729	98822	93629	72477
17	52257	64895	96218	45817	93951	30547	93632	21510	17326	95743
18	27531	76301	89645	24680	93157	56419	92677	05539	81408	37221
19	17406	68465	66526	13785	92655	25101	95658	54255	07336	17904
20	87810	83955	12467	83985	39484	80179	96878	67468	16173	29937
21	01109	37024	09219	04303	65058	07201	50126	56572	97194	99595
22	67362	79269	61078	70412	89414	45697	17368	48025	41999	45286
23	38002	58000	50220	34603	73647	06894	84712	52922	73303	22802
24	60044	14258	82451	24551	14223	77858	61729	69565	62211	90630
25	55818	55177	80015	88181	96369	57150	37206	02369	18457	29621
26	82646	47169	71375	65259	13194	59086	81076	08421	47402	25764
27	47133	75669	28424	83710	21907	46183	21782	04475	88099	33155
28	62065	06444	34797	56543	90176	41665	53588	71810	26557	83977
29	52765	89407	17693	33927	97348	72061	14231	12340	44493	64194
30	68651	84960	60535	51369	08459	97693	31991	37836	37247	50762
31	74437	48122	89309	16025	06062	10840	22809	28746	30682	48082
32	49051	14405	76357	57632	46511	00666	09647	61493	66875	29164
33	95023	70370	60841	58975	63641	71478	48327	82378	17689	49232
34	19358	28765	57897	93980	61832	10202	79416	40162	85205	87337
35	95489	73778	86660	39424	89005	68527	85534	77132	95116	65790
36	07758	15002	18281	35417	07440	56681	31392	91160	85337	79306
37	27602	69590	13299	50384	25829	85184	89773	97149	16399	41287
38	75864	68804	37205	39021	67019	38964	62848	40359	22254	54700
39	47313	78390	64495	14918	97584	73636	55745	33592	16050	86578
40	13406	80860	65073	73149	74121	97974	60190	50744	52846	91673

If a random variate X from the normal distribution $N(\mu, \sigma^2)$ with mean μ and variance σ^2 is needed, it is constructed from the standard normal value by

$$X = \mu + Z\sigma$$

BOOTSTRAPPING

Drawing a bootstrap sample from a data set of size n is equivalent to drawing n integers from the set $1, 2, \ldots, n$ with replacement. This can be achieved by taking the product of n times each of a set of numbers drawn at random on the unit interval and rounding up to the nearest integer. If twelve random numbers 0.240, 0.827, 0.950, 0.240, 0.066, 0.855, 0.239, 0.416, 0.207, 0.239, 0.417, 0.450 are drawn, then a bootstrap sample of size 12 would consist of items 3, 10, 12, 3, 1, 11, 3, 5, 3, 3, 6, 6.

RANDOM SEQUENCES

Testing hypotheses about DNA sequences often makes use of random sequences. Two possibilities are those of ***representative*** and ***shuffled*** sequences (Fitch 1983). A representative sequence is one drawn at random from a population that has the same characteristics as the sequence at hand, while a shuffled sequence is just a permutation of the elements in that sequence. There may be very little difference in results based on the two approaches for long sequences.

To draw a representative sequence of length m that has the same expected base frequencies as those in the actual sequence, an infinite population of bases is imagined. This has the same base frequencies p_i, $i = 1, 2, 3, 4$ for bases `A, C, G, T` as does the actual sequence. These enable the unit interval to be partitioned into four segments, with boundaries

$$0, \; p_1, \; p_1 + p_2, \; p_1 + p_2 + p_3, \; 1$$

These segments therefore have lengths p_1, p_2, p_3, p_4 respectively. For the mth site in the random sequence, a random number is drawn from the unit interval. If this number lies in the ith segment, then the site is filled with base i.

The following algorithm for shuffling a sequence was given by Knuth (1981) and depends on the exchange of random pairs of elements in the sequence. The algorithm for a sequence of length m is as follows. Each sequence element is considered in turn and exchanged with another element chosen at random.

1. Set $j = m$.

2. Choose a random number u from the interval [0,1].
   Set $k = [ju] + 1$, i.e., the integral part of $j \times u$ plus 1.

3. Exchange elements k and j.
   Set $j = j - 1$.

4. Return to step 2 if $j > 1$.

To shuffle a sequence of 12 elements, suppose the 12 random numbers 0.240, 0.827, 0.950, 0.240, 0.066, 0.855, 0.239, 0.416, 0.207, 0.239, 0.417, 0.450 are drawn. The numbers k for step 2 of the algorithm are 3, 10, 12, 3, 1, 11, 3, 5, 3, 3, 6, 6. The 12 steps in the shuffling produce the following set of sequences, with the final one being the desired complete shuffle.

1	2	12	4	5	6	7	8	9	10	11	3
1	2	12	4	5	6	7	8	9	11	10	3
1	2	12	4	5	6	7	8	9	3	10	11
1	2	9	4	5	6	7	8	12	3	10	11
8	2	9	4	5	6	7	1	12	3	10	11
8	2	9	4	5	6	10	1	12	3	7	11
8	2	6	4	5	9	10	1	12	3	7	11
8	2	6	4	5	9	10	1	12	3	7	11
8	2	4	6	5	9	10	1	12	3	7	11
8	2	4	6	5	9	10	1	12	3	7	11
8	9	4	6	5	2	10	1	12	3	7	11
2	9	4	6	5	8	10	1	12	3	7	11

Constrained Random Sequences

When sequence statistics, such as the length of the longest direct repeat, are studied by calculating them on a set of random sequences, it may be appropriate to ensure that dinucleotide frequencies are preserved. Such constraints may be built into the algorithms for representative or shuffled sequences (Fitch 1983).

For a representative sequence to preserve dinucleotide frequencies, it is necessary to find these frequencies, p_{ij} for bases ij, in the original sequence in addition to the individual base frequencies p_i. The random sequence is begun by drawing a random base as described in the previous section. If this is of type i, then the next base is selected to be of type j with probability p_{ij}/p_i. In other words the unit interval is partitioned into four segments in a way that depends on frequencies of dinucleotides beginning with base i.

At the third site, the base drawn is of type k with probability p_{jk}/p_j, and so on.

Shufflings with constraints are a little more difficult to construct, and Fitch (1983) describes a method that first counts dinucleotides, including that formed by the last base followed by the first. The initial base i is chosen, without replacement, from the set of m bases in the original sequence. The next base j is chosen, without replacement, from the set of dinucleotides that start with i.

PERMUTATION TESTS

The shuffling procedure of the previous section is also used for permutation tests. To perform a Hardy-Weinberg exact test, with a p-value generated by permutations, the entire set of alleles in the sample is shuffled, and then successive pairs taken to be genotypes. If a sample of six individuals has genotypes

$$A_1A_1 \quad A_1A_1 \quad A_2A_3 \quad A_2A_2 \quad A_1A_3 \quad A_2A_2$$

then the set of alleles can be numbered 1 to 12:

1	2	3	4	5	6	7	8	9	10	11	12
A_1	A_1	A_1	A_1	A_2	A_3	A_2	A_2	A_1	A_3	A_2	A_2

and the shuffling illustrated in the previous section produces

2	9	4	6	5	8	10	1	12	3	7	11
A_1	A_1	A_1	A_3	A_2	A_2	A_3	A_1	A_2	A_1	A_2	A_2

The new sample is, therefore,

$$A_1A_1 \quad A_1A_3 \quad A_2A_2 \quad A_1A_3 \quad A_1A_2 \quad A_2A_2$$

Appendix C

Answers to Exercises

CHAPTER 1

Exercise 1.2

Mendel's upper and lower quartiles are less far apart than are those from the normal distribution.

Exercise 1.3

The expected numbers for round or wrinkled seeds are three-quarters or one-quarter the sample size (round plus wrinkled). The chi-square test statistic is given in Equation 1.1 and leads to the following results:

	round		wrinkled		
Plant	*o*	*e*	*o*	*e*	X^2
1	45	42.75	12	14.25	0.47
2	27	26.25	8	8.75	0.09
3	24	23.25	7	7.75	0.10
4	19	21.75	10	7.25	1.39
5	32	32.25	11	10.75	0.01
6	26	24.00	6	8.00	0.67
7	88	84.00	24	28.00	0.76
8	22	24.00	10	8.00	0.67
9	28	25.50	6	8.50	0.98
10	25	24.00	7	8.00	0.17

Even if the hypothesis is true, one in 10 tests are expected to case rejection simply by chance if a 10% significance level is used. None of these 10 tests cause rejection, reinforcing the impression of a better-than-expected fit of the data to the hypothesis.

CHAPTER 2

Exercise 2.1

a. The sample allele frequencies are

Generation	$\tilde{p}_{1.6}$	$\tilde{p}_{2.7}$	$\tilde{p}_{3.9}$
5	0.0491	0.9189	0.0320
15	0.0769	0.8417	0.0814
25	0.1592	0.7724	0.0684

Note that these are not maximum likelihood estimates since the non-zero inbreeding and associated departure from Hardy-Weinberg has been ignored.

b. The estimated variances and covariances of allele frequencies are

$$\begin{aligned}
\text{Var}(\tilde{p}_{1.6}) &= (0.0491 + 0.0485 - 0.0048)/2972 = 3.03 \times 10^{-5} \\
\text{Var}(\tilde{p}_{2.7}) &= (0.9189 + 0.9125 - 1.6888)/2972 = 4.80 \times 10^{-5} \\
\text{Var}(\tilde{p}_{3.9}) &= (0.0320 + 0.0283 - 0.0020)/2972 = 1.96 \times 10^{-5} \\
\text{Cov}(\tilde{p}_{1.6}, \tilde{p}_{2.7}) &= (0.0061 - 0.1805)/5944 = -2.94 \times 10^{-5} \\
\text{Cov}(\tilde{p}_{1.6}, \tilde{p}_{3.9}) &= (0.0007 - 0.0063)/5944 = -0.094 \times 10^{-5} \\
\text{Cov}(\tilde{p}_{2.7}, \tilde{p}_{3.9}) &= (0.0067 - 0.1176)/5944 = -1.86 \times 10^{-5}
\end{aligned}$$

c. A single inbreeding coefficient f is estimated from Equation 2.28 as

Generation	$\tilde{H} = 1 - \sum_u \tilde{P}_{uu}$	$\tilde{h} = 1 - \sum_u \tilde{p}_u^2$	$\hat{f}$
5	0.0135	0.1522	0.9116
10	0.0436	0.2790	0.8437
15	0.0106	0.3734	0.9717

These estimates are virtually the same as the initial values for the iterative scheme of Hill et al. (1995). The final values for that scheme are also very close to the initial values.

If an estimate was to be obtained for each allele, the method of moments would take note of:

$$\begin{aligned}
\mathcal{E}(\tilde{p}_u) &= p_u \\
\mathcal{E}(\tilde{P}_{uu}) &= p_u^2 + p_u(1 - p_u)f \\
\mathcal{E}(\tilde{p}_u^2) &= p_u^2 + p_u(1 - p_u)f/2n
\end{aligned}$$

to arrive at the estimate

$$\hat{f}_u = \frac{(\tilde{P}_{uu} - \tilde{p}_u^2) + \frac{1}{2n}(\tilde{p}_u - \tilde{P}_{uu})}{(\tilde{p}_u - \tilde{p}_u^2) - \frac{1}{2n}(\tilde{p}_u - \tilde{P}_{uu})}$$

The single estimate for the locus, Equation 2.28, is the ratio of the sums over all alleles u of the numerator and denominator of this expression.

Exercise 2.2

With equal initial values, the first 10 iterates of the Newton-Raphson iteration approach are:

Iterate	M	S	F	O
1	0.7406	0.1145	0.0702	0.0747
2	0.7402	0.1155	0.0701	0.0741
3	0.7406	0.1156	0.0701	0.0738
4	0.7408	0.1156	0.0701	0.0735
5	0.7410	0.1156	0.0701	0.0733
6	0.7411	0.1156	0.0701	0.0732
7	0.7412	0.1156	0.0701	0.0731
8	0.7412	0.1156	0.0701	0.0731
9	0.7413	0.1156	0.0701	0.0730
10	0.7413	0.1156	0.0701	0.0730

With equal initial values, the first 40 iterates of the EM algorithm approach are:

Iterate	M	S	F	O	Iterate	M	S	F	O
1	0.5330	0.1359	0.0981	0.2330	2	0.6600	0.1196	0.0757	0.1447
3	0.6908	0.1163	0.0718	0.1210	4	0.7055	0.1153	0.0705	0.1086
5	0.7141	0.1150	0.0701	0.1008	6	0.7197	0.1149	0.0699	0.0954
7	0.7237	0.1149	0.0699	0.0915	8	0.7267	0.1150	0.0698	0.0884
9	0.7290	0.1150	0.0699	0.0861	10	0.7309	0.1151	0.0699	0.0842
11	0.7323	0.1152	0.0699	0.0826	12	0.7336	0.1152	0.0699	0.0813
13	0.7346	0.1153	0.0700	0.0802	14	0.7355	0.1153	0.0700	0.0793
15	0.7362	0.1153	0.0700	0.0785	16	0.7368	0.1154	0.0700	0.0778
17	0.7374	0.1154	0.0700	0.0772	18	0.7378	0.1154	0.0700	0.0767
19	0.7383	0.1154	0.0701	0.0763	20	0.7386	0.1155	0.0701	0.0759
21	0.7389	0.1155	0.0701	0.0755	22	0.7392	0.1155	0.0701	0.0752
23	0.7394	0.1155	0.0701	0.0750	24	0.7397	0.1155	0.0701	0.0747
25	0.7399	0.1155	0.0701	0.0745	26	0.7400	0.1155	0.0701	0.0744
27	0.7402	0.1155	0.0701	0.0742	28	0.7403	0.1155	0.0701	0.0741
29	0.7404	0.1155	0.0701	0.0739	30	0.7405	0.1156	0.0701	0.0738
31	0.7406	0.1156	0.0701	0.0737	32	0.7407	0.1156	0.0701	0.0736
33	0.7408	0.1156	0.0701	0.0735	34	0.7408	0.1156	0.0701	0.0735
35	0.7409	0.1156	0.0701	0.0734	36	0.7409	0.1156	0.0701	0.0734
37	0.7410	0.1156	0.0701	0.0733	38	0.7410	0.1156	0.0701	0.0733
39	0.7411	0.1156	0.0701	0.0732	40	0.7411	0.1156	0.0701	0.0732

The matrix of estimated variances and covariances of the first three alleles, from the information matrix, is

	S	M	F
S	.00001909	−.00000834	−.00000515
M	−.00000834	.00001002	−.00000085
F	−.00000515	−.00000085	.00000644

Exercise 2.3

a. For the first set of genotypic counts, there is one real solution to the cubic equation described before Equation 2.27. This root is 0.7227.

b. There are three real roots, 0.40, 0.20, and 0.30 to the cubic equation, and these have log-likelihoods of −224.00, −224.00, and −225.34. The third root therefore gives the maximum likelihood.

c. There are three real roots, 0.37, 0.22, and 0.27 to the cubic equation, and these have log-likelihoods of −226.38, −227.20, and −227.36. The third root therefore gives the maximum likelihood.

Exercise 2.4

Write

$$\hat{f} = \frac{\sum_u \tilde{P}_{uu} - \sum_u \tilde{p}_u^2}{1 - \sum_u \tilde{p}_u^2} = \frac{\tilde{X} - \tilde{Y}}{1 - \tilde{Y}}$$

where

$$\tilde{X} = \sum_u \tilde{P}_{uu} \quad , \quad \tilde{Y} = \sum_u \tilde{p}_u^2$$

and

$$X = \sum_u P_{uu} \quad , \quad Y = \sum_u p_u^2$$

The derivatives of $\hat{f}$ with respect to any variable z may be found from

$$\begin{aligned} \frac{\partial \hat{f}}{\partial z} &= \frac{\partial \hat{f}}{\partial \tilde{X}} \frac{\partial \tilde{X}}{\partial z} + \frac{\partial \hat{f}}{\partial \tilde{Y}} \frac{\partial \tilde{Y}}{\partial z} \\ &= \frac{1}{1 - \tilde{Y}} \frac{\partial \tilde{X}}{\partial z} + \frac{\tilde{X} - 1}{(1 - Y)^2} \frac{\partial Y}{\partial z} \end{aligned}$$

Taking derivatives with respect to homozygote counts, heterozygote counts, and total sample size, and evaluating them at expected values of counts:

$$\frac{\partial \tilde{X}}{\partial n_{uu}} = \frac{1}{n} \quad , \quad \frac{\partial \tilde{Y}}{\partial n_{uu}} = \frac{2p_u}{n}$$
$$\frac{\partial \tilde{X}}{\partial n_{uv}} = 0 \quad , \quad \frac{\partial \tilde{Y}}{\partial n_{uv}} = \frac{p_u + p_v}{n}, \ v \neq u$$
$$\frac{\tilde{X}}{\partial n} = -\frac{X}{n} \quad , \quad \frac{\tilde{Y}}{\partial n} = -\frac{2Y}{n}$$

Putting these into the derivatives for $\hat{f}$:

$$\begin{aligned}
\frac{\partial \hat{f}}{\partial n_{uu}} &= \frac{1}{1-Y}\frac{1}{n} + \frac{X-1}{(1-Y)^2}\frac{2p_u}{n} \\
\frac{\partial \hat{f}}{\partial n_{uv}} &= \frac{1}{1-Y}0 + \frac{X-1}{(1-Y)^2}\frac{p_u+p_v}{n}, \quad v \neq u \\
\frac{\partial \hat{f}}{\partial n} &= \frac{1}{1-Y}\frac{-X}{n} + \frac{X-1}{(1-Y)^2}\frac{-2Y}{n}
\end{aligned}$$

As a special case, suppose there is Hardy-Weinberg equilibrium, $f = 0$ and $X = Y$ (this is not saying that $\hat{f} = 0$ or that $\tilde{X} = \tilde{Y}$). Then

$$\begin{aligned}
\frac{\partial \hat{f}}{\partial n_{uu}} &= \frac{1-2p_u}{n(1-Y)} \\
\frac{\partial \hat{f}}{\partial n_{uv}} &= -\frac{p_u+p_v}{n(1-Y)} \\
\frac{\partial \hat{f}}{\partial n} &= \frac{Y}{n(1-Y)}
\end{aligned}$$

Putting these derivatives into Fisher's approximate variance formula:

$$\begin{aligned}
n\text{Var}(\hat{\text{f}}) &= \sum_u \left(\frac{1-2p_u}{1-Y}\right)^2 p_u^2 + \sum_u \sum_{v \neq u} \left(\frac{p_u+p_v}{1-Y}\right)^2 p_u p_v - \left(\frac{Y}{1-Y}\right)^2 \\
&= \frac{1}{1-\sum_u p_u^2}\left(\sum_u p_u^2 - 2\sum_u p_u^3 + (\sum_u p_u^2)^2\right)
\end{aligned}$$

CHAPTER 3

Exercise 3.1

The sample allele frequencies are $\tilde{p}_A = 0.75, \tilde{p}_a = 0.25$, allowing the calculation of expected genotypic counts:

	AA	Aa	aa	Total
Observed	6	3	1	10
Expected	5.625	3.750	0.625	10
Obs. − Exp.	0.375	−0.750	0.375	0.00

The chi-square test statistic is

$$\begin{aligned} X^2 &= \frac{(0.375)^2}{5.625} + \frac{(-0.750)^2}{3.750} + \frac{(0.375)^2}{0.625} \\ &= 0.40 \end{aligned}$$

The hypothesis of Hardy-Weinberg frequencies is not rejected, although the chi-square test should not be applied for expected numbers less than 1.

A better test is provided by an exact test. There are three possible genotypic arrays with the same allelic counts:

AA	Aa	aa	Probability
5	5	0	0.5201
6	3	1	0.4334
7	1	2	0.0464

The probability of the observed data, or a less likely dataset, when Hardy-Weinberg holds, is therefore 0.4798 and the hypothesis, once again, is not rejected.

Exercise 3.2

For a sample of $2n$ gametes with gametic counts $n_{AB}, n_{Ab}, n_{aB}, n_{ab}$, the unconstrained likelihood is

$$L_1 = \frac{n!}{n_{AB}!n_{Ab}!n_{aB}!n_{ab}!}(p_{AB})^{n_{AB}}(p_{Ab})^{n_{Ab}}(p_{aB})^{n_{aB}}(p_{ab})^{n_{ab}}$$

and the likelihood constrained by the hypothesis of linkage equilibrium is

$$\begin{aligned} L_0 &= \frac{n!}{n_{AB}!n_{Ab}!n_{aB}!n_{ab}!}(p_A p_B)^{n_{AB}}(p_A p_b)^{n_{Ab}}(p_a p_B)^{n_{aB}}(p_a p_b)^{n_{ab}} \\ &= \frac{n!}{n_{AB}!n_{Ab}!n_{aB}!n_{ab}!}(p_A)^{n_A}(p_a)^{n_a}(p_B)^{n_B}(p_b)^{n_b} \end{aligned}$$

where $n_A = n_{AB} + n_{Ab}$ etc. From multinomial theory, these likelihoods are maximized when gamete or allele frequencies are set to their observed values, and the ratio of these maximum likelihoods is

$$\frac{L_0}{L_1} = \frac{(n_A)^{n_A}(n_a)^{n_a}(n_B)^{n_B}(n_b)^{n_b}}{(2n)^{2n}(n_{AB})^{n_{AB}}(n_{Ab})^{n_{Ab}}(n_{aB})^{n_{Ab}}(n_{ab})^{n_{ab}}}$$

For the *Bam*HI and *Xho*I data in Table 1.6, this statistic becomes

$$\frac{L_0}{L_1} = \frac{(11)^{11}(6)^6(11)^{11}(6)^6}{(17)^{17}(5)^5(6)^6(6)^6(0)^0} = \frac{(11)^{22}}{(5)^5(17)^{17}}$$

and taking minus twice the logarithm gives a one degree of freedom chi-square statistic:

$$-2\ln\left(\frac{L_0}{L_1}\right) = 6.92$$

The likelihood ratio test therefore leads to rejection of the hypothesis, as did the exact and chi-square goodness-of-fit tests in Table 3.4.

Exercise 3.3

Assuming Hardy-Weinberg, gametic frequencies may be estimated from the EM algorithm. The sample allele frequencies are $\tilde{p}_A = \tilde{p}_B = 0.75$, the sample gametic frequency $\tilde{p}_{AB}$ is 0.6147, and the estimated linkage disequilibrium coefficient is $\tilde{D}_{AB} = 0.0522$. The corresponding chi-square test statistic is

$$\begin{aligned} X^2_{AB} &= \frac{40 \times (0.0522)^2}{0.75 \times 0.25 \times 0.75 \times 0.25} \\ &= 3.10 \end{aligned}$$

and the hypothesis of linkage equilibrium is not rejected.

If Hardy-Weinberg is not assumed, the composite disequilibrium coefficient is

$$\begin{aligned} \tilde{\Delta}_{AB} &= \frac{18+2+3+1}{20} - 2 \times 0.75 \times 0.75 \\ &= 0.075 \end{aligned}$$

The one-locus disequilibrium coefficients are

$$\begin{aligned} \tilde{D}_A &= \frac{12}{20} - (0.75)^2 = 0.0375 \\ \tilde{D}_B &= \frac{13}{20} - (0.75)^2 = 0.0875 \end{aligned}$$

and the test statistic is

$$\begin{aligned} X^2_{AB} &= \frac{20 \times (0.075)^2}{(0.75 \times 0.25 + 0.0375)(0.75 \times 0.25 + 0.0875)} \\ &= 1.82 \end{aligned}$$

The hypothesis of equilibrium is not rejected.

Note that it appears that locus B is not in Hardy-Weinberg equilibrium, so the two estimates (0.052 and 0.0750) are not of the same quantity.

Exercise 3.4

For any allele, $0 \le P_{ii} \le p_i$. If $P_{ii} = p_i^2 + D_{ii}$ this means that

$$-p_i^2 \le D_{ii} \le p_i(1-p_i)$$

The maximum value of D_{ii} is $p_i(1-p_i)$.

For three equally frequent alleles, $p_i = 1/3$ and half the maximum disequilibrium is $1/9$. The six genotype frequencies are, therefore:

Observed	Expected	Obs. – Exp.
$P_{11} = 2/9$	$p_1^2 = 1/9$	$P_{11} - p_1^2 = 1/9$
$P_{22} = 2/9$	$p_2^2 = 1/9$	$P_{22} - p_2^2 = 1/9$
$P_{33} = 2/9$	$p_3^2 = 1/9$	$P_{33} - p_3^2 = 1/9$
$P_{12} = 1/9$	$2p_1p_2 = 2/9$	$P_{12} - 2p_1p_2 = -1/9$
$P_{13} = 1/9$	$2p_1p_3 = 2/9$	$P_{13} - 2p_1p_3 = -1/9$
$P_{23} = 1/9$	$2p_2p_3 = 2/9$	$P_{23} - 2p_2p_3 = -1/9$

Using these population frequencies, the form of the chi-square test statistic becomes

$$\begin{aligned} X^2 &= n[3\frac{(1/9)^2}{1/9} + 3\frac{(-1/9)^2}{2/9}] \\ &= n/2 \end{aligned}$$

For 3 degrees of freedom, the noncentrality parameter is 14.2 for 90% power and 5% significance level. Hence

$$\begin{aligned} n/2 &\geq 14.2 \\ n &\geq 28.4 \end{aligned}$$

Exercise 3.5

The sample correlation coefficient r_{AB} is

$$\begin{aligned} r_{AB} &= \frac{\sum_i(x_iy_i) - \frac{1}{2n}\sum_i x_i \sum_i y_i}{\sqrt{[\sum_i x_i^2 - \frac{1}{2n}(\sum_i x_i)^2][\sum_i y_i^2 - \frac{1}{2n}(\sum_i y_i)^2]}} \\ &= \frac{\tilde{p}_{AB} - \tilde{p}_A\tilde{p}_B}{\sqrt{\tilde{p}_A(1-\tilde{p}_A)\tilde{p}_B(1-\tilde{p}_B)}} \\ &= \frac{\tilde{D}_{AB}}{\sqrt{\tilde{p}_A(1-\tilde{p}_A)\tilde{p}_B(1-\tilde{p}_B)}} \end{aligned}$$

From Equation 3.10, the chi-square test statistic for linkage disequilibrium is, therefore,

$$X_{AB}^2 = (2n)r_{AB}^2$$

CHAPTER 4

Exercise 4.1

	Locus 1	Locus 2	Locus 3	Average
Population 1	$\tilde{H}_{11} = 0.2$	$\tilde{H}_{12} = 0.6$	$\tilde{H}_{13} = 0.0$	$\tilde{H}_{1.} = 0.2667$
Population 2	$\tilde{H}_{21} = 0.6$	$\tilde{H}_{22} = 0.6$	$\tilde{H}_{23} = 0.0$	$\tilde{H}_{2.} = 0.4000$

l	$\tilde{H}_{il}$	$\tilde{H}_{il} - \tilde{H}_{i.}$	l, l'	$\tilde{H}_{i:l,l'}$
Population $i = 1$				
1	0.2	−0.0667	1, 2	0.2
2	0.6	0.3333	1, 3	0.0
3	0.0	−0.2667	2, 3	0.0
Population $i = 2$				
1	0.6	0.2000	1, 2	0.2
2	0.6	0.2000	1, 3	0.0
3	0.0	−0.4000	2, 3	0.0

In population 1,

$$\begin{aligned}
\mathrm{Var}(\tilde{H}_{1.}) &\hat{=} \frac{1}{45}[(0.16 + 0.24 + 0) + 2(0.08 + 0.0 + 0.0)] \\
&= 0.0124 \\
s_H^2 &= \frac{1}{2}(0.0044 + 0.1111 + 0.0711) \\
&= 0.0933 \\
s_H^2/3 &= 0.0311
\end{aligned}$$

In population 2,

$$\begin{aligned}
\mathrm{Var}(\tilde{H}_{2.}) &\hat{=} \frac{1}{45}[(0.24 + 0.24 + 0) + 2(-0.08 + 0.0 + 0.0)] \\
&= 0.0071 \\
s_H^2 &= \frac{1}{2}(0.04 + 0.04 + 0.16) \\
&= 0.12 \\
s_H^2/3 &= 0.04
\end{aligned}$$

For the analysis of variance:

$$\begin{aligned}
SS_1 &= 3.4667 \\
SS_2 &= 4.6667 \\
SS_3 &= 5.2000 \\
SS_4 &= 5.6000 \\
SS_5 &= 10.0000 \\
C &= 3.3333
\end{aligned}$$

so that

Source	df	SS	MS
Populations	1	0.1333	0.1333
Individuals/Popns	8	1.2000	0.1500
Loci	2	1.8667	0.9333
Loci by Popns	2	0.2667	0.1333
Loci by Indivs/Popns	16	3.2000	0.2000

Only the variance component for loci by individuals within populations is non-negative. The estimated variance of average heterozygosity is the mean square for populations divided by $mn = 15$, or 0.0089.

Exercise 4.2

The various gene diversities are:

	$l = 1$	$l = 2$	$l = 3$	Total
$i = 1$	$p_B = 0.7$	$p_A = 0.7$	$p_A = 1.0$	
	$p_D = 0.3$	$p_B = 0.2$		
		$p_C = 0.1$		
	$\tilde{D}_{11} = 0.42$	$\tilde{D}_{12} = 0.46$	$\tilde{D}_{13} = 0.00$	$\tilde{D}_{1.} = 0.88$
$i = 2$	$p_B = 0.6$	$p_A = 0.5$	$p_A = 1.0$	
	$p_D = 0.2$	$p_B = 0.3$		
	$p_E = 0.2$	$p_C = 0.1$		
		$p_D = 0.1$		
	$\tilde{D}_{21} = 0.56$	$\tilde{D}_{22} = 0.64$	$\tilde{D}_{23} = 0.00$	$\tilde{D}_{2.} = 1.20$
Total	$\tilde{D}_{.1} = 0.98$	$\tilde{D}_{.2} = 1.10$	$\tilde{D}_{.3} = 0.00$	$\tilde{D}_{..} = 2.08$

leading to the analysis of variance

Source	df	SS	MS
Populations	1	0.7381−0.7210=0.0171	0.0171
Loci	2	1.0852−0.7210=0.3642	0.1821
Loci by Popns.	2	1.1112−0.7381−1.0852+0.7210=0.0089	0.0045

The estimated variance component for a population average diversity is $\sigma_p^2 + \frac{1}{m}\sigma_{lp}^2$. Either by estimating each component separately, or by dividing the population mean square by $m = 3$, the estimate is 0.0057.

Taking the variance over single-locus diversities, in population 1:

$$s_D^2/3 = 0.0216$$

and for population 2:

$$s_D^2/3 = 0.0405$$

Exercise 4.3

Suppose n_l individuals are scored at locus l, and that $n_{ll'}$ individuals are scored at both loci l and l'. Then the covariance of the two single-locus heterozygosities is modified from Equation 4.2 to

$$\mathrm{Cov}(\tilde{H}_l, \tilde{H}_{l'}) = \frac{n_{ll'}}{n_l n_{l'}}(H_{ll'} - H_l H_{l'})$$

because

$$\begin{aligned}
\tilde{H}_l &= \frac{1}{n_l}\sum_j x_{jl} \\
\tilde{H}_{l'} &= \frac{1}{n_{l'}}\sum_j x_{jl'} \\
\mathcal{E}\sum_j x_{jl}x_{jl'} &= n_{ll'}H_{ll'}
\end{aligned}$$

Similarly, Equation 4.3 becomes

$$\begin{aligned}
\mathrm{Var}(\tilde{H}) &= \frac{1}{m^2}\sum_l \frac{H_l(1-H_l)}{n_l} \\
&+ \frac{1}{m^2}\sum_l \sum_{l' \neq l} \frac{n_{ll'}}{n_l n_{l'}}(H_{ll'} - H_l H_{l'})
\end{aligned}$$

CHAPTER 5

Exercise 5.1

The contingency tables of observed (and expected) allele counts are:

Locus B	
Allele 3	Allele 4
1 (0.9)	9 (9.1)
2 (0.7)	6 (7.3)
1 (0.5)	5 (5.5)
0 (0.5)	6 (5.5)
0 (0.7)	8 (7.3)
0 (0.6)	7 (6.4)

Locus C		
Allele 2	Allele 3	Allele 4
1 (0.7)	6 (3.5)	1 (3.7)
1 (0.7)	2 (3.5)	5 (3.7)
0 (0.6)	2 (2.7)	4 (2.8)
0 (0.6)	4 (2.7)	2 (2.8)
1 (0.7)	3 (3.5)	4 (3.7)
1 (0.7)	2 (3.1)	4 (3.3)

Locus D

Allele 1	Allele 2	Allele 3	Allele 4
0 (0.9)	0 (0.7)	6 (5.0)	4 (3.4)
0 (0.7)	2 (0.5)	3 (4.0)	3 (2.7)
1 (0.5)	0 (0.4)	3 (3.0)	2 (2.0)
0 (0.5)	0 (0.4)	3 (3.0)	3 (2.0)
1 (0.6)	0 (0.5)	5 (3.5)	1 (2.4)
2 (0.6)	1 (0.5)	2 (3.5)	2 (2.4)

The three chi-square test statistics are: locus B, 5.08 (5 d.f.), locus C, 8.63 (10 d.f.) and locus D, 15.30 (15 d.f.). None of these values are significant at the 5% level.

For a random-populations analysis, for locus B Table 5.3 becomes

Source	d.f.	S.S.	$\mathcal{E}$(MS)
Between	5	0.4111	0.0822
Within	39	3.2333	0.0829

and $n_c = 7.45$. From Equation 5.2, $\hat{\theta} = -0.0011$. Note that this is not a multiple of the chi-square goodness-of-fit statistic because Equation 5.2 provides an unbiased estimate and the relation between chi-square and $\hat{\theta}$ holds only for large numbers of large samples.

Estimates for each of the loci are:

Locus	Allele	$\hat{\theta}$
B	3	−0.0011
	4	−0.0011
	All	−0.0011
C	2	−0.1069
	3	0.0465
	4	0.0364
	All	0.0202
D	1	0.0345
	2	0.0537
	3	−0.0503
	4	−0.0901
	All	−0.0424
All	All	−0.0107

Amalgamation of estimates over alleles and loci was shown in Equation 5.3. Loci that are not polymorphic do not contribute to the overall estimate.

Exercise 5.2

There are four ways of partitioning the three "3" alleles between the two samples: (3,0), (2,1), (1,2), and (0,3). From Equation 5.1, the probabilities of these four events, conditional on the fact that there were three "3"s and 15 "4"s, are 5/34, 15/34, 35/102 and 7/102. None of the outcomes would cause rejection of the hypothesis of equal allele frequencies.

Exercise 5.3

Euclidean distances:

$$d_{12} = 0.1414 \quad , \quad d_{34} = 0.1414$$

Distance takes into account the differences of allele frequencies, so both pairs of populations are equally distant.

Nei's distances:

$$D_{12} = 0.0200 \quad , \quad D_{34} = 0.0075$$

Distance reflects actions of drift and mutation. Mutation is a more rapid force for frequencies near 1.0 than for frequencies near 0.5, so populations 3 and 4 are much closer than populations 1 and 2.

Coancestry distance:

$$\theta_{12} = 0.0198 \quad , \quad \theta_{34} = 0.0541$$

Distance reflects action of drift only. Drift is a more rapid force for frequencies near 0.5 than for frequencies near 1.0, so populations 1 and 2 are much closer than populations 3 and 4.

Exercise 5.4

For locus 4:

allele	$\hat{F}$	$\hat{\theta}$	$\hat{f}$	$\hat{\sigma^2}_P$	$\hat{\sigma^2}_I$	$\hat{\sigma^2}_G$
1	−.0082	.0633	−.0764	.0073	−.0083	.1167
2	−.0082	.0633	−.0764	.0073	−.0083	.1167
Total	−.0082	.0633	−.0764			

For locus 5:

allele	$\hat{F}$	$\hat{\theta}$	$\hat{f}$	$\hat{\sigma^2}_P$	$\hat{\sigma^2}_I$	$\hat{\sigma^2}_G$
1	−.0226	−.0107	−.0118	−.0002	−.0003	.0222
2	−.0226	−.0107	−.0118	−.0002	−.0003	.0222
Total	−.0226	−.0107	−.0118			

For locus 6:

allele	$\hat{F}$	$\hat{\theta}$	$\hat{f}$	$\hat{\sigma^2}_P$	$\hat{\sigma^2}_I$	$\hat{\sigma^2}_G$
1	.5659	.2922	.3868	.0860	.0806	.1278
2	.0058	.0450	−.0411	.0086	−.0075	.1889
3	.0056	.0898	−.0925	.0105	−.0099	.1167
4	−.1497	.0493	−.2093	.0067	−.0269	.1556
Total	.2009	.1517	.0581			

For locus 7:

allele	$\hat{F}$	$\hat{\theta}$	$\hat{f}$	$\hat{\sigma^2}_P$	$\hat{\sigma^2}_I$	$\hat{\sigma^2}_G$
1	.1111	.0527	.0617	.0102	.0113	.1722
2	−.0645	−.0031	−.0612	−.0004	−.0074	.1278
3	.3877	.1720	.2605	.0187	.0235	.0667
Total	.1325	.0676	.0696			

Over all loci:

$\hat{F}$	$\hat{\theta}$	$\hat{f}$
.1403	.1077	.0365

Bootstrapping over loci:

	$\hat{F}$	$\hat{\theta}$	$\hat{f}$
Lower 95% limit	−.0144	.0316	−.0764
Upper 95% limit	.2009	.1517	.0656

CHAPTER 6

Exercise 6.1

Uncle/niece Suppose X and Y are full sibs, with parents G and H. Individuals Y and Z have an offspring W, so that X and W are related as uncle and niece. Assuming no coancestry between Z and X, the coancestry calculations are

$$\begin{aligned}
\theta_{XW} &= \frac{1}{2}(\theta_{XY} + \theta_{XZ}) \\
&= \frac{1}{4}(\theta_{XG} + \theta_{XH}) \\
&= \frac{1}{8}(\theta_{GG} + 2\theta_{GH} + \theta_{HH}) \\
&= \frac{1}{16}(2 + F_G + F_H) \\
&= \frac{1}{8}
\end{aligned}$$

Parent/offspring Suppose X and Y have offspring Z. Then the coancestry of parent X and offspring Z is

$$\begin{aligned}
\theta_{XZ} &= \frac{1}{2}(\theta_{XX} + \theta_{XY}) \\
&= \frac{1}{4}(1 + F_X) \\
&= \frac{1}{4}
\end{aligned}$$

First cousins Now suppose individuals A and B have offspring C and D. Individuals C and G have offspring X, and individuals D and H have offspring Y. Then X and Y are first cousins. If G, H are unrelated to C, D, then

$$\begin{aligned}
\theta_{XY} &= \frac{1}{4}(\theta_{GD} + \theta_{GH} + \theta_{CD} + \theta_{CH}) \\
&= \frac{1}{4}\theta_{CD} \\
&= \frac{1}{16}(\theta_{AA} + \theta_{AB} + \theta_{BA} + \theta_{BB}) \\
&= \frac{1}{32}(2 + F_A + F_B) \\
&= \frac{1}{16}
\end{aligned}$$

Exercise 6.2

There are three nonzero δ measures for uncle $X(a, b)$ and nephew $Y(c, d)$. If c is the gene Y received from the sib of X:

$$\delta_0 = 1/2, \ \ \delta_{ac} = 1/4, \ \ \delta_{bc} = 1/4$$

Putting these into Table 6.2 provides the following frequencies for the uncle/nephew pair:

$$\begin{aligned}
AA, AA &: p_A^3(1 + p_A)/2 \\
AA, Aa &: p_A^2 p_a(1 + 2p_A) \\
AA, aa &: 2p_A^2 p_a^2 \\
Aa, Aa &: p_A p_a(p_A + p_a + 4p_A p_a)/2 \\
Aa, aa &: p_A p_a^2(1 + 2p_a) \\
aa, aa &: p_a^3(1 + p_a)/2
\end{aligned}$$

It may be easier to follow these calculations by setting out all the intermediate steps:

ab	cd	$\delta_0 = \frac{1}{2}$	$\delta_{ac} = \frac{1}{4}$	$\delta_{bc} = \frac{1}{4}$
AA	AA	p_A^4	p_A^3	p_A^3
AA	Aa	$p_A^3 p_a$	$p_A^2 p_a$	$p_A^2 p_a$
AA	aA	$p_A^3 p_a$	–	–
AA	aa	$p_A^2 p_a^2$	–	–
Aa	AA	$p_A^3 p_a$	$p_A^2 p_a$	–
Aa	Aa	$p_A^2 p_a^2$	$p_A p_a^2$	–
Aa	aA	$p_A^2 p_a^2$	–	$p_A^2 p_a$
Aa	aa	$p_A p_a^3$	–	$p_A p_a^2$
aA	AA	$p_A^3 p_a$	–	$p_A^2 p_a$
aA	Aa	$p_A^2 p_a^2$	–	$p_A p_a^2$
aA	aA	$p_A^2 p_a^2$	$p_A^2 p_a$	–
aA	aa	$p_A p_a^3$	$p_A p_a^2$	–
aa	AA	$p_A^2 p_a^2$	–	–
aa	Aa	$p_A p_a^3$	–	–
aa	aA	$p_A p_a^3$	$p_A p_a^2$	$p_A p_a^2$
aa	aa	p_a^4	p_a^3	p_a^3

Exercise 6.3

If independence between all 14 alleles in the profile is assumed, the conditional frequencies are

Population	$\theta = 0.00$	$\theta = 0.03$
Afr. Am.	4.6×10^{-8}	5.3×10^{-5}
Caucasian	1.3×10^{-5}	4.2×10^{-3}

Exercise 6.4

When the alleged father AF of type A_jA_k is a relative of the true father R, the chance that R would contribute the obligate paternal allele A_j to the child is

$$P = \Pr(R \to A_j | AF = A_jA_k) = \Pr(R \to A_j, AF = A_jA_k)/\Pr(AF = A_jA_k)$$

If $X = 1/\Pr(AF = A_jA_k)$, this probability becomes

$$\begin{aligned} P &= X\Pr(R = A_jA_j, AF = A_jA_k) + \frac{X}{2}\Pr(R = A_jA_k, AF = A_jA_k) \\ &\quad + \frac{X}{2}\sum_{m\neq j,k}\Pr(R = A_jA_m, AF = A_jA_k) \end{aligned}$$

The joint frequencies for uncle/nephew were given in Exercise 6.2. Therefore

$$\begin{aligned} P &= Xp_j^2p_k(1+2p_j) + \frac{X}{4}p_jp_k(p_j+p_k+4p_jp_k) \\ &\quad + \frac{X}{4}p_jp_k(1-p_j-p_k)(1+4p_j) \\ &= \frac{p_j(2+p_j)}{2} \end{aligned}$$

and the paternity index is

$$L = \frac{\frac{1}{2}\times\frac{1}{2}}{\frac{1}{2}\times\frac{p_j(2+p_j)}{2}} = \frac{1}{p_j(2+p_j)}$$

For first cousins, AF(a,b) and R(c,d) where a,c are the two alleles that may both have come from the common grandparents of AF, R: $\delta_{ac}=1/4, \delta_0=3/4$ so

$$\begin{aligned} \Pr(R=A_jA_j, AF=A_jA_k) &= \frac{1}{4}p_j^2p_k(1+6p_j) \\ \Pr(R=A_jA_k, AF=A_jA_k) &= \frac{1}{4}p_jp_k(p_j+p_k+12p_jp_k) \\ \Pr(R=A_jA_m, AF=A_jA_k) &= \frac{1}{4}p_jp_kp_m(1+12p_j) \end{aligned}$$

so

$$\begin{aligned} P &= \frac{X}{4}p_j^2p_k(1+6p_j) + \frac{X}{8}p_jp_k(p_j+p_k+12p_jp_k) \\ &\quad + \frac{X}{8}p_jp_k(1-p_j-p_k)(1+12p_j) \\ &= \frac{1+14p_j}{16} \end{aligned}$$

and the paternity index is

$$L = \frac{\frac{1}{2}\times\frac{1}{2}}{\frac{1}{2}\times\frac{1+14p_j}{16}} = \frac{8}{1+14p_j}$$

Exercise 6.5

The suspect is not excluded, on the basis of the DNA evidence, from being the perpetrator of the crime. The two scenarios are

C: Suspect contributed to mixed sample

$\bar{C}$: Suspect did not contribute to mixed sample

In either case, it is known that the victim contributed to the mixed sample. Under C, the mixed sample is certain to occur, but under $\bar{C}$ it is necessary that another person provided allele c. This person may have had genotypes ac, bc, or cc since each of these cases will give the observed profile for the mixture.

The likelihood ratio is

$$\begin{aligned} L &= \frac{\Pr_\phi(\phi|abc)}{\Pr_1(c|abc)} \\ &= \frac{1}{2p_a p_c + 2p_b p_c + p_c^2} \end{aligned}$$

If the system being used is one with "null" alleles n, the unknown perpetrator may also have been of type cn. The frequency of such nulls is unknown, but certainly

$$p_a + p_b + p_c + p_n \leq 1$$

so

$$\begin{aligned} L &= \frac{1}{p_c(2p_a + 2p_b + p_c + 2p_n)} \\ &\geq \frac{1}{p_c(2p_a + 2p_b + 2p_c + 2p_n)} \\ &\geq \frac{1}{2p_c} \end{aligned}$$

A more realistic bound would be obtained when an upper bound on p_n is used.

CHAPTER 7

Exercise 7.1

As an approximation, since the level of recombination is small, assume double heterozygotes are all formed from parental gametes. There are then three classes of F_2 progeny: the 1136 double homozygotes formed from the union of two parental gametes, the 1041 double heterozygotes formed from the union of two parental gametes and the 10 formed from the union of one parental and one recombinant gamete. The likelihood for the recombination fraction c is

$$\begin{aligned} L(c) &\propto \left[\left(\frac{1-c}{2}\right)^2\right]^{1136} \left[2\left(\frac{1-c}{2}\right)^2\right]^{1041} \left[2\left(\frac{c(1-c)}{4}\right)\right]^{10} \\ \ln L(c) &= \text{constant} + 4364\ln(1-c) + 10\ln(c) \\ \frac{\partial L(c)}{\partial c} &= -\frac{4364}{1-c} + \frac{10}{c} \end{aligned}$$

so that

$$\begin{aligned} \hat{c} &= \frac{10}{4374} \\ &= 0.0023 \end{aligned}$$

Exercise 7.2

When the two parental inbred lines are $AABB$ and $aabb$ with A dominant to a and B dominant to b, the gametic array from the F_1 double heterozygotes AB/ab is

$$\frac{1-c}{2}AB + \frac{c}{2}Ab + \frac{c}{2}aB + \frac{1-c}{2}ab$$

so the F_2 phenotypic and genotypic arrays are:

Phenotype	Genotype	Frequency
AB	$AABB$	$\left(\frac{1-c_{AB}}{2}\right)^2$
	$AABb$	$2\left(\frac{1-c_{AB}}{2}\frac{c_{AB}}{2}\right)$
	$AaBB$	$2\left(\frac{1-c_{AB}}{2}\frac{c_{AB}}{2}\right)$
	$AaBb$	$2\left(\frac{1-c_{AB}}{2}\right)^2 + 2\left(\frac{c_{AB}}{2}\right)^2$
Ab	$AAbb$	$\left(\frac{c_{AB}}{2}\right)^2$
	$Aabb$	$2\left(\frac{1-c_{AB}}{2}\frac{c_{AB}}{2}\right)$
aB	$aaBB$	$\left(\frac{c_{AB}}{2}\right)^2$
	$aaBb$	$2\left(\frac{1-c_{AB}}{2}\frac{c_{AB}}{2}\right)$
ab	$aabb$	$\left(\frac{1-c_{AB}}{2}\right)^2$

If the phenotypic array has x, y, z, w of AB, Ab, aB, ab phenotypes, with $x+y+z+w=n$, the likelihood of the recombination fraction c_{AB} is

$$L(c_{AB}) \propto \left(\frac{3-2c_{AB}+c_{AB}^2}{4}\right)^x \left(\frac{c_{AB}(2-c_{AB})}{4}\right)^{y+z} \left(\frac{(1-c_{AB})^2}{4}\right)^w$$

Note that it is not possible to have $c_{AB}=0$ when $y+z>0$. The first derivative of the log-likelihood is

$$\frac{\partial \ln L(c_{AB})}{\partial c_{AB}} = \frac{-2x(1-c_{AB})}{3-2c_{AB}+c_{AB}^2} + \frac{2(y+z)(1-c_{AB})}{c_{AB}(2-c_{AB})} + \frac{-2w}{1-c_{AB}}$$

The maximum likelihood estimate of c_{AB} is that value, in the range $(0, 0.5)$ that maximizes $L(c_{AB})$. It will cause the first derivative to be zero if c_{AB} is strictly between 0 and 0.5.

The algebra can be simplified with the substitution $\theta = (1-c_{AB})^2$. Then the likelihood becomes

$$L(\theta) \propto (2+\theta)^x(1-\theta)^{y+z}\theta^w$$

The first derivative of the log-likelihood is zero when

$$n\theta^2 - (x-2y-2z-w)\theta - 2w = 0$$

and then c_{AB} can be estimated as

$$\hat{c}_{AB} = 1 - \sqrt{\hat{\theta}}$$

For $x = 30, y = 15, z = 14, w = 1$, the quadratic in θ is

$$60\theta^2 + 29\theta - 2 = 0$$

with solution $\hat{\theta} = 0.06121$ and $1 - \hat{c}_{AB} = 0.2474$ or $\hat{c}_{AB} = 0.5$. For valid values of c_{AB} (i.e. $1 \geq \theta \geq 0.25$, the first and second derivatives of the log-likelihood are positive. In this range the likelihood is maximized at $c_{AB} = 0.5$. The usual chi-square test for linkage has the test statistic

$$\begin{aligned} X^2 &= (x - 3y - 3z + 9w)^2/9n \\ &= 4.27 \end{aligned}$$

This is significant, indicating *more* recombination than $c_{AB} = 0.5$. There is no evidence for linkage.

If w is zero, so that no double recessives are seen, the derivative becomes

$$\frac{\partial \ln L(c_{AB})}{\partial c_{AB}} = -2(1 - c_{AB})\left(\frac{x}{3 - 2c_{AB} + c_{AB}^2} - \frac{(n - x)}{c_{AB}(2 - c_{AB})}\right)$$

If $x < 2n/3$, this cannot be zero, the derivative is positive, and the maximum likelihood estimate of c_{AB} is 0.5. If $x = 2n/3$ the derivative is zero when $c_{AB} = 0.5$. In other words, the estimated recombination fraction is

$$\hat{c}_{AB} = \begin{cases} 0.5, & x < 2n/3 \\ 1 - \sqrt{(3x - 2n)/n}, & x \geq 2n/3 \end{cases}$$

For $x = 30, y = 17, z = 13, w = 0$, the estimate is $\hat{c}_{AB} = 0.5$. Note that the chi-square test statistic for linkage between the two loci is

$$\begin{aligned} X^2 &= (x - 3y - 3z + 9w)^2/9n \\ &= 6.67 \end{aligned}$$

This significant result occurs because there is evidence for *more* recombination than expected if $c_{AB} = 0.5$. There is no evidence of linkage.

Exercise 7.3

When the parental inbred lines are $AAbb$ and $aaBB$ with A dominant to a and B dominant to b, the gametic array from the F_1 double heterozygotes Ab/aB is

$$\frac{1 - c_{AB}}{2}Ab + \frac{c_{AB}}{2}AB + \frac{c_{AB}}{2}ab + \frac{1 - c_{AB}}{2}aB$$

so the F_2 phenotypic and genotypic arrays are:

Phenotype	Genotype	Frequency
AB	$AABB$	$\left(\frac{c_{AB}}{2}\right)^2$
	$AABb$	$2\left(\frac{1-c_{AB}}{2}\frac{c_{AB}}{2}\right)$
	$AaBB$	$2\left(\frac{1-c_{AB}}{2}\frac{c_{AB}}{2}\right)$
	$AaBb$	$2\left(\frac{1-c_{AB}}{2}\right)^2 + 2\left(\frac{c_{AB}}{2}\right)^2$
Ab	$AAbb$	$\left(\frac{1-c_{AB}}{2}\right)^2$
	$Aabb$	$2\left(\frac{1-c_{AB}}{2}\frac{c_{AB}}{2}\right)$
aB	$aaBB$	$\left(\frac{1-c_{AB}}{2}\right)^2$
	$aaBb$	$2\left(\frac{1-c_{AB}}{2}\frac{c_{AB}}{2}\right)$
ab	$aabb$	$\left(\frac{c_{AB}}{2}\right)^2$

If the phenotypic array has x, y, z, w of AB, Ab, aB, ab phenotypes, with $x+y+z+w=n$, the likelihood of the recombination fraction is

$$L(c_{AB}) \quad \propto \quad \left(\frac{2+c_{AB}^2}{4}\right)^x \left(\frac{1-c_{AB}^2}{4}\right)^{y+z} \left(\frac{c_{AB}^2}{4}\right)^w$$

Note that it is not possible to have $c_{AB}=0$ when $w>0$. The first derivative of the log-likelihood is

$$\frac{\partial \ln L(c_{AB})}{\partial c_{AB}} \quad = \quad \frac{2xc_{AB}}{2+c_{AB}^2} + \frac{-2(y+z)c_{AB}}{1-c_{AB}^2} + \frac{2w}{c_{AB}}$$

The maximum likelihood estimate of c_{AB} is that value, in the range $(0, 0.5)$ that maximizes $L(c_{AB})$. It will cause the first derivative to be zero if c_{AB} is strictly between 0 and 0.5.

The algebra can be simplified with the substitution $\theta = c_{AB}^2$. Then the likelihood becomes

$$L(\theta) \quad \propto \quad (2+\theta)^x(1-\theta)^{y+z}\theta^w$$

The first derivative of the log-likelihood is zero when

$$n\theta^2 - (x-2y-2z-w)\theta - 2w \quad = \quad 0$$

and then c_{AB} can be estimated as

$$\hat{c}_{AB} \quad = \quad \sqrt{\hat{\theta}}$$

For $x=30, y=15, z=14, w=1$, the quadratic in θ is

$$60\theta^2 + 29\theta - 2 \quad = \quad 0$$

with solution $\hat{\theta} = 0.06121$ and $\hat{c}_{AB} = 0.2474$. The usual chi-square test for linkage has the test statistic

$$\begin{aligned} X^2 &= (x - 3y - 3z + 9w)^2/9n \\ &= 4.27 \end{aligned}$$

This is significant, indicating *less* recombination than $c = 0.5$. There is evidence for linkage.

If w is zero, so that no double recessives are seen, the derivative becomes

$$\frac{\partial \ln L(c_{AB})}{\partial c_{AB}} = 2c_{AB}\left(\frac{x}{2 + c_{AB}^2} - \frac{(n - x)}{1 - c_{AB}^2}\right)$$

which is zero for $c_{AB} = 0$ or $c_{AB} = \sqrt{(3x - 2n)/n}, x > 2n/3$. When $x \leq 2n/3$, the only valid solution is $c_{AB} = 0$. The only individuals known to be recombinants (phenotype ab) are not seen, so there are no obligate recombinants in the sample.

For $x = 30, y = 17, z = 13, w = 0$, the estimated recombination fraction is 0. The chi-square test statistic is 6.67, as before, but now the result is significant because c_{AB} is *less* than 0.5.

Exercise 7.4

From the text:

$$\Pr(T : M_1, NT : M_2 | T : D) = m(1 - m) + (1 - c - m)\mathcal{D}/p$$

and similarly

$$\Pr(T : M_1, NT : M_2 | T : N) = m(1 - m) - (1 - c - m)\mathcal{D}/(1 - p)$$

If $\Pr(T : M_1, NT : M_2)$ is the probability that the M_1M_2 father transmits M_1 to an affected child

$$\begin{aligned} \Pr(T : M_1, NT : M_2) &= \frac{\Pr(T : M_1, NT : M_2, \text{child affected})}{\Pr(\text{child affected})} \\ &= \frac{\Pr(M_1, NT : M_2 | T : D)[f_{DD}p^2 + f_{DN}p(1 - p)]}{f_{DD}p^2 + 2f_{DN}p(1 - p) + f_{NN}(1 - p)^2} \\ &+ \frac{\Pr(M_1, NT : M_2 | T : N)[f_{ND}p(1 - p) + f_{NN}(1 - p)^2]}{f_{DD}p^2 + 2f_{DN}p(1 - p) + f_{NN}(1 - p)^2} \end{aligned}$$

The probability that an M_1M_3 mother also transmits M_1 to the affected child must take into account the genotype of the child. If the child is DD, for example, both parents must have transmitted M_1D. If the child is DN, however, one parent must transmit M_1D and the other M_1N but it is not known which is which. The disease genotypes of the parents and child are not known.

Exercise 7.5

Using the proportions of times a "12" or "21" is seen as a measure of distance for each pair of loci:

	1	2	3	4	5
1	–	.27	.70	.61	.81
2	.27	–	.57	.78	.60
3	.70	.57	–	.43	.23
4	.61	.78	.43	–	.42
5	.81	.60	.23	.42	–

The sar values for all 5!/2=60 possible orders are as follows, and show that the order with smallest sar is 12354.

Order					sar	Order					sar
1	2	3	4	5	1.69	2	3	1	4	5	2.30
1	2	3	5	4	1.49*	2	3	1	5	4	2.50
1	2	4	3	5	1.71	2	3	4	1	5	2.42
1	2	4	5	3	1.70	2	3	5	1	4	2.22
1	2	5	3	4	1.53	2	4	1	3	5	2.32
1	2	5	4	3	1.72	2	4	1	5	3	2.43
1	3	2	4	5	2.47	2	4	3	1	5	2.72
1	3	2	5	4	2.29	2	4	5	1	3	2.71
1	3	4	2	5	2.51	2	5	1	3	4	2.54
1	3	4	5	2	2.15	2	5	1	4	3	2.45
1	3	5	2	4	2.31	2	5	3	1	4	2.14
1	3	5	4	2	2.13	2	5	4	1	3	2.33
1	4	2	3	5	2.19	3	1	2	4	5	2.17
1	4	2	5	3	2.22	3	1	2	5	4	1.99
1	4	3	2	5	2.21	3	1	4	2	5	2.69
1	4	3	5	2	1.87	3	1	5	2	4	2.89
1	4	5	2	3	2.20	3	2	1	4	5	1.87
1	4	5	3	2	1.83	3	2	1	5	4	2.07
1	5	2	3	4	2.41	3	2	4	1	5	2.77
1	5	2	4	3	2.62	3	2	5	1	4	2.59
1	5	3	2	4	2.39	3	4	1	2	5	1.91
1	5	3	4	2	2.25	3	4	2	1	5	2.29
1	5	4	2	3	2.58	3	5	1	2	4	2.09
1	5	4	3	2	2.23	3	5	2	1	4	1.71
2	1	3	4	5	1.82	4	1	2	3	5	1.68
2	1	3	5	4	1.62	4	1	3	2	5	2.48
2	1	4	3	5	1.54	4	2	1	3	5	1.98
2	1	4	5	3	1.53	4	2	3	1	5	2.86
2	1	5	3	4	1.74	4	3	1	2	5	2.00
2	1	5	4	3	1.93	4	3	2	1	5	2.08

Seriation requires only 5 orders, but each one requires four decisions. These orders are as follows, and confirm that 12354 shares the minimum sar.

Start	1st	2nd	3rd	4th	sar
1	12	412	3412	53412	1.54
2	12	123	1235	12354	1.49
3	35	354	2354	12354	1.49
4	45	453	1453	21453	1.53
5	35	354	2354	12354	1.49

CHAPTER 8

Exercise 8.1

From Equation 8.1, $\hat{t}$ must satisfy

$$\frac{6}{\hat{t}} - \frac{523.5065}{1-0.5599\hat{t}} - \frac{136.5984}{1-0.8432\hat{t}} - \frac{49.8650}{1-0.9973\hat{t}} - \frac{85.7428}{1-0.5996\hat{t}}$$

To begin a numerical search for a solution, the value in Equation 8.2 is used:

$$\hat{t} = \frac{6}{523.5065 + 136.5984 + 49.8650 + 85.7428} = 0.0075$$

The following values were found for the left-hand side of the first equation, showing that the maximum likelihood estimate is between 0.00750 and 0.00751.

$\hat{t}$	LHS	$\hat{t}$	LHS
.00741	10.21	.00751	-0.623
.00742	9.114	.00752	-1.691
.00743	8.020	.00753	-2.755
.00744	6.930	.00754	-3.817
.00745	5.842	.00755	-4.876
.00746	4.757	.00756	-5.933
.00747	3.676	.00757	-6.986
.00748	2.597	.00758	-8.037
.00749	1.520	.00759	-9.085
.00750	0.447	.00760	-10.13

Exercise 8.2

The estimated proportions are

$$\hat{\gamma}_{21} = 0.0807$$
$$\hat{\gamma}_{22} = 0.1478$$
$$\hat{\gamma}_{23} = 0.2286$$

so the observed and expected heterozygous offspring counts are

	A_1A_1	A_1A_2	A_2A_2
Observed	65	173	119
Expected	63.03	115.43	178.54

The chi-square statistic is 68.63, which is clearly significant.

CHAPTER 9

Exercise 9.1

The base and dinucleotide counts are

	A	T	C	G	Total
A	28	15	19	30	92
T	7	17	27	57	108
C	32	42	25	4	103
G	25	34	33	42	134

and the trinucleotide counts are

	A	T	C	G	Total
A A	4	6	5	13	28
A T	1	3	5	6	15
A C	4	5	7	2	18
A G	7	7	2	14	30
T A	2	3	2	0	7
T T	2	4	5	6	17
T C	12	7	7	1	27
T G	11	10	16	20	57
C A	11	5	8	8	32
C T	4	5	8	25	42
C C	5	15	5	0	25
C G	1	2	0	1	4
G A	11	1	4	9	25
G T	0	5	9	20	34
G C	11	15	6	1	33
G G	6	14	15	7	42

These results, together with the tetra- and penta-nucleotide counts, lead to the following version of Table 9.7. As expected from the results for the chicken sequence, this rabbit sequence may also be described by a first-order Markov chain.

k	BIC(k)	Log-likelihood(k)	No. parameters	Chi-square (d.f.)
1	1224.27	−603.01	3	
2	1186.52	−556.78	12	46.23*
3	1344.31	−526.29	48	30.49
4	2068.02	−450.78	192	75.51
5	5265.19	−300.55	768	150.23

Exercise 9.2

For the human sequence, the 16 dinucleotide counts, and corresponding base counts (and symmetrized counts) [and odds ratios] are:

	A	T	C	G	Base
A	8 (8) [.263]	6 (7) [.228]	7 (5) [.322]	4 (4) [.225]	50 (44)
T	7 (7) [.228]	7 (8) [.263]	2 (4) [.225]	3 (4) [.258]	38 (44)
C	5 (4) [.258]	3 (4) [.225]	5 (3) [.354]	1 (1) [.118]	29 (24)
G	5 (4) [.225]	3 (5) [.322]	1 (1) [.118]	1 (3) [.354]	19 (24)

From such tables for each of the five sequences, the distance matrix D_δ is found to be

$$D_\delta = \begin{bmatrix} .000 & .014 & .022 & .095 & .058 \\ .014 & .000 & .023 & .104 & .055 \\ .022 & .023 & .000 & .089 & .071 \\ .095 & .104 & .089 & .000 & .131 \\ .058 & .055 & .071 & .131 & .000 \end{bmatrix}$$

CHAPTER 10

Exercise 10.1

The UPGMA tree (Figure C.1) has distances 0.041 from LS4 or LS6 to A, 0.104 from LS7 to B, 0.062 from A to B, 0.161 from LP1 or LA1 to C, 0.200 from LJ1 or LH1 to D, 0.010 from D to E, 0.060 from C to F, and 0.011 from E to F. (PHYLIP)

The neighbor-joining tree (Figure C.2) has distances 0.012 from LS6 to A, 0.071 from LS4 to A, 0.101 from A to B, 0.066 from B to LS7, 0.092 from B to E, 0.010 from E to D, 0.204 from D to LJ1, 0.196 from D to LH1, 0.072 from E to C, 0.156 from C to LA1, and 0.165 from C to LP1. (PHYLIP)

Exercise 10.2

Results are given here just for the serine data, omitting the site where Orangutan has a gap.

The UPGMA tree (Figure C.3) has distances 0.046 from Human or Orangutan to A, 0.050 from Gibbon to B, 0.004 from A to B, 0.065 from Chimpanzee or Gorilla to C, 0.033 from B to D, and 0.018 from C to D. (PHYLIP)

The neighbor-joining tree (Figure C.4) has distances 0.033 from Orangutan to A, 0.059 from Gibbon to A, 0.039 from A to C, 0.016 from C to Human, 0.039 from C to B, 0.056 from B to Chimpanzee and 0.072 from B to Gorilla. (PHYLIP)

The maximum parsimony tree (PHYLIP) has the same structure as in Figure C.4. A total of 18 changes are required: one at sites 9, 10, 11, 12, 17, 25, 28, 29, 32, 36, 39, two at sites 30 and 31, and three at site 40.

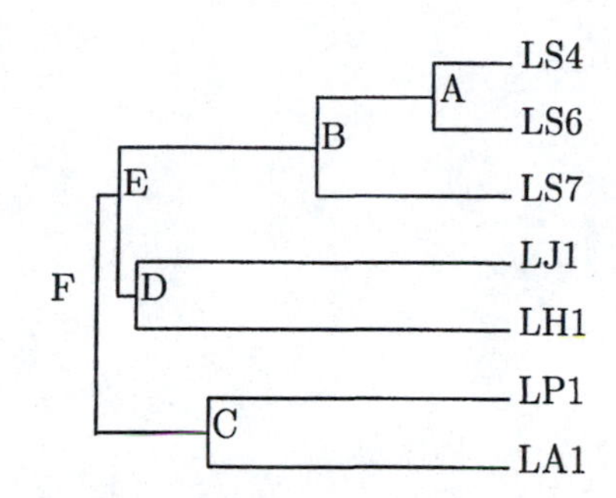

Figure C.1

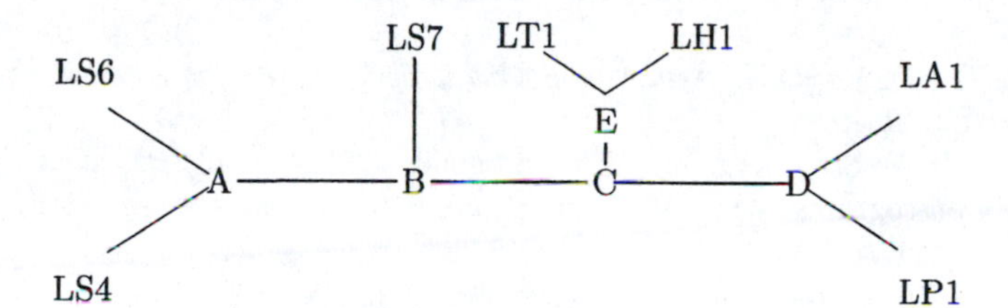

Figure C.2

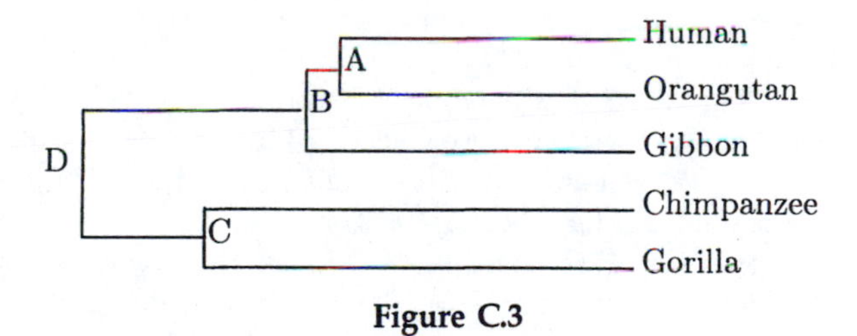

Figure C.3

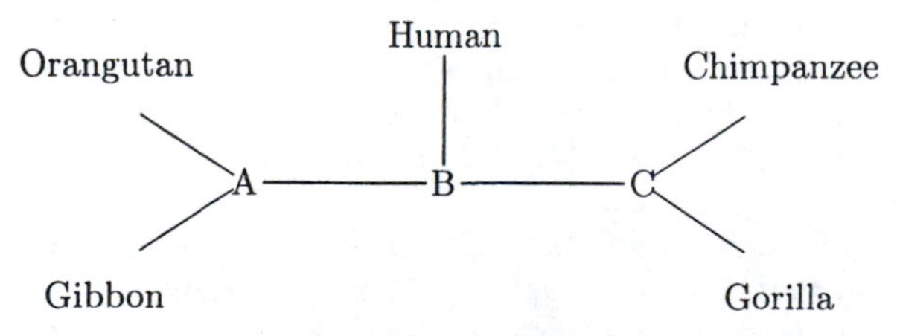

Figure C.4

Bibliography

Abramowitz, M. and I.A. Stegun (Editors). 1970. *Handbook of Mathematical Functions with Formulas, Graphs, and Mathematical Tables.* National Bureau of Standards Applied Math. Series 55. U.S. Government Printing Office, Washington, DC.

Aitken, C.G.G. 1995. *Statistics and the Evaluation of Evidence for Forensic Scientists.* Wiley, New York.

Agard, D.A., R.A. Steinberg and R.M. Stroud. 1981. Quantitative analysis of electrophoretograms: A mathematical approach to super-resolution. Anal. Biochem. 111:257–268.

Allard, R.W., A.L. Kahler and B.S. Weir. 1972. The effect of selection on esterase allozymes in a barley population. Genetics 72:489–503.

Altschul, S.F. 1991. Amino acid substitution matrices from an information theoretic perspective. J. Mol. Biol. 219:555–565.

Altschul, S.F. 1993. A protein alignment scoring system sensitive at all evolutionary distances. J. Mol. Biol. 36:290–300.

Altschul, S.F., M.S. Boguski, W. Gish and J.C. Wootton. 1994. Issues in searching molecular sequence databases. Nature Genet. 6:119–129.

Altschul, S.F., W. Gish, W. Miller, E.W. Myers and D. Lipman. 1990. Basic local alignment search tool. J. Mol. Biol. 215:403–410.

Anderson, S., A.T. Bankier, B.G. Barrell, M.H.L. de Bruijn, A.R. Coulson, J. Drouin, I.C. Eperon, D.P. Nierlich, B.A. Roe, F. Sanger, P.H. Schreier, A.J.H. Smith, R. Staden and I.G. Young. 1981. Sequence and organization of the human mitochondrial genome. Nature 290:457–465.

Arnold, J., A.J. Cuticchia, D.A. Newsome, W.W. Jennings III and R. Ivarie. 1988. Mono- through hexanucleotide composition of the sense strand of yeast DNA: A Markov chain analysis. Nucleic Acids Res. 16:7145–7158.

Arratia, R. and M.S. Waterman. 1985. An Erdös-Rényi law with shifts. Adv. Math. 55:13–23.

Bailey, N.T.J. 1951. Testing the solubility of maximum likelihood equations in the routine application of scoring methods. Biometrics 7:268–274.

Bailey, N.T.J. 1961. *Introduction to the Mathematical Theory of Genetic Linkage.* Oxford Univ. Press, Oxford.

Balakrishnan, V. and L.D. Sanghvi. 1968. Distance between populations on the basis of attribute data. Biometrics 24:859–865.

Baldi, P., Y. Chauvin, T. Hunkapillar and M. McClure. 1994. Hidden markov modles of biological primary sequence information. Proc. Natl. Acad. Sci. USA. 91:1059–1063.

Balding, D.J. and R.A. Nichols. 1994. DNA profile match frequency calculations: how to allow for population stratification, relatedness, database selection and single bands. Forensic Sci. Int. 64:125–140.

Barker, J.S.F., P.D. East and B.S. Weir. 1986. Temporal and microgeographic variation in allozyme frequencies in a natural population of *Drosophila buzzatii.* Genetics 112:577–611.

Bennett, J.H. 1954. On the theory of random mating. Ann. Eugen. 18:311–317.

Bernstein, F. 1925. Zusammenfassende Betrachutungen über die erblichen Blutenstructuren des Menschen. Z. Abstamm. Vererbgsl. 37:237–270.

Bickeböller, H. and F. Clerget-Darpoux. 1995. Statistical properties of the allelic and genotypic transmission/disequilibrium test for multiallelic markers. Genet. Epidem. 12:865–870.

Bishop, D.T., J.A. Williamson and M.H. Skolnick. 1983. A model for restriction fragment length distribution. Am. J. Human Genet. 35:795–815.

Bishop, M.J. and E.A. Thompson. 1986. Maximum likelihood alignment of DNA sequences. J. Mol. Biol. 190:159–165.

Brooks, L.D., B.S. Weir and H.E. Schaffer. 1988. The probabilities of similarities in DNA sequence comparisons. Genomics 3:207–216.

Brown, A.H.D., A.C. Matheson and K.G. Eldridge. 1975. Estimation of the mating system of *Eucalyptus obliqua* L'Hérit. by using allozyme polymorphisms. Aust. J. Bot. 23:931–949.

Brown, W.M., E.M. Prager, A. Wang and A.C. Wilson. 1982. Mitochondrial DNA sequences of primates: Tempo and mode of evolution. J. Mol. Evol. 18:225–239.

Buetow, K.H. and A. Chakravarti. 1987a. Multipoint gene mapping using seriation. I. General methods. Am. J. Hum. Genet. 41:180–188.

Buetow, K.H. and A. Chakravarti. 1987b. Multipoint gene mapping using seriation. I. Analysis of simulated and empirical data. Am. J. Hum. Genet. 41:189–201.

Buri, P. 1956. Gene frequency in small populations of mutant *Drosophila.* Evolution 10:367–402.

Casella, G. 1985. An introduction to empirical Bayes data analysis. Am. Statist. 39:83–87.

Cavalli-Sforza, L.L. and W.F. Bodmer. 1971. *The Genetics of Human Populations.* Freeman, San Francisco.

Cavalli-Sforza, L.L. and A.W.F. Edwards. 1967. Phylogenetic analysis: Models and estimation procedures. Am. J. Human Genet. 19:233–257.

Ceppellini, R., M. Siniscalco and C.A.B. Smith. 1955. The estimation of gene frequencies in a random-mating population. Ann. Human Genet. 20:97–115.

Chakraborty, R. 1977. Estimation of time of divergence from phylogenetic studies. Can. J. Genet. Cytol. 19:217–213.

Chakraborty, R. 1986. Estimation of linkage disequilibrium from conditional haplotype data: Application to β-globin gene cluster in American Blacks. Genet. Epidemiol. 3:323–333.

Chakraborty, R., M. Shaw and W.J. Schull. 1974. Exclusion of probability: The current state of the art. Am. J. Human Genet. 26:477–488.

Chakravarti, A., C.C. Li and K.H. Buetow. 1984. Estimation of the marker gene frequency and linkage disequilibrium from conditional marker data. Am. J. Human Genet. 36:177–186.

Cheliak, W.M., K. Morgan, C. Strobeck, F.C.H. Yeh and B.P. Dancik. 1983. Estimation of mating system parameters in plant populations using the EM algorithm. Theor. Appl. Genet. 65:157–161.

Chimera, J.A., C.R. Harris and M. Litt. 1989. Population genetics of the highly polymorphic locus D16S7 and its use in paternity testing. Am. J. Human Genet. 45:926–931.

Christiansen, F.B. and O. Frydenberg. 1973. Selection component analysis of natural polymorphisms using population samples including mother-offspring combinations. Theor. Pop. Biol. 4:425–445.

Clegg, M.T. and R.W. Allard. 1973. The genetics of electrophoretic variants in *Avena*. II. The esterase E_1, E_2, E_4, E_5, E_6 and anodal peroxidase APX_4 loci in *A. fatua*. J. Hered. 64:2–6.

Clegg, M.T., A.L. Kahler and R.W. Allard. 1978. Estimation of life cycle components of selection in an experimental plant population. Genetics 89:765–792.

Cockerham, C.C. 1967. Group inbreeding and coancestry. Genetics 56:89–104.

Cockerham, C.C. 1969. Variance of gene frequencies. Evolution 23:72–84.

Cockerham, C.C. 1973. Analyses of gene frequencies. Genetics 74:679–700.

Cockerham, C.C. 1984. Drift and mutation with a finite number of allelic states. Proc. Natl. Acad. Sci. U.S.A. 81:530–534.

Cockerham, C.C. and B.S. Weir. 1986. Estimation of inbreeding parameters in stratified populations. Ann. Human Genet. 50:271–281.

Cockerham, C.C. and B.S. Weir. 1987. Correlations, descent measures: Drift with migration and mutation. Proc. Natl. Acad. Sci. U.S.A. 84:8512–8514.

Cockerham, C.C. and B.S. Weir. 1993. Estimation of gene flow from F-statistics. Evolution 47:855–863.

Cochran, W.G. 1954. Some methods for strengthening the common χ^2 tests. Biometrics 10:417–451.

Cox, D.R., M. Burmeister, E.R. Price, S. Kim and R.M. Myers. 1960. Radiation hybrid mapping: a somatic cell genetic method for constructing high resolution maps of mammalian chromosomes. Science 250:245–250.

Crow, J.F. and M. Kimura. 1970. *An Introduction to Population Genetics Theory.* Burgess, Minneapolis,MN.

Cuticchia, A.J., R. Ivarie and J. Arnold. 1992. The application of Markov chain analysis to oligonucleotide frequency prediction and physical mapping of *Drosophila melanogaster*. Nucleic Acids Res. 20:3651–3657.

Dayhoff, M.O. 1979. Survey of new data and computer methods of analysis. Pp. 1–8 in *Atlas of Protein Sequence and Structure.* Volume 5, Supplement 3, 1978. National Biomedical Foundation, Washington, DC.

Dayhoff, M.O., R.M. Schwartz and B.C. Orcutt. 1979. A model of evolutionary change in proteins. Pp. 345–351 in *Atlas of Protein Sequence and Structure*, Volume 5, Supplement 3, 1978. National Biomedical Research Foundation, Washington, DC.

Degens, P.O. 1983. Hierarchical cluster methods as maximum likelihood estimators. Pp. 249–253 in *Numerical Taxonomy*, J. Felsenstein (Editor), Springer-Verlag, Berlin.

Dempster, A.P., N.M. Laird and D.B. Rubin. 1977. Maximum likelihood from incomplete data via the EM algorithm. J. Roy. Stat. Soc. B 39:1–38.

Dobzhansky, T. and H. Levene. 1951. Development of heterosis through natural selection in experimental populations of *Drosophila pseudoobscura*. Am. Nat. 85:247–263.

Dodds, K.G. 1986. *Resampling Methods in Genetics and the Effect of Family Structure in Genetic Data*. Institute of Statistics Mimeo Series 1684T, North Carolina State University, Raleigh, NC.

Doerge, R.W. 1995. Testing for linkage: phase known/unknown. J. Heredity 86:61–62.

Doolittle, R.F. (Editor). 1990. *Molecular Evolution: Computer Analysis of Protein and Nucleic Acid Sequences.* Methods in Enzymology Volume 138. Academic Press, San Diego.

DuMouchel, W.H. and W.W. Anderson. 1968. The analysis of selection in experimental populations. Genetics 58:435–449.

Edwards, A.W.F. 1986. Are Mendel's results really too close? Biol. Rev. 61:295–312.

Edwards, A.W.F. 1992. *Likelihood.* Expanded Edition. Johns Hopkins Press, Baltimore.

Edwards, A.W.F. and L.L. Cavalli-Sforza. 1963. The reconstruction of evolution. Heredity 18:553.

Edwards, A.W.F. and L.L. Cavalli-Sforza. 1964. Reconstruction of evolutionary trees. Pp. 67–76 in *Phenetic and Phylogenetic Classification* (Systematics Association Publication No. 6), V.E. Heywood and J. McNeill (Editors). Systematics Association, London.

Efron, B. 1982. *The Jackknife, the Bootstrap and Other Resampling Plans.* CBMS-NSF Regional Conference Series in Applied Mathematics, Monograph 38. SIAM, Philadelphia.

Efron, B. and R.J. Tibshirani. 1993. *An Introduction to the Bootstrap.* Chapman and Hall, New York.

Ehm, M.G., M. Kimmel and R.W. Cottingham. 1996. Error detection for genetic data, using likelihood methods. Am. J. Hum. Genet. 58:225–234.

Elder, J.K. and E.M. Southern. 1987. Computer-aided analysis of one-dimensional restriction fragment gels. Pp. 165–172 in *Nucleic Acid and Protein Sequence Analysis,* M.J. Bishop and C.J. Rawlings (Editors), IRL Press, Oxford.

Elston. R.C. 1986a. Probability and paternity testing. Am. J. Human Genet. 39:112–122.

Elston, R.C. 1986b. Some fallacious thinking about the paternity index: Comments.

Am. J. Human Genet. 39:670–672.

Elston, R.C. and J. Stewart. 1971. A general method for the analysis of pedigree data. Human Hered. 21:523–542.

Endler, J. 1986. *Natural Selection in the Wild.* Princeton Univ. Press, Princeton, NJ.

Erlich, H.A. (Editor). 1989. *PCR Technology: Principles and Applications for DNA Amplification.* Stockton, New York.

Estabrook, G.F., C.S. Johnson and F.R. McMorris. 1975. An idealized concept of the true cladistic character. Math. Biosci. 23:263–272.

Evett, I.W. 1987. Bayesian inference and forensic science: problems and perspectives. The Statistician 36:99–105.

Evett, I.W. 1990. The theory of interpreting scientific transfer evidence. Pp. 141–180 in *Forensic Science Progress* 4, A. Maehley and R.L. Willams (Editors), Springer-Verlag, Berlin.

Evett, I.W. 1992. DNA statistics: putting the problems into perspective. Jurimetrics J. 33:139–145.

Evett, I.W. 1995. Avoiding the transposed conditional. Science and Justice 35:127–131.

Ewens, W.J. 1983. The role of models in the analysis of molecular genetic data, with particular reference to restriction fragment data. Pp. 45–73 in *Statistical Analysis of DNA Sequence Data,* B.S. Weir (Editor). Dekker, New York.

Felsenstein, J. 1978a. The number of evolutionary trees. Syst. Zool. 27:27–33.

Felsenstein, J. 1978b. Cases in which parsimony or compatibility methods will be positively misleading. Syst. Zool. 27:401–410.

Felsenstein, J. 1981. Evolutionary trees from DNA sequences: A maximum likelihood approach. J. Mol. Evol. 17:368–376.

Felsenstein, J. 1983. Methods for inferring phylogenies: A statistical view. Pp. 315–334 in *Numerical Taxonomy*, J. Felsenstein (Editor). Springer-Verlag, Berlin.

Felsenstein, J. 1985. Confidence limits on phylogenies: An approach using the bootstrap. Evolution 39:783–791.

Felsenstein, J. 1988. Phylogenies from molecular sequences: Inference and reliability. Annu. Rev. Genet. 22:521–565.

Felsenstein, J. 1993. *PHYLIP (Phylogeny Inference Package)*, version 3.5c. Department of Genetics, University of Washington, Seattle.

Finney, D.J., R. Latscha, B.M. Bennett and P. Hsu. 1963. *Tables for Testing Significance in a 2 × 2 Contingency Table.* Cambridge Univ. Press, Cambridge.

Fisher, R.A. 1925. *Statistical Methods for Research Workers*, 13th Edition, Hafner, New York.

Fisher, R.A. 1935. The logic of inductive inference. J. Roy. Stat. Soc. 98:39–54.

Fisher, R.A. 1936a. Has Mendel's work been rediscovered? Annals of Science 1:115–137.

Fisher, R.A. 1936. "The coefficient of racial likeness" and the future of craniometry. J. Roy. Anthrop. Inst. 66:57–63.

Fisher, R.A. and B. Balmakund. 1928. The estimation of linkage from the offspring of selfed heterozygotes. J. Genet. 20:79–92.

Fitch, W.M. 1983. Random sequences. J. Mol. Biol. 163:171–176.

Fitch, W.M. 1971. Toward defining the course of evolution: Minimum change for a specific tree topology. Syst. Zool. 20:406–416.

Fitch, W.M. and M. Margoliash. 1967. Construction of phylogenetic trees. Science 155:279–284.

Fleischmann, R.D., M.D. Adams, O. White et al. 1995. Whole-genome random sequencing and assembly of *Haemophilus influenzae* Rd. Science 269:496–512.

Fu, Y.X. and J. Arnold. 1990. A table of exact sample sizes for use with Fisher's exact test for 2 × 2 tables. Biometrics 48:1103–1112.

Fyfe, J.L. and N.T.J. Bailey. 1951. Plant breeding studies in leguminous forage crops. I. Natural cross-breeding in winter beans. J. Agric. Sci. 41:371–378.

Gaut, B.S. and B.S. Weir. 1994. Detecting substitution-rate heterogeneity among regions of a nucleotide sequence. Mol. Biol. Evol. 11:620–629.

Gelfand, A.E. 1971. Seriation. Pp. 186–201 in *Mathematics in the Archaelogical and Historical Sciences*, F.R. Hodson, D.G. Kendall and P. Tauta (Editors), Edinburgh Univ. Press, Edinburgh.

Gnedenko, B.V. 1963. *The Theory of Probability.* Chelsea, New York.

Goldman, N. 1990. Maximum likelihood inference of phylogenetic trees, with special reference to a Poisson process model of DNA substitution and to parsimony analyses. Syst. Zool. 39:345–361.

Good, P. 1994. *Permutation Tests.* Springer-Verlag, New York.

Goodman, L.A. 1960. On the exact variance of products. J. Am. Stat. Assoc. 57:54–60.

Goodman, M.M. 1972. Distance analysis in biology. Syst. Zool. 21:174–286.

Goodman, M.M., C.W. Stuber, K. Newton and H.H. Weissinger. 1980. Linkage relationships of 19 enzyme loci in maize. Genetics 96:697–710.

Gotoh, O. 1982. An improved algorithm for matching biological sequences. J. Mol. Biol. 162:705–708.

Green, A.G., A.H.D. Brown and R.O. Oram. 1980. Determination of outcrossing rate in a breeding population of *Lupinus albus* L. (White lupin). Z. Pflanzenzüchtg. 84:181–191.

Gunel, E. and S. Wearden. 1995. Bayesian estimation and testing of gene frequencies. Theor. Appl. Genet. 91:534–543.

Guo, S.W. and E.A. Thompson. 1992. Performing the exact test of Hardy-Weinberg proportion for multiple alleles. Biometrics 48:361–372.

Gürtler, H. 1956. Principles of blood group statistical evaluation of paternity cases at the University Institute of Forensic Medicine, Copenhagen. Acta Med. Leg. Soc., Liège 9:83–93.

Gyapay, G., J. Morissette, A. Vignal, C. Dib, C. Fizames, P. Millasseau, S. Marc, G. Bernadi, M. Lathrop and J. Weissenbach. 1994. The 1993-94 Généthon human genetic linkage map. Nature Genet. 7:246–339.

Haldane, J.B.S. 1919. The combination of linkage values and the calculation of distances between the loci of linked factors. J. Genet. 8:299–309.

Haldane, J.B.S. 1954. An exact test for randomness of mating. J. Genet. 52:631–635.

Haldane, J.B.S. 1956. The estimation of viabilities. J. Genet. 54:294–296.

Haldane, J.B.S. and C.A.B. Smith. 1947. A new estimate of the linkage between the genes for colour-blindness and haemophilia in man. Ann. Eugen. 14:10–31.

Handt, O., M. Richards, M. Trommsdorff, C. Kilger, J. Simanainen, O. Georgiev, K. Bauer, A. Stone, R. Hedges, W. Schaffner, G. Utermann, B. Sykes and S. Paäbo. 1994. Molecular genetic analysis of the Tyrolean ice man. Science 264:1775–1778.

Hartl, D.L. and A.G. Clark. 1989. *Principles of Population Genetics,* Second Edition. Sinauer, Sunderland, MA.

Hasegawa, M., H. Kishino and T. Yano. 1985. Dating of the human- ape splitting by a molecular clock of mitochondrial DNA. J. Mol. Evol. 22:160–174.

Haynam, G.E., Z. Govindarajulu and F.C. Leone. 1970. Tables of the cumulative non-central chi-square distribution. Pp. 1–78 in *Selected Tables in Mathematical Statistics,* Volume 1, H.L. Harter and D.B. Owen (Editors). Am. Math. Society, Providence, RI.

Hedges, S.B., S. Kumar, K. Tamura and M. Stoneking. 1992. Human origins and analysis of mitchondrial DNA sequences. Science 255:737–739.

Henikoff, S. and J.G. Henikoff. 1992. Amino acid substitution matrices from protein blocks. Proc. Natl. Acad. Sci. USA 89:10915–10919.

Henikoff, S. and J.G. Henikoff. 1994. Protein family classification based on searching a database of blocks. Genomics 19:97–107.

Hernández, J.L. and B.S. Weir. 1989. A disequilibrium coefficient approach to Hardy-Weinberg testing. Biometrics 45:53–70.

Hill, W.G., H.A. Babiker, L.C. Ranford-Cartwright and D. Walliker. 1995. Estimation of inbeeding coefficients from genotypic data on multiple alleles, and application to estimation of clonality in malaria parasites. Genet. Res. 65:53–61.

Hill, W.G. and B.S. Weir. 1988. Variances and covariances of squared linkage disequilibria in finite populations. Theor. Pop. Biol. 33:54–78.

Hill, W.G. and B.S. Weir. 1994. Maximum-likelihood estimation of gene location by linkage disequilibrium. Am. J. Hum. Genet. 54:705–714.

Hillis, D.M., C. Moritz and B.K. Mable. (Editors) 1996. *Molecular Systematics*, 2nd Edition, Sinauer, Sunderland, MA.

Hudson, R.R., M. Slatkin and W.P. Maddison. 1992. Estimation of levels of gene flow from DNA sequence data. Genetics 132:583–589.

Hudson, T.J. and 49 others. 1995. An STS-based map of the human genome. Science 270:1945–1954.

Hunt, L.T. and M.O. Dayhoff. 1982. Evolution of chromosomal proteins. Pp. 193–239 in *Macromolecular Sequences in Systematic and Evolutionary Biology*, M. Goodman (Editor). Plenum, New York.

Immer, F.R. 1930. Formulae and tables for calculating linkage intensities. Genetics 15:81–98.

Jeffreys, A.J., V. Wilson and S.L. Thein. 1985. Individual- specific "fingerprints" of human DNA. Nature (London) 316:76–79.

Jiang,C. 1987. Estimation of F-statistics in subdivided populations. Ph.D. Thesis, North Carolina State University.

Jones, D.T., W.R. Taylor and J.M. Thornton. 1992. The rapid generation of mutation data matrices from protein sequences. Comp. Appl. Biosci. 8:275–282.

Jorde, L.B. 1995. Linkage disequilibrium as a gene-mapping tool. Am. J. Hum. Genet. 56:11–14.

Jukes, T.H. and C.R. Cantor. 1969. Evolution in protein molecules. Pp. 21–123 in *Mammalian Protein Metabolism,* H.N. Munro (Editor). Academic Press, New York.

Kan, Y.W. and A.M. Dozy. 1978. Polymorphism of DNA sequence adjacent to the human β-globin structural gene: Relationship of sickle cell anemia. Proc. Natl. Acad. Sci. U.S.A. 75:5631–5635.

Kaplan, N.L. and B.S. Weir. 1992. Expected behavior of conditional linkage disequilibrium. Am. J. Hum. Genet. 51:333–343.

Kaplan, N.L. and B.S. Weir. 1995. Are moment bounds on the recombination fraction between a marker and a disease locus too good to be true? Allelic association mapping revisited for simple genetic diseases in the Finnish population. Am. J. Hum. Genet. 57:1486–1498.

Kaplan, N.L., W.G. Hill and B.S. Weir. 1995. Likelihood methods for locating disease genes in nonequilibrium populations. Am. J. Hum. Genet. 56:18–32.

Karlin, S. and S.F. Altschul. 1990. Methods for assessing the significance of molecular sequence features by using general scoring schemes. Proc. Natl. Acad. Sci. USA 87:2264–2268.

Karlin, S. and C. Burge. 1995. Dinucleotide relative abundance extremes: a genomic significance. Trends in Genetics 11:283–290.

Karlin, S. and L.R. Cardon. 1994. Computational DNA sequence analysis. Ann. Rev. Microbiol. 48:619–654.

Karlin, S., G. Ghandour, F. Ost, S. Tavaré and L.J. Korn. 1983. New approaches for computer analysis of nucleic acid sequences. Proc. Natl. Acad. Sci. U.S.A. 80:5660–5664.

Karlin, S. and I. Ladunga. 1994. Comparisons of eukaryotic genomic sequences. Proc. Natl. Acad. Sci. USA 91:12832–12836.

Karlin, S., I. Ladunga and B.E. Blaisdell. 1994. Heterogeneity of genomes: measures and values. Proc. Natl. Acad. Sci. USA 91:12837–12841.

Katz, R.W. 1981. On some criteria for estimating the order of a Markov chain. Technometrics 23:243–249.

Kendall, M.G. and A. Stuart. 1979. *The Advanced Theory of Statistics.* Volume 2. *Inference and Relationship.* Hafner, New York.

Kim, J. and M.A. Burgman. 1988. Accuracy of phylogenetic-estimation methods under unequal evolutionary rates. Evolution 42:596–602.

Kimura, M. 1980. A simple method for estimating evolutionary rates of base substitutions through comparative studies of nucleotide sequences. J. Mol. Evol. 16:111–120.

Kimura, M. and J.F. Crow. 1964. The number of alleles that can be maintained in a finite population. Genetics 49:725–738.

Kirkpatrick, S., C.D. Gelatt and M.P. Vecchi. 1983. Optimization by simulated annealing. Science 220:671–680.

Knuth, D.E. 1981. *The Art of Computer Programming.* Volume 2. *Seminumerical Algorithms*, Second Edition. Addison-Wesley, Reading, MA.

Kosambi, D.D. 1944. The estimation of map distances from recombination values. Ann. Eugen. 12:172–175.

Kreitman, M. 1983. Nucleotide polymorphism at the alcohol dehydrogenase locus of *Drosophila melanogaster.* Nature (London) 304:412–417.

Krogh, A., M. Brown, I.S. Mian, K. Sjölander and D. Haussler. 1994. Hidden Markov models in computational biology: applications to protein modelling. J. Mol. biol. 235:1501–1531.

Kruskal, J.B. 1983. An overview of sequence comparison. Pp. 1–44 in *Time Warps, String Edits, and Macromolecules.* D. Sankoff and J.B. Kruskal, (Editors). Addison-Wesley, Reading, MA.

Kuhner, M.K. and J. Felsenstein. 1994. A simulation comparison of phylogeny algorithms under equal and unequal evolutionary rates. Mol. Biol. Evol. 11:459–468.

Lander, E.S. 1989. Analysis with restriction enzymes. Pp. 35–51 in *Mathematical Methods for DNA Sequences*, M.S. Waterman (Editor). CRC Press, Boca Raton, FL.

Lange, K. 1995. Applications of the Dirichlet distribution to forensic match probabilities. Genetica 96:107–117.

Lange, K. and R.C. Elston. 1975. Extensions to pedigree analysis. I. Likelihood calculations for simple and complex pedigrees. Human Hered. 25:95–105.

Langley, C.H., Y.N. Tobari and K.I. Kojima. 1974. Linkage disequilibrium in natural populations of *Drosophila melanogaster.* Genetics 78:921–936.

Langley, C.H., E. Montgomery and W.F. Quattlebaum. 1982. Restriction map variation in the *Adh* region of *Drosophila.* Proc. Natl. Acad. Sci. U.S.A. 79:5631–5635.

Laurie-Ahlberg, C.C. and B.S. Weir. 1979. Allozyme variation and linkage disequilibrium in some laboratory populations of *Drosophila melanogaster.* Genetics 92:1295–1314.

Lawrence, C.E., S.E. Altschul, M.S. Boguski, J.S. Liu, A.F. Neuwald and J.C. Wootton. 1993. Detecting subtle sequence signals: a Gibbs sampling strategy for multiple alignment. Science 262:208–214.

Lempert, R. 1991. Some caveats concerning DNA as criminal evidence: with tnaks to the Reverend Bayes, Cardozo Law Rev. 13:303–341.

Lewin, B. 1994. *Genes V.* Wiley, New York.

Lewis, P.A.W., A.S. Goodman and J.M. Miller. 1969. A pseudo- random number generator for the system/360. IBM Syst. J. 8:136–146.

Lewontin, R.C. 1974. *The Genetic Basis of Evolutionary Change.* Columbia Univ. Press, New York.

Lewontin, R.C. 1988. On measures of gametic disequilibrium. Genetics 120:849–852.

Lewontin, R.C. and C.C. Cockerham. 1959. The goodness-of-fit test for detecting natural selection in random mating populations. Evolution 13:561–564.

Li, C.C. 1988. Pseudo-random mating populations. In celebration of the 80th anniversary of the Hardy-Weinberg law. Genetics 119:731–737.

Li, W-H. and D. Graur. 1991. *Fundamentals of Molecular Evolution.* Sinauer, Sunderland, MA.

Lincoln, S.E. and E.S. Lander. 1992. Systematic detection of errors in genetic linkage data. Genomics 14:604–610.

Lipman, D.J. and W.R. Pearson. 1985. Science 227:1435–1441.

Lockhart, P.J., M.A. Steel, M.D. Hendy and D. Penny. 1994. Recovering evolutionary trees under a more realistic model of sequence evolution. Mol. Biol. Evol. 11:605–612.

Macpherson, J.N., B.S. Weir and A.J. Leigh-Brown. 1990. Extensive linkage disequilibrium in the achaete-scute complex of *Drosophila melanogaster.* 126:121–129.

Maiste, P.J. 1993. *Comparison of Statistical Tests for Independence at Genetic Loci with many Alleles.* Ph.D. Thesis, North Carolina State University, Raleigh, NC.

Maiste, P.J. and B.S. Weir. 1992. Estimating linkage disequilibrium from conditional data. Am. J. Hum. Genet. 50:1139–1140.

Maiste, P.J. and B.S. Weir. 1995. A comparison of tests for independence in the FBI RFLP data bases. Genetica 96:125–138.

Maizel, J.V. Jr. and R.P. Lenk. 1981. Enhanced graphic matrix analysis of nucleic acid and protein sequences. Proc. Natl. Acad. Sci. U.S.A. 78:7665–7669

Majumder, P.P. and M. Nei. 1983. A note on the positive identification of paternity by using genetic markers. Human Hered. 33:29–35.

Manly, B.F.J. 1985. *The Statistics of Natural Selection on Animal Populations.* Chapman and Hall, London.

Manly, B.F.J. 1991. *Randomization and Monte Carlo Methods in Biology.* Chapman and Hall, New York.

Margush, T. and F.R. McMorris. 1981. Consensus n-trees. Bull. Math. Biol. 43:239–244.

Mather, K. 1951. *The Measurement of Linkage in Heredity,* Second Edition. Methuen, London.

Meagher, R.B., M.D. McLean and J. Arnold. 1988. Recombination within a subclass of restriction fragment length polymorphisms may help link classical and molecular genetics. Genetics 120:809–818.

Meagher, T.R. 1986. Analysis of paternity within a natural population of *Chamaelirium luteum.* I. Identification of most-likely male parents. Am. Nat. 128:199-215.

Mendel, G. 1866. Versuche über Pflanzen-Hybriden. Verhandlungen des naturforschenden Vereines in Brünn 4,3-47. (English translation by Royal Horticultural Society of London, reprinted in Peters, J.A. 1959. *Classic Papers in Genetics.* Prentice-Hall, Englewood Cliffs, NJ.)

Miyamoto, M.M. and J. Cracraft (Editors). 1991. *Phylogenetic Analysis of DNA Sequences.* Oxford Univ. Press, New York.

Morton, N.E. 1955. Sequential tests for the detection of linkage. Am. J. Human Genet. 7:277–318.

Morton, N.E. 1964. Genetic studies in northeastern Brazil. Cold Spring Harbor Symp. Quant. Biol. 29:69–79.

Mosimann, J.E. 1962. On the compound multinomial distribution, the multivariate beta distribution, and correlations among proportions. Biometrika 49:65–82.

Mott, R.F., T.B.L. Kirkwood and R.N. Curnow. 1989. A test for the statistical significance of DNA sequence similarities for application in databank searches. CABIOS 5:123–131.

Mourant, A.E., A.C. Kopec and K. Domaniewska-Sobczak. 1976. *The Distribution of the Human Blood Groups and Other Polymorphisms*, Second Edition. Oxford Univ. Press, Oxford.

Mukai, T., R.A. Cardellino, T.K. Watanabe and J.F. Crow. 1974. The genetic variance for viability and its components in a local population of *Drosphila melanogaster*. Genetics 78:1195–1208.

Mullis, K.B. and F. Faloona. 1987. Specific synthesis of DNA *in vitro* via a polymerase catalyzed chain reaction. Meth. Enzymol. 155:335–350.

Muse, S.V. 1995. Estimating synonymous and nonsynonymous substitution rates. Mol. Biol. Evol. 13:105–114.

Muse, S.V. and B.S. Gaut. 1994. A likelihood approach for comparing synonymous and nonsynonymous nucleotide substitution rates, with application to the chloroplast genome. Mol. Biol. Evol. 11:715–724.

Muse, S.V. and B.S. Weir. 1992. Testing for equality of evolutionary rates. Genetics 132:269–276.

Needleman, S.B. and C.D. Wunsch. 1970. A general method applicable to the search for similarities in the amino acid sequence of two proteins. J. Mol. Biol. 48:443–453.

Nei, M. 1972. Genetic distance between populations. Am. Nat. 106:283–292.

Nei, M. 1978. Estimation of average heterozygosity and genetic distance from a small number of individuals. Genetics 89:583–590.

Nei, M. 1987. *Molecular Evolutionary Genetics.* Columbia Univ. Press, New York.

Nei, M. and W-H. Li. 1980. Non-random association between electromporhs and inversion chromosomes in finite populations. Genet. Res. 35:65–83.

Nei, M., F. Tajima and Y. Tateno. 1983. Accuracy of estimated phylogenetic trees from molecular data. II. Gene frequency data. J. Mol. Evol. 19:153–170.

Novitski, C.E. 1995. Another look at some of Mendel's results. J. Heredity 86:62–66.

Nussinov, R. 1984. Doublet frequencies in evolutionary distinct groups. Nucleic Acids Res. 12:1749–1763.

Nussinov, R. 1987. Nucleotide quartets in the vicinity of eukaryotic transcriptional initiation sites: Some DNA and chromatin structural implications. DNA 6:613–622.

O'Hagan, A. 1994. *Kendall's Advanced Theory of Statistics.* Volume 2b. *Bayesian Inference.* Halstead Press, New York.

Oliver, S.G., Q.J.M. van der Art, M.L. Agostoni-Carbone et al. 1992. The complete sequence of yeast chromosome III. Nature 357:38–46.

Ott, J. 1974. Estimation of the recombination fraction in human pedigrees: Efficient computation of the likelihood for human linkage studies. Am. J. Human Genet. 34:630–649.

Ott, J. 1991. *Analysis of Human Genetic Linkage.* Second Edition. Johns Hopkins Press, Baltimore.

Ott, J. 1993. Detecting marker inconsistencies in human gene mapping. Hum. Hered. 43:25–30.

Pamilo, P. and M. Nei. 1988. Relationships between gene trees and species trees. Mol. Biol. Evol. 5:568–583.

Parker, R.C., R.M. Watson, and J. Vinograd. 1977. Mapping of closed circular DNAs by cleavage with restriction endonucleases and calibration by agarose gel electrophoresis. Proc. Natl. Acad. Sci. U.S.A. 74:851-855.

Pearson, W.R. 1982. Automatic construction of restriction site maps. Nucleic Acids Res. 10:217–227.

Pearson, W.R. 1990. Rapid and sensitive sequence comparison with FASTP and FASTA. Meth. Enzymology 183:63–98.

Pearson, W.R. and D.J. Lipman. 1988. Improved tools for biological sequence comparison. Proc. Natl. Acad. Sci. USA 85:2444–2448.

Piegorsch, W.W. 1983. The questions of fit in the Gregor Mendel controversy. Commun. Statist. Theor. Meth. 12:2289–2304.

Piegorsch, W.W. 1986. The Gregor Mendel controversy: Early issues of goodness-of-fit and recent issues of genetic linkage. Hist. Sci. 24:173–182.

Posten, H.O. 1989. An effective algorithm for the noncentral chi-squared distribution function. Amer. Statist. 43:261–263.

Prout, T. 1965. The estimation of fitness from genotypic frequencies. Evolution 19:546–551.

Queen, C.L. and L.J. Korn. 1980. Computer analysis of nucleic acids and proteins. Pp. 595–609 in *Methods in Enzymology.* Volume 65, L. Grossman and K. Moldave (Editors). Academic Press, New York.

Quenouille, M. 1956. Notes on bias in estimation. Biometrika 43:253–260.

Race, R.R., R. Sanger, S.D. Lawler and D. Bertinshaw. 1949. The inheritance of the MNS blood groups: A second series of families. Heredity 3:205–213.

Rao, D.C., N.E. Morton, J. Lindsten, M. Hultén and S. Yee. 1977. A mapping function for man. Human Hered. 27:99–104.

Raymond, M. and F. Rousset. 1995. An exact test for population differentiation. Evolution 49:1280–1283.

Reynolds, J. 1981. *Genetic Distance and Coancestry.* Institute of Statistics Mimeo Series 1341, North Carolina State University, Raleigh, NC.

Reynolds, J., B.S. Weir and C.C. Cockerham. 1983. Estimation of the coancestry coefficient: Basis for a short-term genetic distance. Genetics 105:767–779.

Roeder, K. 1994. DNA fingerprinting: a review of the controversy (with discussion). Statistical Sci. 9:222–278.

Roeder, K., M. Escobar, J.B. Kadane and I. Balazs. 1994. Measuring hetrogeenity in forensic databases using hierarchical Bayes models. Technical Report, Carnegie Mellon University Department of Statistics, Pittsburgh, USA.

Roff, D.A. and P. Bentzen. 1989. The statistical analysis of mitochondrial DNA polymorphisms: χ^2 and the problem of small sample sizes. Mol. Biol. Evol. 6:539–545.

Rogers, J.S. 1972. Measures of genetic similarity and genetic distance. In *Studies in Genetics VII.* University of Texas Publication 7213, Austin.

Rohlf, F.J. and R.R. Sokal. 1995. *Statistical Tables,* Third Edition, Freeman, San Francisco.

Rzhetsky, A. and M. Nei. 1992. A simple method for estimating and testing minimum-evolution trees. Mol. Biol. Evol. 9:945–967.

Saiki, R.K., S. Scharf, F. Faloona, K.B. Mullis, G.T. Horn, H.A. Erlich and N. Arnheim. 1985. Enzymatic amplification of β-globin genomic sequences and restriction site analysis for diagnosis of sickle cell anemia. Science 230:1350.

Saitou, N. and M. Nei. 1987. The neighbor-joining method: a new method for reconstrucing phylogenetic trees. Mol. Biol. Evol. 4:406–425.

Sanger, F., S. Nocklen and A.R. Coulson. 1977. DNA sequencing with chain-terminating inhibitors. Proc. Natl. Acad. Sci. USA. 74:560–564.

Sankoff, D., C. Morel and R.J. Cedergren. 1973. Evolution of 5S RNA and the non-randomness of base replacement. Nature New Biol. 245:232–234.

Santibánez-Koref, M. and J.G. Reich. 1986. Dinucleotide frequencies in different reading frame positions of coding mammalian DNA sequences. Biomed. Biochim. Acta 45:737–748.

Sarich, V.M. and A.C. Wilson. 1973. Generation time and genomic evolution in primates. Science 179:1144–1147.

Schaffer, H.E. 1983. Determination of DNA fragment size from gel electrophoresis mobility. Pp. 1–14 in *Statistical Analysis of DNA Sequence Data.* B.S. Weir (Editor). Dekker, New York.

Schrage, L. 1979. A more portable Fortran random number generator. ACM Trans. Math. Software 5:132–138.

Schroeder, J.L. and F.B. Blattner. 1978. Least-squares method for restriction mapping. Gene 4:167–174.

Schwartz, R.M. and M.O. Dayhoff. 1979. Matrices for detecting distant relationships. Pp. 353-358 in *Atlas of Protein Sequence and Structure,* Volume 5, Supplement 3, 1978. National Biomedical Research Foundation, Washington, DC.

Shaw, D.V., A.L. Kahler and R.W. Allard. 1981. A multilocus estimator of mating system parameters in plant populations. Proc. Natl. Acad. Sci. U.S.A. 78:1298–1302.

Shen, S., J.L. Slightom and O. Smithies. 1981. A history of the human fetal globin gene duplication. Cell 26:191–203.

Slatkin, M. 1985. Rare alleles as indicators of gene flow. Evolution 39:53–65.

Slatkin, M. 1990. Gene flow and the genetic structure of populations. Pp. 105–122 in *Population Biology of Genes and Molecules*, N. Takahata and J.F. Crow (Editors). Baifukan, Tokyo. 39:53–65.

Smith, C.A.B. 1957. Counting methods in genetical statistics. Ann. Human Genet. 21:254–276.

Smith, C.A.B. 1977. A note on genetic distance. Ann. Human Genet. 21:254–276.

Smith, T.F. and M.S. Waterman. 1981. The identification of common molecular subsequences. J. Mol. Biol. 147:195–197.

Smith, T.F. and M.S. Waterman. 1992. The continuing case of the Florida dentist. Science 256:1165–1171.

Smith, T.F., M.S. Waterman and J.R. Sadler. 1983. Statistical characterization of nucleic acid sequence functional domains. Nucleic Acids Res. 11:2205–2220.

Sneath, P.H.A. and R.R. Sokal. 1973. *Numerical Taxonomy.* Freeman, San Francisco.

Sokal, R.R. and F.J. Rohlf. 1995. *Biometry,* Third Edition. Freeman, San Francisco.

Spielman, R.S., R.E. McGinnis and W.J. Ewens. 1993. Transmission test for linkage disequilibrium: the insulin gene region and insulin-dependent diabetes mellitus (IDDM). Am. J. Hum. Genet. 52:506–516.

Spielman, R.S., J.V. Neel and F.H.F. Li. 1977. Inbreeding estimation from population data: models, procedures and implications. Genetics 85:355–371.

Steel, R.G.D. and J.H. Torrie. 1980. *Principles and Procedures of Statistics,* Second Edition. McGraw-Hill, New York.

Stephens, J.C., D. Briscoe and S.J. O'Brien. 1994. Mapping by admixture linkage disequilibrium in human populations: limits and guidelines. Am. J. Hum. Genet. 55:809–824.

Stuart, A. and J.K. Ord. 1987. *Kendall's Advanced Theory of Statistics.* Fifth Edition. Oxford Univ. Press, New York.

Swofford, D.L., G.J. Olsen, P.J. Waddell and D.M. Hillis. 1996. Phylogeny reconstruction. Pp. 407–514 in *Molecular Systematics*, 2nd edition, D.M. Hillis, C. Moritz and B.K. Mabel (Editors). Sinauer, Sunderland, MA.

Swofford,D.L., J.L. Thorne, J. Felsenstein and B.M. Wiegmann. 1996. The topology-dependent permutation test for monophyly does not test for monophyly. Syst. Biol. 45:575–579.

Sytsma, K.J. and B.A. Schaal. 1985. Genetic variation, differentiation, and evolution in a species of complex tropical shrubs based on isozymic data. Evolution 39:582–593.

Tateno, Y., M. Nei and F. Tajima. 1982. Accuracy of estimated phylogenetic trees from molecular data. J. Mol. Evol. 18:387–404.

Tavaré, S. and B.W. Giddings. 1989. Some statistical aspects of the primary structure of nucleotide sequences. Pp. 117–132 in *Mathematical Methods for DNA Sequences,* M.S. Waterman (Editor). CRC Press, Boca Raton, FL.

Templeton, A.R. 1992. Human origins and analysis of mitochondrial DNA sequences. Science 253:1503–1507.

Terwilliger, J.D. and J. Ott. 1994. *Handbook of Human Genetic Linkage.* Johns Hopkins Press, Baltimore.

Thomson, G. and M.P. Baur. 1984. Third order linkage disequilibrium. Tissue Antigens 24:250–255.

Thompson, E.A. 1987. Crossover counts and likelihood in multipoint linkage analysis. IMA J. Math. Applied Med. Biol. 4:93–108.

Thompson, J.D., D.G. Higgins and T.J. Gibson. 1994. CLUSTAL W: improving the sensitivity of pogressive multiple sequence alignment through sequence weighting, position-specific gap penalties and weight matrix choice. Nucleic Acids Res. 22:4673–4680.

Thorne, J.L., H. Kishino and J. Felsenstein. 1991. An evolutionary model for maximum likelihood alignment of DNA sequences. J. Mol. Biol. 33:114–121.

Thorne, J.L., H. Kishino and J. Felsenstein. 1992. Inching toward reality: an improved likelihood model of sequence evolution. J. Mol. Biol. 34:3–16.

Tukey, J. 1958. Bias and confidence in not quite large samples. Ann. Math. Stat. 29:614.

Vigilant, L., M. Stoneking, H. Harpending, K. Hawkes and A.C. Wilson. 1991. African populations and the evolution of human mitochondrial DNA. Science 253:1503–1507.

Vithayasai, C. 1973. Exact critical values of the Hardy-Weinberg test statistic for two alleles. Commun. Stat. 1:229–242.

Vogel, F. and A.G. Motulsky. 1986. *Human Genetics,* Second Edition. Springer-Verlag, New York.

Waterman, M.S. 1995. *Introduction to Computational Biology.* Chapman and Hall, London.

Weeks, D.E. 1991. Human linkage analysis: strategies for locus ordering. Pp. 297–330 in *Advanced Techniques in Chromosome Research.* K.W. Adolph (Editor). Dekker, New York.

Weir, B.S. 1979. Inferences about linkage disequilibrium. Biometrics 25:235–254.

Weir, B.S. 1989a. Sampling properties of gene diversity. Pp. 23–42 in *Plant Population Genetics, Breeding and Genetic Resources,* A.H.D. Brown, M.T. Clegg, A.L. Kahler and B.S. Weir (Editors). Sinauer, Sunderland, MA.

Weir, B.S. 1989b. Locating the cystic fibrosis gene on the basis of linkage disequilibrium with markers? Pp. 81–86 in *Multipoint Mapping and Linkage Based upon Affected Pedigree Members: Genetic Analysis Workshop 6.* R.C. Elston, M.A. Spence, S.E. Hodge and J.W. MacCluer (Editors), Liss, New York.

Weir, B.S. 1993. Analysis of DNA sequences. Stat. Meth. Med. Res. 2:225–239.

Weir, B.S. 1994. The effects of inbreeding on forensic calculations. Ann. Rev. Genet. 28:597–621.

Weir, B.S. (Editor). 1995. *Human Identification: The Use of DNA Markers.* Kluwer, Dordrecht.

Weir, B.S., R.W. Allard and A.L. Kahler. 1972. Analysis of complex allozyme polymorphisms in a barley population. Genetics 72:505–523.

Weir, B.S., R.W. Allard and A.L. Kahler. 1974. Further analysis of complex allozyme polymorphisms in a barley population. Genetics 78:911–919.

Weir, B.S. and C.J. Basten. 1990. Sampling strategies for DNA sequence distances. Biometrics 46:551–582.

Weir, B.S. and L.D. Brooks. 1986. Disequilibrium on human chromosome 11p. Genet. Epidemiol. Suppl. 1:177–183

Weir, B.S. and C.C. Cockerham. 1978. Testing hypotheses about linkage disequilibrium with multiple alleles. Genetics 88:633–642.

Weir, B.S. and C.C. Cockerham. 1979. Estimation of linkage disequilibrium in randomly mating populations. Heredity 42:105–111.

Weir, B.S. and C.C. Cockerham. 1984. Estimating F-statistics for the analysis of population structure. Evolution 38:1358–1370.

Weir, B.S. and C.C. Cockerham. 1989. Complete characterization of disequilibrium at two loci. Pp. 86–110 in *Mathematical Evolutionary Theory*, M.E. Feldman (Editor). Princeton Univ. Press, Princeton.

Weir, B.S. and W.G. Hill. 1986. Nonuniform recombination within the human β-globin gene cluster. Am. J. Hum. Genet. 38:776—778.

Weir, B.S., J. Reynolds and K.G. Dodds. 1990. The variance of sample heterozygosity. Theor. Pop. Biol. 37:235–253.

Weir, B.S., C.M. Triggs, L. Starling, L.I. Stowell, K.A.J. Walsh and J.S. Buckleton. 1997. Interpreting DNA mixtures. J. Forensic Sciences 42:213–222.

Weir, B.S. and S.R. Wilson. 1986. Log-linear models for linked loci. Biometrics 42:665–670.

Welsh, J. and M. McClelland. 1990. Fingerprinting genomes using PCRT with arbitrary primers. Nucleic Acids Res. 18:7213–7218.

Williams, C.J. and S. Evarts. 1989. The estimation of concurrent paternity probabilities in natural populations. Theor. Pop. Biol. 35:90–112.

Williams, C.J., W.W. Anderson and J. Arnold. 1990. Generalized linear modeling methods for selection component experiments. Theor. Pop. Biol. 37:398–423.

Williams, J.G.K., A.R. Kubelik, K.J. Livak, J.A. Rafalski and S.V. Tingey. 1990. DNA polymorphism amplified by arbitrary primers are useful as genetic markers. Nucleic Acids Res. 18:6531–6535.

Workman, P.L. and S.K. Jain. 1966. Zygotic selection under mixed random mating and self-fertilization: theory and problems of estimation. Genetics 54:159–171.

Wright,S. 1945. The differential equation of the distribution of gene frequencies. Proc. Natl. Acad. Sci. USA 31:383–389.

Wright, S. 1951. The genetical structure of populations. Ann. Eugen. 15:323–354.

Wu, C-I. and W-H. Li. 1985. Evidence for higher rates of nucleotide substitution in rodents than in man. Proc. Natl. Acad. Sci. U.S.A. 82:1741–1745.

Yasuda, N. and M. Kimura. 1968. A gene-counting method of maximum likelihood for estimating gene frequencies in ABO and ABO-like systems. Ann. Human Genet. 31:409–420.

Yates, F. 1934. Contingency tables involving small numbers and the X^2 test. J. Roy. Stat. Soc. Suppl. 1:217–235.

Yates, F. 1984. Tests of significance for 2 × 2 contingency tables. J. Roy. Stat. Soc. A 147:426–463.

Yeh, F.C. and K. Morgan. 1987. Mating system and multilocus associations in a natural population of *Pseudotsuga menziesii* (Mirb.) Franco. Theor. Appl. Genet. 73:799–808.

Zaykin, D., L. Zhivotovsky and B.S. Weir. 1995. Exact tests for association between alleles at arbitrary numbers of loci. Genetica 96:169–178.

Zuker, M. 1989. The use of dynamic programming algorithms in RNA secondary structure prediction. Pp. 159–184 in *Mathematical Methods for DNA Sequences,* M. Waterman (Editor). CRC Press, Boca Raton, FL.

Author Index

Subject Index